SPECTROSCOPIC PROPERTIES OF NATURAL FLAVONOIDS

SPECTROSCOPIC PROPERTIES OF NATURAL FLAVONOIDS

Goutam Brahmachari
Visva-Bharati University, India

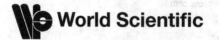
World Scientific

NEW JERSEY · LONDON · SINGAPORE · BEIJING · SHANGHAI · HONG KONG · TAIPEI · CHENNAI · TOKYO

Published by

World Scientific Publishing Co. Pte. Ltd.

5 Toh Tuck Link, Singapore 596224

USA office: 27 Warren Street, Suite 401-402, Hackensack, NJ 07601

UK office: 57 Shelton Street, Covent Garden, London WC2H 9HE

British Library Cataloguing-in-Publication Data
A catalogue record for this book is available from the British Library.

SPECTROSCOPIC PROPERTIES OF NATURAL FLAVONOIDS

ISBN 978-981-3275-68-3

For any available supplementary material, please visit
https://www.worldscientific.com/worldscibooks/10.1142/11140#t=suppl

Typeset by Stallion Press
Email: enquiries@stallionpress.com

Printed in Singapore

Dedicated to

Ranjit Chattopadhyay
Malati Chattopadhyay

Abbreviations

UV	Ultraviolet Spectrocopy
FT-IR	Fourier Transform Infrared Spectroscopy
^{1}H NMR	Proton (^{1}H) Nuclear Magnetic Resonance Spectrocopy
^{13}C NMR	Carbon (^{13}C) Nuclear Magnetic Resonance Spectrocopy
COSY	Correlated Spectroscopy (2D-NMR)
NOSEY	Nuclear Overhauser Effect Spectroscopy
HMQC	Heteronuclear Multiple Quantum Coherence Spectroscopy (2D-NMR)
HMBC	Heteronuclear Multiple Bond Coherence Spectroscopy (2D-NMR)
EI-MS	Electron Impact Mass Spectroscopy
CI-MS	Chemical Ionization Mass Spectroscopy
HR-MS	High Resolution Mass Spectroscopy
TOF-MS	Time-Of-Flight Mass Spectroscopy
ESI-MS	Electronspray Ionization Mass Spectroscopy
MeOH	Methanol
EtOH	Ethanol
$(CH_3)_2CO$	Acetone
$CHCl_3$	Chloroform
$CDCl_3$	Chloroform-d (deuterochloroform)
DMSO-d_6	Dimethyl sulfoxide-d_6 (Hexadeuterodimethyl sulfoxide)
Api	β-Apiofuranosyl
Arb	α-L-Arabinopyranosyl

Gal	β-Galactopyranosyl
Glc	β-D-Glucopyranosyl
Rha	α-L-Rhamnopyranosyl
Xyl	β-D-Xylopyranosyl

Foreword

The birth of *'Chemistry for Human Health'* started with the discovery that *Mother Nature* has provided us with the cure for many diseases in form of natural products, or the so called secondary metabolites. Plant and microbial world offer a wide diversity of compounds that can serve as exciting new pharmacophores with novel mechanisms of action for controlling disease processes. Later on marine natural products added greater diversity to the structural architect. The challenge to understand nature's magical chemistry is continuously stimulating much of our intellectual thoughts, scientific discoveries, and advances in technology. Understanding the structures of the secondary metabolites is arguably the starting point to make progress in pharmaceutical research. Until and unless one knows the structure of a bioactive molecule, further advancement to bring its activity to a higher level and also to enhance the efficiencies of other pharmaceutical parameters is not possible. The chemistry of penicillin remains a classic example. Once the structure was confirmed by X-ray and synthesis, new derivatives could be made with broader activity profile.

It is a pleasure for me to write a foreword on this book on *Spectroscopic Properties of Natural Flavonoids* written by Professor Goutam Brahmachari. Flavonoids have emerged as an important class of secondary metabolites possessing wide array of biological activities. Antioxidant, anti-mutagenic, anti-carcinogenic, anti-proliferative and anti-inflammatory are just to name a few. This class of natural products has already played an indispensible privileged skeleton for the development of new therapeutic agents. Due to their wide range of activities, correlation of structural motifs with the bioactivity has become a subject of immense interest. So having a book rich with information on physical characteristics and spectral data of 150 selected natural flavonoids will certainly aid to our research in quest for new natural products of this class. Mentioning the biological activities wherever reported and NMR spectral data assignment are added attraction of this book. It reminded me of our earlier days, when those working on natural product chemistry that included terpenes, shikimates etc. heavily relied on the famous handbook by Scott and Devon.

Vital role of flavonoids in a variety of nutraceutical, pharmaceutical, medicinal and cosmetic applications has stimulated current trends of research on isolation, identification, characterization and functions of flavonoids and eventually their beneficial effect on health. I have no doubt that this book will readily become attractive and helpful to practicing natural product chemists. I hope that Prof. Brahmachari's next effort will be to cover other natural products of similar importance and expand the present version.

Congratulations and many thanks to Professor Brahmachari for his outstanding effort to come out with this important handbook.

26th July, 2018   Amit Basak
                          Indian Institute of Technology, Kharagpur
                                                        India

Preface

When studying science, the examples are more useful than the rules

— Sir Isaac Newton

Structural elucidation of organic molecules has always been a challenge in the field of organic chemistry, and it is obvious that the task is more challenging when one deals with natural products characterized by unexpected and unprecedented scaffolds. The scope of such investigations with natural products is diverse and broad. It may be recalled in this connection that more than 20,000 and more than 30,000 new marine-derived and higher plant-derived compounds, respectively, were isolated and structurally investigated in the past 10–12 years, with the help of modern spectroscopic techniques. Identification and in-depth understanding of the chemical structure of a newly isolated compound from any natural source is the very first and key step prior to advancing into further developments.

Bioactive natural products are a rich source of novel therapeutics, and are of great interest and promise in the present day research directed towards drug design and discovery. Many naturally occurring bioactive compounds and/or their derivatives have become drugs of central importance, and represent a high percentage of the drugs used today. At present, plants, microorganisms, and marine invertebrates represent major sources of natural products for discovering new and novel drugs. Medicinal chemistry of such bioactive compounds encompasses a vast area that primarily includes their isolation from natural sources, followed by their structural elucidation. Among diverse natural products, flavonoids comprise a group of phenolic secondary plant metabolites that are widespread in nature. Major flavonoids that have well categorized structures as well as structure

function-relationships are flavans, flavanones, flavones, isoflavones, fla-
vonols, flavanols, flavanonols, catechins, chalcones, dihydrochalcones,
aurones, anthocyanidins and anthocyanins. Bio-flavonoids are well-
known for their multi-directional pharmacological potentials that include
antioxidant, anticancer, antitumor, cytotoxic, enzyme inhibitory, anti-
inflammatory, antimicrobial, anti-HIV, anti-diabetic, anti-platelet aggre-
gation, and neuroprotective activity. Hence, this group of natural
polyphenols is now-a-days regarded as promising and significantly attractive natural substances to enrich the current therapy options against a
variety of diseases. Under this purview, researchers from different fields
are actively engaged with natural flavonoids.

Spectroscopic Properties of Natural Flavonoids — Volume 1 offers
detailed physical characteristics and spectral data of 150 selected compounds from this important class of natural products, arranged according
to their chemical structures in various subclasses. Natural source, molecular formula, chemical structure, IUPAC name, physical characteristics
(state, melting point, molecular weight, and specific rotation) and detailed
spectral data (UV, FT-IR, ^{1}H-NMR, ^{13}C-NMR, 2D-NMR, Mass) along
with their assignments for each compound are documented. Being a first-
time comprehensive reference guide on natural flavonoids, which presents
the actual spectral data rather than just references to the data, this book
will prove to be of great benefit to advanced organic chemistry students,
organic chemists, natural product chemists, medicinal chemists and
pharmacologists.

I would like express my sincere thanks to Professor Amit Basak,
Indian Institute of Technology, Kharagpur, India, for his keen interest in
the manuscript and for writing the foreword to the book.

I would also like to express my deep sense of appreciation to all of the
editorial and publishing staff members associated with World Scientific
Publishing Co. Pte. Ltd., Singapore, for their keen interest in publishing
the work as well as their all-round help so as to ensure that the highest
standards of publication have been maintained in bringing out this book.
My effort will be successful only when it is found helpful to the readers
at large. Every step has been taken to make the manuscript error-free; in
spite of that, some errors might have crept in. Any remaining error is, of

course, of my own. Constructive comments on the approach of the book from the readers will be highly appreciated.

Finally, I should thank my wife and my son for their well understanding and allowing me enough time throughout the entire period of writing; without their support, this work would not have been successful.

Goutam Brahmachari
Chemistry Department
Visva-Bharati (a Central University)
Santiniketan, India
July 2018

Contents

xv

Acacetin 7-*O*-(3-acetyl-β-D-glucopyranoside)

IUPAC name: (2*S*,3*R*,4*S*,5*R*,6*R*)-3,5-dihydroxy-2-((5-hydroxy-2-(4-methoxyphenyl)-4-oxo-4*H*-chromen-7-yl)oxy)-6-(hydroxymethyl) tetrahydro-2*H*-pyran-4-yl acetate

Sub-class: Flavone glycoside

Chemical structure

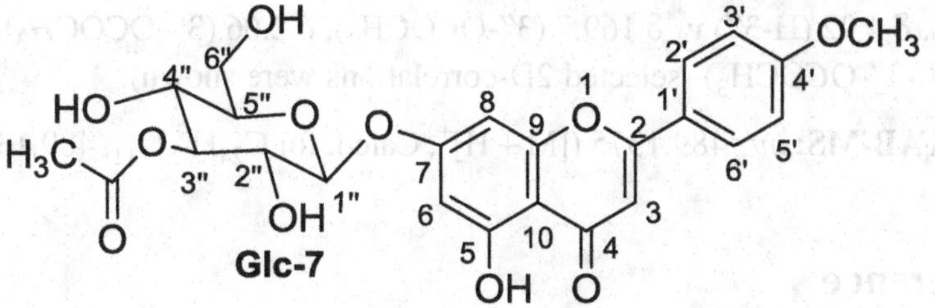

Source: *Chrysanthemum sinense* Sabine (Family: Asteraceae); flowers

Molecular formula: $C_{24}H_{24}O_{11}$

Molecular weight: 488

State: Pale yellow amorphous solid

Bioactivity studied: Xanthine oxidase inhibitor (IC_{50} 80.3 μM)

Specific rotation: $[\alpha]^{20}_{D}$ −35.6o (MeOH, *c* 0.25)

IR (KBr): ν_{max} 3400 (OH), 1660 (α,β-unsaturated carbonyl), 1605, 1450 cm^{-1}

1

^{1}H-NMR (DMSO-d_6, 400 MHz): δ 6.95 (1H, s, H-3), 6.87 (1H, d, J = 2.0 Hz, H-6), 6.47 (1H, d, J = 2.0 Hz, H-8), 8.06 (2H, d, J = 8.8 Hz, H-2′ and H-6′), 7.13 (2H, d, J = 8.8 Hz, H-3′ and H-5′), 12.92 (1H, s, 5-OH), 3.87 (3H, s, 4′-OCH_3), 5.23 (1H, d, J = 7.8 Hz, H-1″), 3.41 (1H, dd, J = 9.5, 7.8 Hz, H-2″), 4.92 (1H, t, J = 9.5 Hz, H-3″), 3.41 (1H, dd, J = 9.5, 8.8 Hz, H-4″), 3.53 (1H, m, H-5″), 3.71 (1H, dd, J = 11.7, 6.1 Hz, H-6″a), 3.46 (1H, dd, J = 11.7, 3.9 Hz, H-6″b), 2.06 (3H, s, 3″-OCOCH_3).

^{13}C-NMR (DMSO-d_6, 100 MHz): δ 163.8 (C-2), 103.8 (C-3), 182.0 (C-4), 156.9 (C-5), 94.9 (C-6), 162.7 (C-7), 99.6 (C-8), 161.1 (C-9), 105.5 (C-10), 122.6 (C-1′), 128.4 (C-2′ and C-6′), 114.6 (C-3′ and C-5′), 162.5 (C-4′), 55.6 (4′-OCH_3), 99.4 (C-1″), 71.0 (C-2″), 77.4 (C-3″), 67.3 (C-4″), 76.7 (C-5″), 60.2 (C-6″), 169.7 (3″-OCOCH$_3$), 21.1 (3″-OCOCH$_3$).

HMBC: δ 6.95 (H-3) *vs* δ 163.8 (C-2), 103.8 (C-3), 182.0 (C-4), 105.5 (C-10) and 122.6 (C-1′), δ 6.87 (H-6) *vs* δ 156.9 (C-5), 162.7 (C-7) and 105.5 (C-10), δ 6.47 (H-8) *vs* δ 162.7 (C-7), 161.1 (C-9) and 105.5 (C-10), δ 12.92 (5-OH) *vs* δ 156.9 (C-5), 94.9 (C-6) and 105.5 (C-10), δ 8.06 (H-2′/H-6′) *vs* δ 163.8 (C-2) and 162.5 (C-4′), δ 7.13 (H-3′/H-5′) *vs* δ 122.6 (C-1′), δ 3.87 (4′-OCH_3) *vs* δ 162.5 (C-4′), δ 5.23 (H-1″) *vs* δ 162.7 (C-7), δ 4.92 (H-3″) *vs* δ 169.7 (3″-OCOCH$_3$), δ 2.06 (3″-OCOCH_3) *vs* δ 169.7 (3″-OCOCH$_3$) (selected 2D-correlations were shown).

HR-FAB-MS: m/z 489.1375 ([M + H]$^+$, Calcd. for $C_{24}H_{25}O_{11}$, 489.1397).

Reference

Nguyen MTT, Awale S, Tezuka Y, Ueda J-Y, Tran Q L, Kadota S. (2006). Xanthine oxidase inhibitors from flowers of *Chrysanthemum sinense*. *Planta Med* **72**: 46–51.

6'''-O-Acetyl amurensin

IUPAC name: ((2R,3S,4S,5R,6S)-6-((3,5-dihydroxy-2-(4-hydroxyphenyl)-8-(3-methylbut-2-en-1-yl)-4-oxo-4H-chromen-7-yl)oxy)-3,4,5-trihydroxytetrahydro-2H-pyran-2-yl)methyl acetate

Sub-class: Flavonol glycoside

Chemical structure

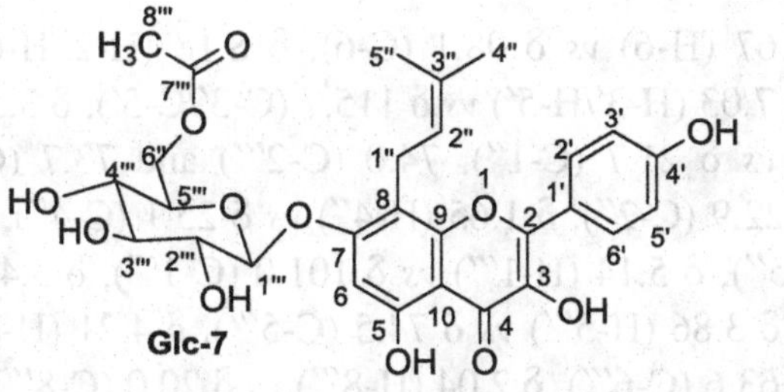

Glc-7

Source: *Phellodendron japonicum* Maxim. (Family: Rutaceae); leaves

Molecular formula: $C_{28}H_{30}O_{12}$

Molecular weight: 558

State: Yellow powder

Melting point: 235–237 °C

Specific rotation: $[\alpha]^{25}_{D}$ −91.3° (MeOH, c 0.05)

UV (MeOH): λ_{max} (log ε) 271 (4.46), 327 (4.20), 373 (4.36) nm

IR (KBr): ν_{max} 3361 (OH), 2923, 1719 (ester carbonyl), 1646 (α,β-unsaturated carbonyl), 1599, 1259, 1081 cm^{-1}

^{1}H-NMR (Acetone-d_6, 400 MHz): δ 6.67 (1H, s, H-6), 8.18 (2H, d, J = 8.8 Hz, H-2' and H-6'), 7.03 (2H, d, J = 8.8 Hz, H-3' and H-5'), 3.52-3.74 (4H, m, H-1″, H-2‴, H-3‴), 5.28 (1H, br t, J = 6.8 Hz, H-2″), 1.65 (3H, s, H-4″), 1.82 (3H, s, H-5″), 5.14 (1H, d, J = 7.6 Hz, H-1‴), 3.45 (1H, m, H-4‴), 3.86 (1H, td, J = 8.6, 2.0 Hz, H-5‴), 4.21 (1H, m, H-6‴a), 4.46 (1H, d, J = 9.6 Hz, H-6‴b), 2.04 (3H, s, H-8‴), 8.07 (1H, br, 3-OH), 12.14 (1H, s, 5-OH), 9.05 (1H, br s, 4'-OH), 4.51 (1H, d, J = 4.4 Hz, 2″-OH), 4.55 (1H, br s, 3‴-OH), 4.65 (1H, br s, 4‴-OH).

^{13}C-NMR (Acetone-d_6, 100 MHz): δ 147.8 (C-2), 136.1 (C-3), 176.4 (C-4), 159.3 (C-5), 98.1 (C-6), 160.8 (C-7), 109.0 (C-8), 153.5 (C-9), 105.1 (C-10), 122.6 (C-1'), 129.9 (C-2' and C-6'), 115.7 (C-3' and C-5'), 159.6 (C-4'), 21.7 (C-1″), 122.9 (C-2″), 131.4 (C-3″), 25.1 (C-4″), 17.5 (C-5″), 101.0 (C-1‴), 74.0 (C-2‴), 73.7 (C-3‴), 70.7 (C-4‴), 74.5 (C-5‴), 63.6 (C-6‴), 170.3 (C-7‴), 20.0 (C-8‴).

HMQC: δ 6.67 (H-6) *vs* δ 98.1 (C-6), δ 8.18 (H-2'/H-6') *vs* δ 129.9 (C-2'/C-6'), δ 7.03 (H-3'/H-5') *vs* δ 115.7 (C-3'/C-5'), δ 3.52-3.74 (H-1″/H-2‴/H-3‴) *vs* δ 21.7 (C-1″), 74.0 (C-2‴) and 73.7 (C-3‴), δ 5.28 (H-2″) *vs* δ 122.9 (C-2″), δ 1.65 (H-4″) *vs* δ 25.1 (C-4″), δ 1.82 (H-5″) *vs* δ 17.5 (C-5″), δ 5.14 (H-1‴) *vs* δ 101.0 (C-1‴), δ 3.45 (H-4‴) *vs* δ 70.7 (C-4‴), δ 3.86 (H-5‴) *vs* δ 74.5 (C-5‴), δ 4.21 (H-6‴a) and 4.46 (H-6‴b) *vs* δ 63.6 (C-6‴), δ 2.04 (H-8‴) *vs* δ 20.0 (C-8‴).

FAB-MS: m/z (%rel): 559 [M + H]$^+$ (17), 355 (24), 299 (16), 185 (100), 149 (20).

HR-FABMS: m/z 559.1819 ([M + H]$^+$, Calcd. for $C_{28}H_{31}O_{12}$, 559.1816).

Reference

Chiu C-Y, Li C-Y, Chiu C-C, Niwa M, Kitanaka S, Damu A G, Lee E-J, Wu T-S. (2005). Constituents of leaves of *Phellodendron japonicum* Maxim. and their antioxidant activity. *Chem Pharm Bull* **53**: 1118–1121.

(2*R*,3*R*)-2″-Acetyl astilbin [(2*R*,3*R*)-5,7,3′,4′-tetrahydroxyflavanonol 2″-acetylrhamnoside]

IUPAC name: (2*R*,3*S*,4*R*,5*R*,6*S*)-2-(((2*R*,3*R*)-2-(3,4-dihydroxyphenyl)-5,7-dihydroxy-4-oxochroman-3-yl)oxy)-4,5-dihydroxy-6-methyltetrahydro-2*H*-pyran-3-yl acetate

Sub-class: Flavanone *C*-glycoside

Chemical structure

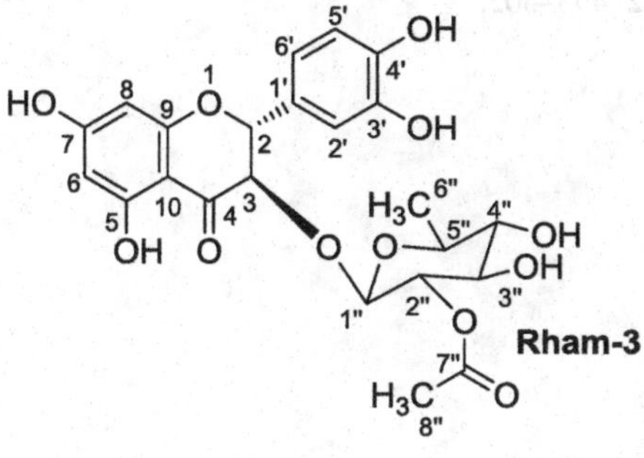

Source: *Smilax corbularia* Kunth. (Family: Smilacaceae); rhizomes

Molecular formula: $C_{23}H_{24}O_{12}$

Molecular weight: 492

State: Pale brown amorphous powder

Specific rotation: $[\alpha]^{25}_D$ +19.6° (MeOH, *c* 0.3)

5

UV (MeOH): λ_{max} (log ε) 220 (4.49), 230 (4.39), 290 (4.35), 330 (3.88) nm

^{1}H-NMR (CD$_3$OD, 400 MHz): δ 5.11 (1H, d, J = 10.5 Hz, H-2β), 4.53 (1H, d, J = 10.5 Hz, H-3α), 5.90 (1H, d, J = 2.0 Hz, H-6), 5.92 (1H, d, J = 2.0 Hz, H-8), 6.93 (1H, d, J = 2.0 Hz, H-2'), 6.76 (1H, d, J = 8.0 Hz, H-5'), 6.79 (1H, dd, J = 8.2 Hz, H-6'), 3.97 (1H, d, J = 1.5 Hz, H-1''), 4.87 (1H, dd, J = 3.0, 1.5 Hz, H-2''), 3.82 (1H, dd, J = 10.0, 3.5 Hz, H-3''), 3.25 (1H, t, J = 10.0 Hz, H-4''), 4.17 (1H, m, H-5''), 1.18 (3H, d, J = 6.0, 1.5 Hz, H-6''), 1.97 (3H, s, H-8'').

^{13}C-NMR (CD$_3$OD, 100 MHz): δ 83.7 (C-2), 78.3 (C-3), 195.3 (C-4), 165.6 (C-5), 96.4 (C-6), 168.8 (C-7), 97.5 (C-8), 164.1 (C-9), 102.5 (C-10), 128.9 (C-1'), 115.1 (C-2'), 146.7 (C-3'), 147.4 (C-4'), 116.5 (C-5'), 120.4 (C-6'), 99.3 (C-1''), 73.2 (C-2''), 70.5 (C-3''), 74.1 (C-4''), 70.6 (C-5''), 17.8 (C-6''), 171.5 (C-7''), 21.7 (C-8'').

HRFABMS: m/z 493.1355 [M + H]$^+$ (Calcd. for $C_{23}H_{25}O_{12}$, 493.1346).

Reference

Wungsintaweekul B, Umehara K, Miyase T, Noguchi H. (2011). Estrogenic and anti-estrogenic compounds from the Thai medicinal plant, *Smilax corbularia* (Smilacaceae). *Phytochemistry* **72**: 495–502.

Aciculatinone

IUPAC name: 5-Hydroxy-8-((2*R*,5*R*,6*R*)-5-hydroxy-6-methyl-4-oxotetrahydro-2*H*-pyran-2-yl)-2-(4-hydroxyphenyl)-7-methoxy-4*H*-chromen-4-one

Sub-class: Flavone *C*-glycoside

Chemical structure

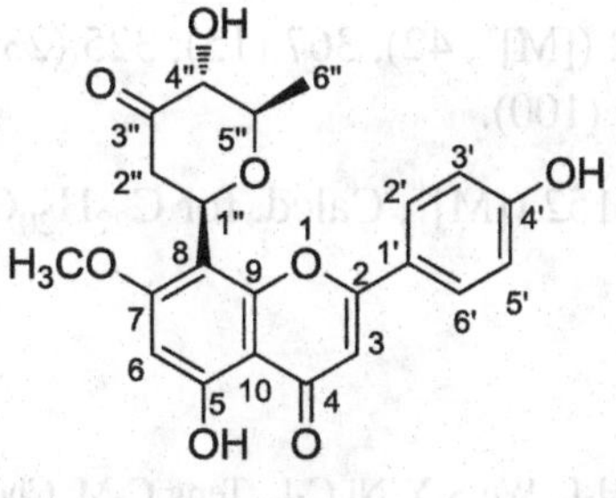

Source: *Chrysopogon aciculatis* (Family: Poaceae); whole grass

Molecular formula: $C_{20}H_{20}O_8$

Molecular weight: 412

State: Light yellow prisms (MeOH-H_2O, 3:1 v/v)

Melting point: 146–148 °C

Bioactivity studied: Cytotoxic [IC_{50} values of 18.36, 18.06, 11.59 and 13.33 µM, respectively against MCF-7, H450, HT-29 and CEM cancer cell lines]

Specific rotation: $[\alpha]_D$ +62° (MeOH, c 0.27)

UV (MeOH): λ_{max} (logε) 214 (4.36), 268 (4.15), 335 (4.31) nm

IR (KBr): ν_{max} 3500 (OH), 1718 (oxo), 1651 (γ-pyrone carbonyl), 1603, 1500, 1455, (aromatic unsaturaion), 1362, 1243, 1180 cm^{-1}

^{1}H-NMR (Acetone-d_6, 600 MHz): δ 6.69 (1H, s, H-3), 6.48 (1H, s, H-6), 8.08 (2H, d, J = 8.4 Hz, H-2′ and H-6′), 7.06 (2H, d, J = 8.4 Hz, H-3′ and H-5′), 5.40 (1H, dd, J = 12.0, 3.0 Hz, H-1″), 3.55 (1H, dd, J = 13.8, 12.0 Hz, H-2″a), 2.51 (1H, dd, J = 13.8, 3.0 Hz, H-2″b), 4.09 (1H, d, J = 9.6 Hz, H-4″), 3.60 (1H, qd, J = 9.6, 6.0 Hz, H-5″), 1.48 (3H, d, J = 6.0 Hz, H-6″), 13.41 (1H, s, 5-OH), 3.97 (3H, s, 7-OCH_3).

^{13}C-NMR (Acetone-d_6, 150 MHz): δ 165.5 (C-2), 103.9 (C-3), 183.5 (C-4), 163.6 (C-5), 95.9 (C-6), 163.3 (C-7), 106.7 (C-8), 156.1 (C-9), 105.8 (C-10), 123.4 (C-1′), 129.6 (C-2′ and C-6′), 116.9 (C-3′ and C-5′), 162.1 (C-4′), 72.2 (C-1″), 45.6 (C-2″), 207.0 (C-3″), 79.9 (C-4″), 80.3 (C-5″), 19.7 (C-6″), 57.0 (7-OCH_3).

EIMS: m/z (%rel) 412 ([M]$^+$, 42), 367 (15), 325 (23), 296 (42), 295 (45), 117 (90), 189 (36), 58 (100).

HR-EIMS: m/z 412.1152 ([M]$^+$, Calcd. for $C_{22}H_{20}O_8$, 412.1146).

Reference

Shen C-C, Cheng J-J, Lay H-L, Wu S-Y, Ni C-L, Teng C-M, Chen C-C. (2012). Cytotoxic apigenin derivatives from *Chrysopogon aciculatis*. *J Nat Prod* **75**: 198–201.

Afzelin A
[6,7-(2″,2″-dimethylpyrano)-3,5, 4′-trihydroxyflavanone]

IUPAC name: 3,5-Dihydroxy-2-(4-hydroxyphenyl)-8,8-dimethyl-2,3-dihydropyrano[3,2-g]chromen-4(8H)-one

Sub-class: Flavanone

Chemical structure

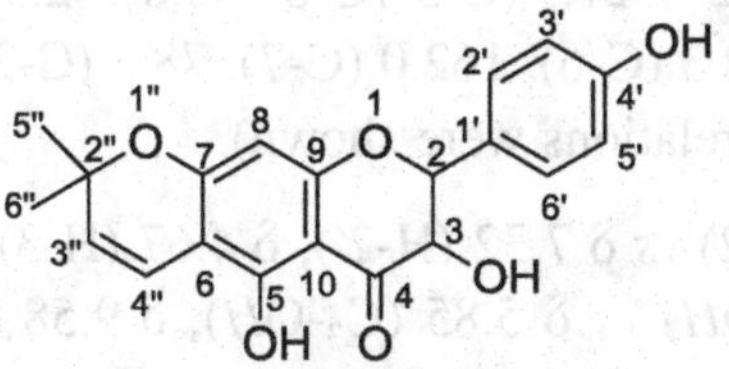

Source: *Hymenostegia afzelii* (Family: Caesalpiniaceae); stem barks and leaves

Molecular formula: $C_{20}H_{18}O_6$

Molecular weight: 354

State: Yellow needles

Melting point: 192.2–192.7 °C

Specific rotation: $[\alpha]^{20}_D$ +23.5° (MeOH, c 0.9)

Bioactivity studied: Cytotoxic

IR (KBr): v_{max} 3321 (OH), 1608 cm^{-1}

^{1}H-NMR (DMSO-d_6, 500 MHz): δ 5.11 (1H, d, J = 11.5 Hz, H-2), 4.67 (1H, dd, J = 11.5, 6.0 Hz, H-3), 5.91 (1H, s, H-8), 7.32 (2H, d, J = 8.5 Hz, H-2′ and H-6′), 6.79 (2H, d, J = 8.5 Hz, H-3′ and H-5′), 5.67 (1H, d, J = 10.0 Hz, H-3″), 6.53 (1H, d, J = 10.0 Hz, H-4″), 1.39 (6H, s, 2 × 2″-CH_3), 5.85 (1H, d, J = 6.0 Hz, C_3-OH, D_2O exchangeable), 12.23 (1H, s, C_5-OH, D_2O exchangeable), 9.58 (1H, s, $C_4′$-OH, D_2O exchangeable).

^{13}C-NMR (DMSO-d_6, 125 MHz): δ 83.0 (C-2), 71.5 (C-3), 198.9 (C-4), 157.9 (C-5), 102.3 (C-6), 162.0 (C-7), 95.7 (C-8), 161.3 (C-9), 101.3 (C-10), 127.4 (C-1′), 129.6 (C-2′ and C-6′), 115.0 (C-3′ and C-5′), 157.3 (C-4′), 78.3 (C-2″), 127.1 (C-3″), 114.5 (C-4″), 27.9 (C-5″ and C-6″).

HMQC: δ 5.11 (H-2) *vs* δ 83.0 (C-2), δ 4.67 (H-3) *vs* δ 71.5 (C-3), δ 5.91 (H-8) *vs* δ 95.7 (C-8), 7.32 (H-2′/H-6′) *vs* δ 129.6 (C-2′/C-6′), δ 6.79 (H-3′/ H-5′) *vs* δ 115.0 (C-3′/C-5′), δ 5.67 (H-3″) *vs* δ 127.1 (C-3″), δ 6.53 (H-4″) *vs* δ 114.5 (C-4″), δ 1.39 (2″-CH_3) *vs* δ 27.9 (C-5″/C-6″).

HMBC: δ 5.11 (H-2) *vs* δ 71.5 (C-3), 198.9 (C-4) and 129.6 (C-6′), δ 4.67 (H-3) *vs* δ 83.0 (C-2), 198.9 (C-4) and 127.4 (C-1′), δ 5.67 (H-3″) *vs* δ 78.3 (C-2″), 78.3 (C-2″), 27.9 (C-5″/C-6″) and 102.3 (C-6), δ 6.53 (H-4″) *vs* δ 157.9 (C-5), 102.3 (C-6), 162.0 (C-7), 78.3 (C-2″) and 127.1 (C-3″) (selected HMBC correlations were shown).

NOESY: δ 5.03 (H-2) *vs* δ 7.32 (H-2′), δ 4.67 (H-3) *vs* δ 7.32 (H-2′ and H-6′), δ 12.23 (C_5-OH) *vs* δ 5.85 (C_3-OH), δ 9.58 ($C_4′$-OH) *vs* δ 6.79 (H-3′ and H-5′), δ 5.67 (H-3″) *vs* δ 6.53 (H-4″).

HR-EIMS: *m/z* 354.10940 ([M]$^+$, Calcd. for $C_{20}H_{18}O_6$, 354.11034).

Reference

Awantu A F, Lenta B N, Donfack E V, Wansi J D, Neumann B, Stammler H-G, Noungoue D T, Tsamo E, Sewald N. (2011). Flavonoids and other constituents of *Hymenostegia afzelii* (Caesalpiniaceae). *Phytochemistry Lett* **4**: 315–319.

Afzelin C
[7,6-(2″,2″-dimethylpyrano)-5-hydroxy-3,4′-(3,3-dimethylallyloxy) flavone]

IUPAC name: 3,5-Dihydroxy-2-(4-hydroxyphenyl)-8,8-dimethyl-2,3-dihydropyrano[3,2-g]chromen-4(8H)-one

Sub-class: Flavone

Chemical structure

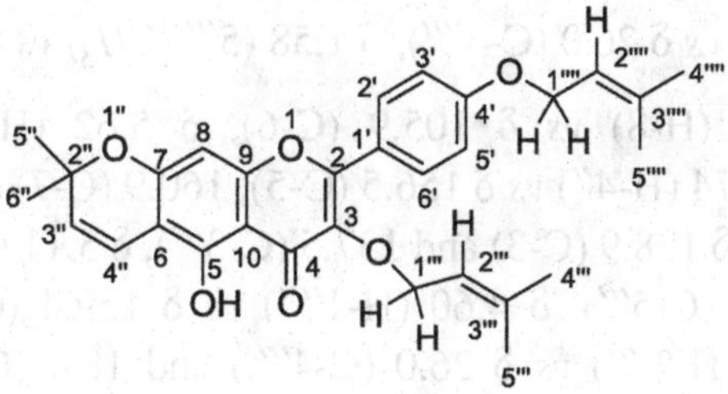

Source: *Hymenostegia afzelii* (Family: Caesalpiniaceae); stem barks and leaves

Molecular formula: $C_{30}H_{32}O_6$

Molecular weight: 488

State: Yellow oil

Bioactivity studied: Cytotoxic

IR (KBr): v_{max} 3281 (OH), 1655, 1588 cm^{-1}

^{1}H-NMR (DMSO-d_6, 500 MHz): δ 6.36 (1H, s, H-8), 8.07 (2H, d, $J = 9.0$ Hz, H-2′ and H-6′), 7.01 (2H, d, $J = 9.0$ Hz, H-3′ and H-5′), 5.62 (1H, d, $J = 10.0$ Hz, H-3″), 6.74 (1H, d, $J = 10.0$ Hz, H-4″), 1.46 (6H, s, 2 × 2″-CH$_3$), 4.55 (2H, d, $J = 7.5$ Hz, H-1‴), 5.11 (1H, t, $J = 7.5$ Hz, H-2‴), 1.82 (3H, s, 4‴-CH$_3$), 1.78 (3H, s, 5‴-CH$_3$), 4.60 (2H, d, $J = 6.5$ Hz, H-1⁗), 5.39 (1H, t, $J = 6.5$ Hz, H-2⁗), 1.61 (3H, s, 4⁗-CH$_3$), 1.58 (3H, s, 5⁗-CH$_3$), 13.06 (1H, s, C$_5$-OH, D$_2$O exchangeable).

^{13}C-NMR (DMSO-d_6, 125 MHz): δ 156.3 (C-2), 138.9 (C-3), 179.2 (C-4), 156.5 (C-5), 105.9 (C-6), 160.9 (C-7), 94.9 (C-8), 156.5 (C-9), 105.2 (C-10), 123.3 (C-1′), 130.4 (C-2′ and C-6′), 114.7 (C-3′ and C-5′), 159.4 (C-4′), 78.1 (C-2″), 128.2 (C-3″), 115.7 (C-4″), 28.4 (C-5″ and C-6″), 77.2 (C-1‴), 119.9 (C-2‴), 137.5 (C-3‴), 18.2 (C-4‴), 25.9 (C-5‴), 69.1 (C-1⁗), 119.3 (C-2⁗), 139.5 (C-3⁗), 26.0 (C-4⁗), 18.4 (C-5⁗).

HMQC: δ 6.36 (H-8) *vs* δ 94.9 (C-8), δ 8.07 (H-2′/H-6′) *vs* δ 130.4 (C-2′/C-6′), δ 7.01 (H-3′/ H-5′) *vs* δ 114.7 (C-3′/C-5′), δ 5.62 (H-3″) *vs* δ 128.2 (C-3″), δ 6.74 (H-4″) *vs* δ 115.7 (C-4″), δ 1.46 (2″-CH$_3$) *vs* δ 28.4 (C-5″/C-6″), δ 4.55 (H-1‴) *vs* δ 77.2 (C-1‴), δ 5.11 (H-2‴) *vs* δ 119.9 (C-2‴), δ 1.82 (4‴-CH$_3$) *vs* δ 18.2 (C-4‴), δ 1.78 (5‴-CH$_3$) *vs* δ 25.9 (C-5‴), 4.60 (H-1⁗) *vs* δ 69.1 (C-1⁗), δ 5.39 (H-2⁗) *vs* δ 119.3 (C-2⁗), δ 1.61 (4⁗-CH$_3$) *vs* δ 26.0 (C-4⁗), δ 1.58 (5⁗-CH$_3$) *vs* δ 18.4 (C-5⁗).

HMBC: δ 6.36 (H-8) *vs* δ 105.9 (C-6), δ 5.62 (H-3″) *vs* δ 28.4 (C-5″/C-6″), δ 6.74 (H-4″) *vs* δ 156.5 (C-5), 160.9 (C-7) and 78.1 (C-2″), δ 4.55 (H-1‴) *vs* δ 138.9 (C-3) and 137.5 (C-3‴), δ 5.11 (H-2‴) *vs* δ 18.2 (C-4‴) and 25.9 (C-5‴), δ 4.60 (H-1⁗) *vs* δ 159.4 (C-4′) and 139.5 (C-3⁗), δ 5.39 (H-2⁗) *vs* δ 26.0 (C-4⁗) and 18.4 (C-5⁗) (selected HMBC correlations were shown).

HR-ESIMS: *m/z* 489.22707 ([M + H]$^+$, Calcd. for C$_{30}$H$_{33}$O$_6$, 489.22717).

Reference

Awantu A F, Lenta B N, Donfack E V, Wansi J D, Neumann B, Stammler H-G, Noungoue D T, Tsamo E, Sewald N. (2011). Flavonoids and other constituents of *Hymenostegia afzelii* (Caesalpiniaceae). *Phytochemistry Lett* **4**: 315–319.

Amyrisin B [5,7-dihydroxy-2-(4-((2-hydroxy-3-methylbut-3-en-1-yl)oxy)phenyl)-6-methoxy-4*H*-chromen-4-one]

IUPAC name: 5,7-Dihydroxy-2-(4-((2-hydroxy-3-methylbut-3-en-1-yl)oxy)phenyl)-6-methoxy-4*H*-chromen-4-one

Sub-class: *O*-Prenylated flavone

Chemical structure

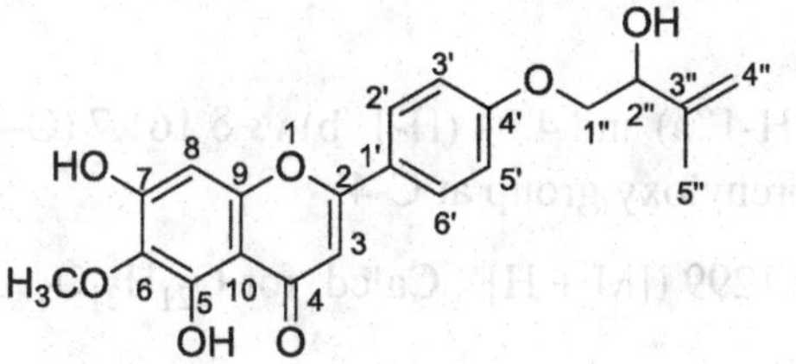

Source: *Amyris madrensis* S. Watson (Family: Rutaceae); leaves and twigs

Molecular formula: $C_{21}H_{20}O_7$

Molecular weight: 384

State: Yellow powder

Bioactivity studied: Anticancerous [IC_{50} = (17.5 ± 4.5) µM against PC-3 cancer cells]

Specific rotation: [α]$^{20}_D$ +6.3° (MeOH, c 0.03)

UV (MeCN-H2O): λ_{max} 274, 335 nm

^{1}H-NMR (CDCl$_3$, 500 MHz): δ 6.58 (1H, s, H-3), 6.60 (1H, s, H-8), 7.84 (2H, d, J = 9.0 Hz, H-2′ and H-6′), 7.05 (2H, d, J = 9.0 Hz, H-3′ and H-5′), 4.14 (1H, dd, J = 9.5, 3.2 Hz, H-1″a), 4.04 (1H, t, J = 9.2 Hz, H-1″b), 4.53 (1H, m, H-2″), 5.19 (1H, s, H-4″a), 5.06 (1H, s, H-4″b), 1.86 (3H, s, H-5″), 4.05 (3H, s, 6-OCH_3), 13.01 (1H, s, 5-OH), 6.49 (1H, s, 7-OH).

^{13}C-NMR (CDCl$_3$, 125 MHz): δ 164.1 (C-2), 103.9 (C-3), 182.7 (C-4), 152.3 (C-5), 130.4 (C-6), 155.1 (C-7), 93.4 (C-8), 153.8 (C-9), 105.7 (C-10), 124.5 (C-1′), 128.3 (C-2′, C-6′), 115.1 (C-3′, C-5′), 161.7 (C-4′), 71.9 (C-1″), 74.1 (C-2″), 143.3 (C-3″), 113.3 (C-4″), 18.7 (C-5″), 60.7 (C$_6$-OCH_3).

HSQC: δ 6.58 (H-3) *vs* δ 103.9 (C-3), δ 6.60 (H-8) *vs* δ 93.4 (C-8), δ 7.84 (H-2′/H-6′) *vs* δ 128.3 (C-2′/C-6′), δ 7.05 (H-3′/ H-5′) *vs* δ 115.1 (C-3′/C-5′), δ 4.14 (H-1″a) and 4.04 (H-1″b) *vs* δ 71.9 (C-1″), δ 4.53 (H-2″) *vs* δ 74.1 (C-2″), δ 5.19 (H-4″a) and 5.06 (H-4″b) *vs* δ 113.3 (C-4″), δ 1.86 (H-5″) *vs* δ 18.7 (C-5″), δ 4.05 (6-OCH_3) *vs* δ 60.7 (C$_6$-OCH_3).

HMBC: δ 4.14 (H-1″a) and 4.04 (H-1″b) *vs* δ 161.7 (C-4′) indicating the presence of the prenyloxy group at C-4′.

HRMS: *m/z* 385.1299 ([M + H]$^+$, Calcd. for $C_{21}H_{21}O_7$, 385.1287).

Reference

Peng J, Hartley R M, Fest G A, Mooberry S L. (2012). Amyrisins A-C, *O*-prenylated fla-vonoids from Amyris madrensis. *J Nat Prod* 75: 494–496.

Amyrisin C [5,7-Dihydroxy-6-methoxy-2-(3-methoxy-4-((3-methylbut-2-en-1-yl)oxy)phenyl)-4*H*-chromen-4-one]

IUPAC name: 5,7-Dihydroxy-6-methoxy-2-(3-methoxy-4-((3-methylbut-2-en-1-yl)oxy)phenyl)-4*H*-chromen-4-one

Sub-class: *O*-Prenylated flavone

Chemical structure

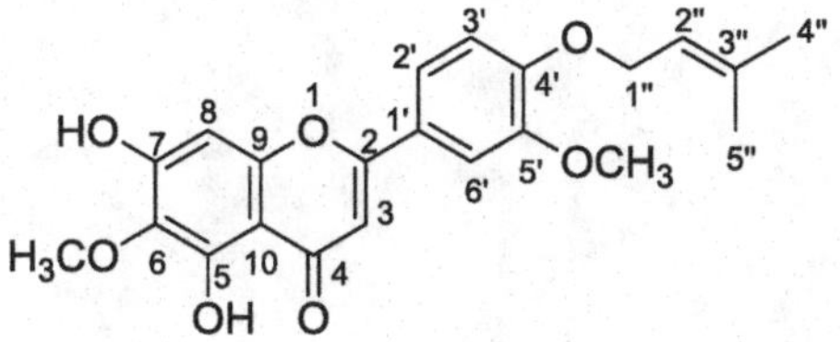

Source: *Amyris madrensis* S. Watson (Family: Rutaceae); leaves and twigs

Molecular formula: $C_{22}H_{22}O_7$

Molecular weight: 398

State: Yellow powder

Bioactivity studied: Anticancerous [IC_{50} = (23.05 ± 5.3) µM against PC-3 cancer cells]

15

UV (MeCN-H$_2$O): λ_{max} 275, 342 nm

^{1}H-NMR (CDCl$_3$, 500 MHz): δ 6.96 (1H, s, H-3), 6.98 (1H, s, H-8), 7.224 (1H, d, J = 2.1 Hz, H-2′), 6.97 (1H, d, J = 8.6 Hz, H-5′), 7.49 (1H, dd, J = 8.5, 2.1 Hz, H-6′), 4.67 (2H, d, J = 6.6 Hz, H-1″), 5.52 (1H, t, J = 6.6 Hz, H-2″), 1.77 (3H, s, H-4″), 1.80 (3H, s, H-5″), 4.05 (3H, s, 6-OCH_3), 3.96 (3H, s, 3′-OCH_3), 13.09 (1H, s, 5-OH).

^{13}C-NMR (CDCl$_3$, 125 MHz): δ 164.1 (C-2), 103.9 (C-3), 182.7 (C-4), not detected (C-5), 130.1 (C-6), 155.0 (C-7), 93.4 (C-8), 153.1 (C-9), 105.8 (C-10), 123.5 (C-1′), 108.8 (C-2′), 149.6 (C-3′), 151.6 (C-4′), 112.7 (C-5′), 120.1 (C-6′), 66.1 (C-1″), 119.3 (C-2″), 138.7 (C-3″), 18.0 (C-4″), 25.5 (C-5″), 61.0 (C$_6$-OCH$_3$), 56.3 (C$_{3′}$-OCH$_3$).

HMBC: δ 4.67 (H-1″) *vs* δ 151.6 (C-4′), δ 3.96 (3′-OCH$_3$) *vs* δ 149.6 (C-3′) (selected HMBC correlations were shown).

HRMS: *m/z* 399.1447 ([M + H]$^+$, Calcd. for C$_{22}$H$_{23}$O$_7$, 399.1444).

Reference

Peng J, Hartley RM, Fest GA, Mooberry SL. (2012). Amyrisins A-C, *O*-prenylated flavonoids from *Amyris madrensis. J Nat Prod* **75**: 494–496.

Angepubebisin

IUPAC name: (*E*)-(*R*)-(7-methoxy-2-oxo-2H-chromen-6-yl)((*S*)-2,2,5,5-tetramethyl-1,3-dioxolan-4-yl)methyl 2-methylbut-2-enoate

Sub-class: Coumarin

Chemical structure

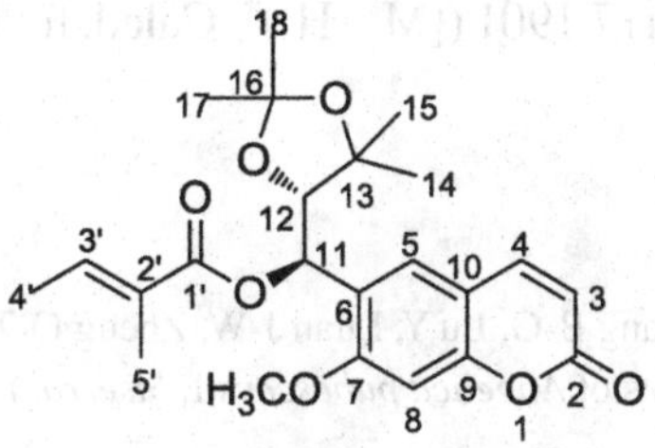

Source: *Angelica pubescens* Maxim. f. *biserrata* Shan et Yuan (Family: Umbelliferae); roots

Molecular formula: $C_{23}H_{28}O_7$

Molecular weight: 416

State: Amorphous powder

Specific rotation: $[\alpha]^{20}_D$ −59.0° (CHCl$_3$, *c* 0.09)

UV (MeOH): λ_{max} (log ε) 221 (4.60), 245 (1.32), 296 (4.11), 322 (4.32) nm

FT-IR (KBr): ν_{max} 2980, 1734 (lactone carbonyl), 1704 (α,β-unsaturated ester carbonyl), 1621, 1566, 1384, 1365 (aromatic unsaturation), 1257, 1206, 1141, 1118, 1051, 1015, 912, 822, 733, 476 cm^{-1}

^{1}**H-NMR (CDCl$_3$, 500 MHz):** δ 6.27 (1H, d, J = 9.6 Hz, H-3), 7.64 (1H, d, J = 9.6 Hz, H-4), 7.46 (1H, s, H-5), 6.83 (1H, s, H-8), 6.19 (1H, d, J = 5.7 Hz, H-11), 4.17 (1H, d, J = 5.7 Hz, H-12), 1.04 (3H, s, H-14), 1.18 (3H, s, H-15), 1.35 (3H, s, H-16), 1.50 (3H, s, H-17), 6.97 (1H, dq, H-3′), 1.82 (3H, d, J = 7.8 Hz, H-4′), 1.84 (3H, s, H-5′), 3.95 (3H, s, 7-OCH_3).

^{13}**C-NMR (CDCl$_3$, 125 MHz):** δ 160.9 (C-2), 113.5 (C-3), 143.4 (C-4), 128.0 (C-5), 124.0 (C-6), 159.6 (C-7), 99.4 (C-8), 155.7 (C-9), 112.2 (C-10), 67.9 (C-11), 83.7 (C-12), 79.9 (C-13), 26.9 (C-14), 23.3 (C-15), 107.2 (C-16), 27.2 (C-17), 28.5 (C-18), 166.8 (C-1′), 128.6 (C-2′), 138.2 (C-3′), 14.5 (C-4′), 12.1 (C-5′), 56.2 (7-OCH$_3$).

EI-MS: m/z 417 [M + H]$^+$, 401 [M – CH$_3$]$^+$, 341 [M– 5CH$_3$]$^+$, 288, 259, 230, 205, 189, 158, 129, 83, 71, 59, 55.

ESI-TOF-MS: m/z 439 [M + Na]$^+$, 417 [M + H]$^+$

HR-FTICR-MS: m/z 417.1901 ([M + H])$^+$, Calcd. for C$_{23}$H$_{29}$O$_7$, 417.1907).

Reference

Yang X-W, Zhang C-Y, Zhang B-G, Lu Y, Luan J-W, Zheng Q-T. (2009). Novel coumarin and furan from the roots of *Angelica pubescens* f. *biserrata*. *J Asian Nat Prod Res* **11**: 698–703.

Apigenin-4'-*O*-(2''-*O*-*p*-coumaroyl)-β-D-glucopyranoside

IUPAC name: (*E*)-(2*S*,3*R*,4*S*,5*S*,6*R*)-2-(4-(5,7-dihydroxy-4-oxo-4*H*-chromen-2-yl)phenoxy)-4,5-dihydroxy-6-(hydroxymethyl)tetrahydro-2*H*-pyran-3-yl 3-(4-hydroxyphenyl)acrylate

Sub-class: Flavone glycoside

Chemical structure

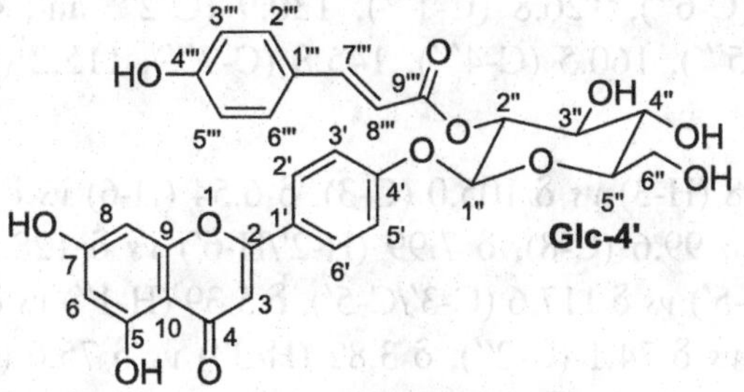

Source: *Palhinhaea cernua* (Family: Lycopodiaceae); whole plants

Molecular formula: $C_{30}H_{26}O_{12}$

Molecular weight: 578

State: Light yellow amorphous powder

Bioactivity studied: Xanthine oxidase inhibitor (IC$_{50}$ 23.95 ± 0.43 μM)

Specific rotation: $[\alpha]^{20}_{D}$ +0.17° (MeOH, *c* 0.16)

UV: λ_{max} (MeOH) 222 (sh, 2.89), 273 (2.87), 275 (2.86), 314 (3.00) nm

IR (KBr): ν_{max} 3361.8 (OH), 3249.4, 2946.7, 2899.5, 1714 (ester carbonyl), 1704.9, 1659 (γ-pyrone carbonyl), 1622.5, 1607.9, 1586.4, 1576.8, 1510, 1455, 1434.7 (aromatic unsaturaion), 1394.2 cm^{-1}

^{1}H-NMR (Acetone-d_6, 500 MHz): δ 6.68 (1H, s, H-3), 6.54 (1H, d, J = 2.1 Hz, H-6), 6.25 (1H, d, J = 2.1 Hz, H-8), 7.99 (2H, d, J = 8.7 Hz, H-2′ and H-6′), 7.20 (2H, d, J = 8.7 Hz, H-3′ and H-5′), 5.39 (1H, d, J = 8.1 Hz, H-1″), 5.17 (1H, dd, J = 9.4, 8.1 Hz, H-2″), 3.82 (1H, t, J = 9.0 Hz, H-3″), 3.61 (1H, t, J = 9.0 Hz, H-4″), 3.71 (1H, m, H-5″), 3.96 (1H, dd, J = 12.1, 1.6 Hz, H-6″a), 3.76 (1H, dd, J = 12.1, 5.8 Hz, H-6″b), 7.54 (2H, d, J = 8.5 Hz, H-2‴ and H-6‴), 6.88 (2H, d, J = 8.5 Hz, H-3‴ and H-5‴), 7.66 (1H, br d, J = 15.9 Hz, H-7‴), 6.37 (1H, br d, J = 15.9 Hz, H-8‴).

^{13}C-NMR (Acetone-d_6, 125 MHz): δ 164.2 (C-2), 105.0 (C-3), 182.7 (C-4), 158.7 (C-5), 94.7 (C-6), 164.8 (C-7), 99.6 (C-8), 163.2 (C-9), 105.2 (C-10), 126.0 (C-1′), 128.9 (C-2′ and C-6′), 117.6 (C-3′ and C-5′), 161.0 (C-4′), 99.4 (C-1″), 74.1 (C-2″), 75.6 (C-3″), 71.2 (C-4″), 78.1 (C-5″), 62.2 (C-6″), 126.8 (C-1‴), 130.9 (C-2‴ and C-6‴), 116.5 (C-3‴ and C-5‴), 160.5 (C-4‴), 145.8 (C-7‴), 115.2 (C-8‴), 166.5 (C-9‴).

HMQC: δ 6.68 (H-3) *vs* δ 105.0 (C-3), δ 6.54 (H-6) *vs* δ 94.7 (C-6), δ 6.25 (H-8) *vs* δ 99.6 (C-8), δ 7.99 (H-2′/H-6′) *vs* δ 128.9 (C-2′/C-6′), δ 7.20 (H-3′/H-5′) *vs* δ 117.6 (C-3′/C-5′), δ 5.39 (H-1″) *vs* δ 99.4 (C-1″), δ 5.17 (H-2″) *vs* δ 74.1 (C-2″), δ 3.82 (H-3″) *vs* δ 75.6 (C-3″), δ 3.61 (H-4″) *vs* δ 71.2 (C-4″), δ 3.71 (H-5″) *vs* δ 78.1 (C-5″), δ 3.96 (H-6″a) and 3.76 (H-6″b) *vs* δ 62.2 (C-6″), δ 7.54 (H-2‴/ H-6‴) *vs* δ 130.9 (C-2‴/C-6‴), δ 6.88 (H-3‴/H-5‴) *vs* δ 116.5 (C-3‴/C-5‴), δ 7.66 (H-7‴) *vs* δ 145.8 (C-7‴), δ 6.37 (H-8‴) *vs* δ 115.2 (C-8‴).

HMBC: δ 6.68 (H-3) *vs* δ δ 164.2 (C-2), 105.2 (C-10) and 126.0 (C-1′), δ 6.54 (H-6) *vs* δ 99.6 (C-8) and 105.2 (C-10), δ 6.25 (H-8) *vs* δ 105.2 (C-10), δ 7.99 (H-2′/H-6′) *vs* δ 161.0 (C-4′), δ 7.20 (H-3′/H-5′) *vs* δ 126.0 (C-1′), δ 5.39 (H-1″) *vs* δ 161.0 (C-4′), δ 5.17 (H-2″) *vs* δ 145.8 (C-7‴) and 166.5 (C-9‴), δ 7.54 (H-2‴/ H-6‴) *vs* δ 160.5 (C-4‴), δ 6.88

(H-3'''/H-5''') vs δ 126.8 (C-1'''), δ 7.66 (H-7''') vs δ 126.8 (C-1'''), 130.9 (C-2'''/C-6''') and 166.5 (C-9'''), δ 6.37 (H-8''') vs δ 166.5 (C-9''').

HR-ESI-MS: m/z 579.1200 ([M + H]$^+$, Calcd. for $C_{30}H_{27}O_{12}$, 579.1194).

Reference

Jiao RH, Ge HM, Shi DH, Tan RX. (2006). An apigenin-derived xanthine oxidase inhibitor from *Palhinhaea cernua*. *J Nat Prod* **69**: 1089–1091.

Apigenin 6-*C*-[2″-*O*-(*E*)-feruloyl-β-D-glucopyranosyl]-8-*C*-β-D-glucopyranoside

IUPAC name: (*E*)-(2*S*,3*R*,5*S*,6*R*)-2-(5,7-dihydroxy-2-(4-hydroxyphenyl)-4-oxo-8-((2*R*,3*S*,4*S*,5*R*,6*S*)-3,4,5-trihydroxy-6-(hydroxymethyl)tetrahydro-2*H*-pyran-2-yl)-4*H*-chromen-6-yl)-4,5-dihydroxy-6-(hydroxymethyl)tetrahydro-2*H*-pyran-3-yl 3-(4-hydroxy-3-methoxyphenyl)acrylate

Sub-class: Flavone glycoside

Chemical structure

Source: *Acacia pennata* Willd. (Family: Mimosaceae); leaves

Molecular formula: $C_{37}H_{38}O_{18}$

Molecular weight: 770

State: Yellow amorphous powder

Specific rotation: $[\alpha]^{20}_{D}$ −14.8° (MeOH, c 0.1)

UV: λ_{max} (MeOH) 336, 303 (sh), 269 (sh) nm

IR (KBr): v_{max} 3415 (OH), 2927, 1730 (ester carbonyl), 1654 (γ-pyrone carbonyl), 1585, 1554, 1501 (aromatic unsaturaion), 1384, 1065, 1036, 850 cm^{-1}

^{1}H-NMR (DMSO-d_6, 600 MHz): δ 6.62 (1H, s, H-3), 7.96 (2H, d, J = 8.4 Hz, H-2′ and H-6′), 6.90 (2H, d, J = 8.4 Hz, H-3′ and H-5′), 4.70 (1H, d, J = 7.4 Hz, H-1′′), 4.68* (1H, H-2′′), 3.28* (1H, H-3′′), 3.81* (1H, H-4′′), 3.30* (1H, H-5′′), 3.61* (1H, H-6′′a), 3.65* (1H, H-6′′b), 4.90 (1H, d, J = 7.6 Hz, H-1′′′), 3.85* (1H, H-2′′′), 3.32* (1H, H-3′′′), 3.37* (1H, H-4′′′), 3.26* (1H, H-5′′′), 3.53* (1H, H-6′′′a), 3.75* (1H, H-6′′′b), 3.99 (1H, br s, H-2′′′′), 6.66 (1H, d, J = 8.1 Hz, H-5′′′′), 6.78 (1H, d, J = 8.1 Hz, H-6′′′′), 6.50 (1H, d, J = 15.8 Hz, H-α), 7.54 (1H, d, J = 15.8 Hz, H-β), 3.76 (3H, s, Ar-OCH$_3$) [*multiplicity was not determined due to overlap of the respective peak].

^{13}C-NMR (DMSO-d_6, 150 MHz): δ 162.7 (C-2), 101.7 (C-3), 181.5 (C-4), 161.8 (C-5), 108.4 (C-6), 162.8 (C-7), 105.2 (C-8), 155.8 (C-9), 102.1 (C-10), 122.0 (C-1′), 128.5 (C-2′ and C-6′), 115.8 (C-3′ and C-5′), 161.0 (C-4′), 73.9 (C-1′′), 73.8 (C-2′′), 78.2 (C-3′′), 71.4 (C-4′′), 81.5(C-5′′), 60.0 (C-6′′), 74.0 (C-1′′′), 71.5 (C-2′′′), 79.1 (C-3′′′), 70.8 (C-4′′′), 81.0 (C-5′′′), 61.9 (C-6′′′), 126.5 (C-1′′′′), 111.5 (C-2′′′′), 147.0 (C-3′′′′), 149.9 (C-4′′′′), 114.2 (C-5′′′′), 120.0 (C-6′′′′), 115.6 (C-α), 144.9 (C-β), 166.0 (CO), 55.6 (Ar-OCH$_3$).

HMQC: δ 6.62 (H-3) *vs* δ 162.7 (C-3), δ 7.96 (H-2′/H-6′) *vs* δ 128.5 (C-2′/C-6′), δ 6.90 (H-3′/H-5′) *vs* δ 115.8 (C-3′/C-5′), δ 4.70 (H-1′′) *vs* δ 73.9 (C-1′′), δ 4.68 (H-2′′) *vs* δ 73.8 (C-2′′), δ 3.28 (H-3′′) *vs* δ 78.2 (C-3′′), δ 3.81 (H-4′′) *vs* δ 71.4 (C-4′′), δ 3.30 (H-5′′) *vs* δ 81.5 (C-5′′), δ 3.61 (H-6′′a) and 3.65 (H-6′′b) *vs* δ 60.0 (C-6′′), δ 4.90 (H-1′′′) *vs* δ 74.0 (C-1′′′), δ 3.85 (H-2′′′) *vs* δ 71.5 (C-2′′′), δ 3.32 (H-3′′′) *vs* δ 79.1 (C-3′′′), δ 3.37 (H-4′′′) *vs* δ 70.8 (C-4′′′), δ 3.26 (H-5′′′) *vs* δ 81.0 (C-5′′′), δ 3.53 (H-6′′′a) and 3.75 (H-6′′′b) *vs* δ 61.9 (C-6′′′), δ 3.99 (H-2′′′′) *vs* δ 111.5

(C-2''''), δ 6.66 (H-5'''') *vs* δ 114.2 (C-5''''), δ 6.78 (H-6'''') *vs* δ 120.0 (C-6''''), δ 6.50 (H-α) *vs* δ 115.6 (C-α), δ 7.54 (H-β) *vs* δ 144.9 (C-β), δ 3.76 (Ar-OCH_3) *vs* δ 55.6 (Ar-OCH_3).

HR-ESI-MS: *m/z* 793.6771 ([M + Na]$^+$, Calcd. for $C_{37}H_{38}O_{18}Na$, 793.6766).

Reference

Dongmo AB, Miyamoto T, Yoshikawa K, Arihara S, Lacaille-Dubois M-A. (2007). Flavonoids from *Acacia pennata* and their cyclooxygenase (COX-1 and COX-2) inhibitory activities. *Planta Med* **73**: 1202–1207.

Apigenin 8-C-[α-L-rhamnopyranosyl-(1→4)]-α-D-glucopyranoside

IUPAC name: 8-((2R,3R,4R,5S,6R)-3,4-dihydroxy-6-(hydroxymethyl)-5-((((2S,3R,4R,5R,6S)-3,4,5-trihydroxy-6-methyltetrahydro-2H-pyran-2-yl)oxy)tetrahydro-2H-pyran-2-yl)-5,7-dihydroxy-2-(4-hydroxyphenyl)-4H-chromen-4-one

Sub-class: Flavone C-glycoside

Chemical structure

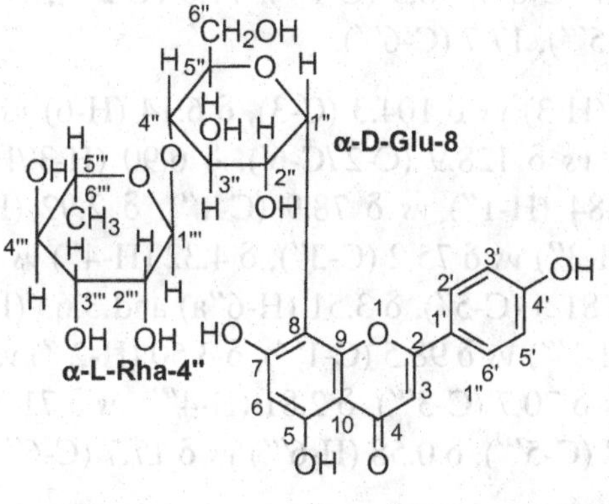

Source: *Diospyros kaki* (Family: Ebenaceae); leaves

Molecular formula: $C_{27}H_{30}O_{14}$

Molecular weight: 578

State: Yellow needles

Melting point: 231–232 °C

Specific rotation: $[\alpha]^{24}_{D}$ +14.5° (MeOH, c 0.20)

UV (MeOH): λ_{max} (log ε) 240 (2.78), 269 (2.95), 310 (3.58) nm

FT-IR (KBr): v_{max} 3396 (OH), 2930, 1678 (α,β-unsaturated carbonyl), 1467 cm^{-1}

^{1}H-NMR (DMSO-d_6, 600 MHz): δ 6.75 (1H, s, H-3), 6.14 (1H, s, H-6), 7.85 (2H, d, J = 8.6 Hz, H-2′ and H-6′), 6.90 (2H, d, J = 8.6 Hz, H-3′ and H-5′), 13.0 (1H, s, 5-OH), 10.5 (2H, br s, 7-OH and 4′-OH), 5.84 (1H, d, J = 2.3 Hz, H-1″), 2.92 (1H, m, H-2″), 3.84 (1H, m, H-3″), 4.32 (1H, m, H-4″), 4.21 (1H, br d, J = 7.2 Hz, H-5″), 3.51 (1H, br s, H-6″a), 3.63 (1H, br d, J = 7.2 Hz, H-6″b), 4.61 (1H, br s, H-1‴), 3.50 (1H, m, H-2‴), 3.41 (1H, m, H-3‴), 2.81 (1H, m, H-4‴), 2.13 (1H, m, H-5‴), 0.58 (3H, br s, H-6‴).

^{13}C-NMR (DMSO-d_6, 150 MHz): δ 161.6 (C-2), 104.3 (C-3), 182.3 (C-4), 160.4 (C-5), 99.5 (C-6), 160.4 (C-7), 103.4 (C-8), 163.4 (C-9), 103.4 (C-10), 121.9 (C-1′), 128.9 (C-2′), 116.5 (C-3′), 164.2 (C-4′), 116.5 (C-5′), 128.9 (C-6′), 78.9 (C-1″), 72.0 (C-2″), 75.2 (C-3″), 80.9 (C-4″), 81.5 (C-5″), 64.4 (C-6″), 98.5 (C-1‴), 71.0 (C-2‴), 70.7 (C-3‴), 71.6 (C-4‴), 68.8 (C-5‴), 17.7 (C-6‴).

HMQC: δ 6.75 (H-3) vs δ 104.3 (C-3), δ 6.14 (H-6) vs δ 99.5 (C-6), δ 7.85 (H-2′/H-6′) vs δ 128.9 (C-2′/C-6′), δ 6.90 (H-3′/H-5′) vs δ 116.5 (C-3′/C-5′), δ 5.84 (H-1″) vs δ 78.9 (C-1″), δ 2.92 (H-2″) vs δ 72.0 (C-2″), δ 3.84 (H-3″) vs δ 75.2 (C-3″), δ 4.32 (H-4″) vs δ 80.9 (C-4″), δ 4.21 (H-5″) vs δ 81.5 (C-5″), δ 3.51 (H-6″a) and 3.63 (H-6″b) vs δ 64.4 (C-6″), δ 4.61 (H-1‴) vs δ 98.5 (C-1‴), δ 3.50 (H-2‴) vs δ 71.0 (C-2‴), δ 3.41 (H-3‴) vs δ 70.7 (C-3‴), δ 2.81 (H-4‴) vs δ 71.6 (C-4‴), δ 2.13 (H-5‴) vs δ 68.8 (C-5‴), δ 0.58 (H-6″) vs δ 17.7 (C-6‴).

HMBC: δ 6.75 (H-3) vs δ 121.9 (C-1′), δ 6.14 (H-6) vs δ 160.4 (C-5), 103.4 (C-8) and 103.4 (C-10), δ 7.85 (H-2′) vs δ 161.6 (C-2) and 164.2 (C-4′), δ 6.90 (H-3′) vs δ 121.9 (C-1′), δ 5.84 (H-1″) vs δ 103.4 (C-8), δ 3.84 (H-3″) vs δ 78.9 (C-1″), δ 4.32 (H-4″) vs δ 72.0 (C-2″), δ 4.21 (H-5″) vs δ 64.4 (C-6″), δ 3.51 (H-6″a) and 3.63 (H-6″b) vs δ 81.5 (C-5″), δ 4.61 (H-1‴) vs δ 80.9 (C-4″), 71.0 (C-2‴) and 70.7 (C-3‴), δ

2.13 (H-5‴) *vs* δ 70.7 (C-3‴) and 17.7 (C-6‴) (selected 2D-correlations were shown).

ESI-MS: *m/z* 579 (100) $[M + H]^+$.

HR-ESI-MS: *m/z* 579.1721 ($[M + H]^+$, Calcd. for $C_{27}H_{31}O_{14}$, 579.1714).

Reference

Chen G, Wei S-H, Huang J, Sun J. (2009). A novel *C*-glycosylflavone from the leaves of *Diospyros kaki*. *J Asian Nat Prod* **11**: 503–507.

Apigenosylide C

IUPAC name: (3'S,5'R,10R)-10-Decyl-5-hydroxy-5'-methyl-2-(4-((((2R,3S,4R,5R,6S)-3,4,5-trihydroxy-6-(hydroxymethyl)tetrahydro-2H-pyran-2-yl)oxy)phenyl)-2'H-spiro[[1,2]dioxino[3,4-h]chromene-9,3'-furan]-2',4,4'(5'H,10H)-trione

Sub-class: Flavone glycoside

Chemical structure

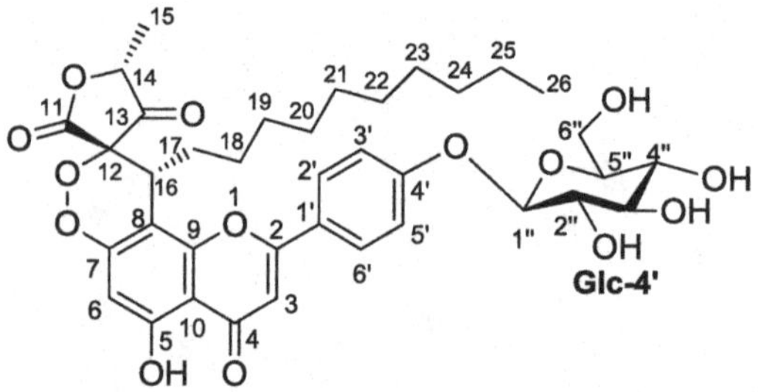

Source: *Machilus japonica* Sieb. & Zucc. Var. *kusanoi* (Hayata) Liao (Family: Lauraceae); leaves

Molecular formula: $C_{37}H_{44}O_{14}$

Molecular weight: 712

State: Yellow solid

Specific rotation: $[\alpha]^{27}_{D}$ −25° (MeOH, *c* 0.2)

UV-vis (MeOH): λ_{max} (log ε) 259 (4.14), 277 (4.37), 341 (4.47) nm

IR (KBr): v_{max} 3418 (OH), 2923, 2852, 1650, 1605, 1507, 1361, 1237, 1184, 1073, 833 cm^{-1}

^{1}H-NMR (CDCl$_3$, 400 MHz): δ 6.64 (1H, s, H-3), 6.22 (1H, s, H-6), 4.50 (1H, q, J = 7.0 Hz, H-14), 1.38 (3H, d, J = 6.6 Hz, H-15), 4.64 (1H, t, J = 8.4 Hz, H-16), 1.9 (2H, m, H-17), 1.20 (14H, m, H-18-24), 1.27 (2H, m, H-25), 0.87 (3H, t, J = 7.0 Hz, H-26), 8.30 (2H, d, J = 8.8 Hz, H-2′ and H-6′), 7.28 (2H, d, J = 8.8 Hz, H-3′ and H-5′), 5.05 (1H, d, J = 7.5 Hz, H-1″), 3.52 (1H, m, H-2″), 3.44 (1H, m, H-3″), 3.65 (1H, m, H-4″), 3.50 (1H, m, H-5″), 3.75 (1H, dd, J = 12.0, 5.1 Hz, H-6″a), 3.88 (1H, m, H-6″b).

^{13}C-NMR (CDCl$_3$, 100 MHz): δ 166.2 (C-2), 103.2 (C-3), 184.5 (C-4), 160.8 (C-5), 103.2 (C-6), 166.2 (C-7), 112.9 (C-8), 156.3 (C-9), 105.3 (C-10), 181.0 (C-11), 97.6 (C-12), 191.3 (C-13), 77.3 (C-14), 18.3 (C-15), 29.1 (C-16), 31.5 (C-17), 29.1 (C-18), 30.0 (C-19), 30.4 (C-20), 30.6 (C-21), 30.64 (C-22), 30.7 (C-23), 33.1 (C-24), 23.7 (C-25), 14.4 (C-26), 126.5 (C-1′), 129.9 (C-2′), 118.0 (C-3′), 162.0 (C-4′), 118.0 (C-5′), 129.9 (C-6′), 101.8 (C-1″), 74.8 (C-2″), 78.0 (C-3″), 71.2 (C-4″), 78.2 (C-5″), 62.4 (6″).

HR-FAB-MS: m/z 712.2728 ([M]$^+$, Calcd. for C$_{37}$H$_{44}$O$_{14}$, 712.2731).

Reference

Lee S-S, Lin Y-S, Chen C-K. (2009). Three adducts of butenolide and apigenin glycoside from the leaves of *Machilus japonica*. *J Nat Prod* **72**: 1249–1252.

Aquisiflavoside [4′,5-dihydroxy-3′, 7-dimethoxyflavone 5-*O*-β-D-xylopyranosyl-(1→6)-β-D-glucopyranoside]

IUPAC name: 2-(4-Hydroxy-3-methoxyphenyl)-7-methoxy-5-(((2*S*,3*R*,4*S*, 5*S*,6*R*)-3,4,5-trihydroxy-6-((((2*R*,3*R*,4*S*,5*R*)-3,4,5-trihydroxytetrahydro-2*H*-pyran-2-yl)oxy)methyl)tetrahydro-2*H*-pyran-2-yl)oxy)-4*H*-chromen-4-one

Sub-class: Flavone gylcoside

Chemical structure

Source: *Aquilaria sinensis* (Lour.) Gilg (Family: Thymelaeaceae); leaves

Molecular formula: $C_{28}H_{32}O_{15}$

Molecular weight: 608

State: Organge amorphous powder

Specific rotation: $[\alpha]^{28}_{D}$ −82.1° (MeOH, *c* 0.5)

Bioactivity studied: Nitric oxide (NO) production inhibitor (IC_{50} 34.95 μM)

UV (MeOH): λ_{max} 262, 328, 402 nm

IR (KBr): ν_{max} 3425 (OH), 2940, 2926, 1633 (α,β-unsaturated carbonyl), 1458, 1280, 1158, 1007, 721, 668 cm^{-1}.

^{1}H-NMR (DMSO-d_6, 300 MHz): δ 6.78 (1H, s, H-3), 6.86 (1H, s, H-6), 7.06 (2H, d, *J* = 2.5 Hz, H-8), 7.53 (1H, d, *J* = 1.5 Hz, H-2′), 6.94 (1H, d, *J* = 8.0 Hz, H-5′), 7.54 (1H, dd, *J* = 8.0, 1.5 Hz, H-6′), 8.41 (1H, s, 4′-OH), 3.89 (3H, s, 7-OCH_3), 3.87 (3H, s, 3′-OCH_3), 4.77 (1H d, *J* = 7.5 Hz, H-1″), 3.29 (1H, dd, *J* = 8.5, 7.5 Hz, H-2″), 3.56 (1H, dd, *J* = 9.0, 8.5 Hz, H-3″), 3.20 (1H, dd, *J* = 9.0, 8.5 Hz, H-4″), 3.35 (1H, dd, *J* = 8.5, 5.5 Hz, H-5″), 3.01 (1H, br d, *J* = 11.0 Hz, H-6″a), 3.68 (1H, dd, *J* = 11.0, 5.5 Hz,, H-6″b), 4.18 (1H, d, *J* = 7.5 Hz, H-1‴), 2.97 (1H, dd, *J* = 8.5, 5.5 Hz, H-2‴), 3.10 (1H, br d, *J* = 8.5 Hz, H-3‴), 3.27 (1H, ddd, *J* = 11.0, 8.5, 5.5 Hz,, H-4‴), 3.64 (1H, br d, *J* = 11.0 Hz,, H-5‴a), 3.97 (1H, dd, *J* = 11.0, 5.0 Hz,, H-5‴b).

^{13}C-NMR (DMSO-d_6, 75 MHz): δ 161.4 (C-2), 106.0 (C-3), 176.9 (C-4), 158.1 (C-5), 102.9 (C-6), 163.6 (C-7), 96.7 (C-8), 158.5 (C-9), 109.2 (C-10), 120.2 (C-1′), 110.0 (C-2′), 148.2 (C-3′), 150.8 (C-4′), 115.8 (C-5′), 120.6 (C-6′), 103.7 (C-1″), 73.4 (C-2″), 75.6 (C-3″), 69.6 (C-4″), 76.0 (C-5″), 68.7 (C-6″), 104.2 (C-1‴), 73.4 (C-2‴), 76.6 (C-3‴), 69.8 (C-4‴), 65.7 (C-5‴), 56.2 (7-OCH_3), 56.0 (3′-OCH_3).

HMBC: δ 6.78 (H-3) *vs* δ 120.2 (C-1′), δ 7.06 (H-8) *vs* δ 102.9 (C-6), 158.5 (C-9) and 109.2 (C-10), δ 3.89 (7-OCH_3) *vs* δ 163.6 (C-7), δ 7.53 (H-2′) *vs* δ 150.8 (C-4′), δ 3.87 (3′-OCH_3) *vs* δ 148.2 (C-3′), δ 6.94 (H-5′) *vs* δ 120.2 (C-1′) and 148.2 (C-3′), δ 7.54 (H-6′) *vs* δ 161.4 (C-2) and 150.8 (C-4′), δ 4.77 (H-1″) *vs* δ 158.1 (C-5), 75.6 (C-3″) and 76.0 (C-5″),

δ 4.18 (H-1‴) *vs* δ 68.7 (C-6″), 76.6 (C-3‴) and 65.7 (C-5‴) (selected 2D-correlations were shown).

HR-ESI-MS (+ve mode): *m/z* 609.1816 ([M + H]$^+$, Calcd. for $C_{28}H_{33}O_{15}$, 609.1814); *m/z* 631.1613 ([M + Na]$^+$, Calcd. for $C_{28}H_{32}O_{15}Na$, 631.1633).

Reference

Yang X-B, Feng J, Yang X-W, Zhao B, Liu J-X. (2012). Aquisiflavoside, a new nitric oxide production inhibitor from the leaves of *Aquilaria sinensis*. *J Asian Nat Prod Res* **114**: 867–872.

Artelasticinol

IUPAC name: 5,7-Dihydroxy-3-(1-hydroxy-3-methylbut-2-en-1-yl)-2-(2-hydroxy-4-methoxyphenyl)-8-(3-methylbut-2-en-1-yl)-4*H*-chromen-4-one

Sub-class: Prenylated flavone

Chemical structure

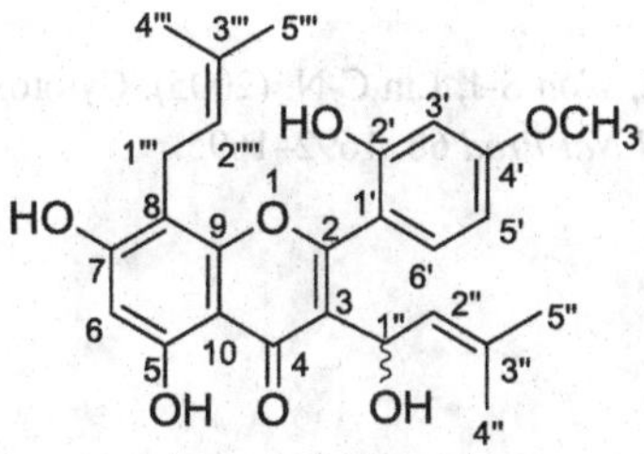

Source: *Artocarpus elasticus* (Family: Moraceae); root bark

Molecular formula: $C_{26}H_{28}O_7$

Molecular weight: 452

State: Orange gum

UV: λ_{max} (MeOH) (log ε): 215 (3.90), 273 (3.82), 295 (sh, 3.32), 370 (3.30) nm

IR (KBr): ν_{max} 3420 (OH), 1651 (α,β-unsaturated carbonyl), 1600 cm^{-1}

^{1}H-NMR (CDCl$_3$, 400 MHz): δ 6.32 (1H, s, H-6), 6.48 (1H, d, J = 2.4 Hz, H-3'), 6.61 (1H, dd, J = 8.8, 2.4 Hz, H-5'), 7.66 (1H, d, J = 8.8 Hz,

33

H-6′), 3.84 (3H, s, 4′-OCH_3), 6.27 (1H, d, J = 9.2 Hz, H-1″), 5.45 (1H, d, J = 9.2 Hz, H-2″), 1.71 (3H, s, H-4″), 1.98 (3H, s, H-5″), 3.57 (2H, d, J = 6.4 Hz, H-1‴), 5.30 (1H, t, J = 6.4 Hz, H-2‴), 1.76 (3H, s, H-4‴), 1.87 (3H, s, H-5‴), 12.78 (1H, s, 5-OH).

^{13}C-NMR (CDCl$_3$, 100 MHz): δ 158.1 (C-2), 109.8 (C-3), 178.8 (C-4), 154.0 (C-5), 99.8 (C-6), 164.5 (C-7), 109.5 (C-8), 160.1 (C-9), 105.5 (C-10), 108.7 (C-1′), 155.3 (C-2′), 102.2 (C-3′), 160.3 (C-4′), 109.1 (C-5′), 139.3 (C-6′), 69.9 (C-1″), 121.0 (C-2″), 134.6 (C-3″), 25.9 (C-4″), 18.6 (C-5″), 21.8 (C-1‴), 124.9 (C-2‴), 134.7 (C-3‴), 25.7 (C-4‴), 18.0 (C-5‴), 55.6 (4′-OCH_3).

EIMS: m/z (%rel.) 452 ([M]$^+$, 100), 435 ([M − H$_2$O]$^+$, 44), 409 ([M − C$_3$H$_6$]$^+$, 78), 395 (87), 353 (48), 219 (19), 165 (73), 69 (88), 55 (86).

HR-EIMS: m/z 452.1830 ([M]$^+$, Calcd. for C$_{26}$H$_{28}$O$_7$, 452.1835).

Reference

Ko H-H, Lu Y-H, Yang S-Z, Won S-J, Lin C-N. (2005). Cytotoxic prenylflavonoids from *Artocarpus elasticus*. *J Nat Prod* **68**: 1692–1695.

Artelastoheterol
[5,2',4',5'-tetrahydroxy-6,7-(2,2-dimethyl-6*H*-pyrano)-8-prenyl-3-(9-hydroxy)prenyl-flavone]

IUPAC name: 5-Hydroxy-3-(1-hydroxy-3-methylbut-2-en-1-yl)-8,8-dimethyl-10-(3-methylbut-2-en-1-yl)-2-(2,4,5-trihydroxyphenyl) pyrano[3,2-*g*]chromen-4(8*H*)-one

Sub-class: Prenylated flavone

Chemical structure

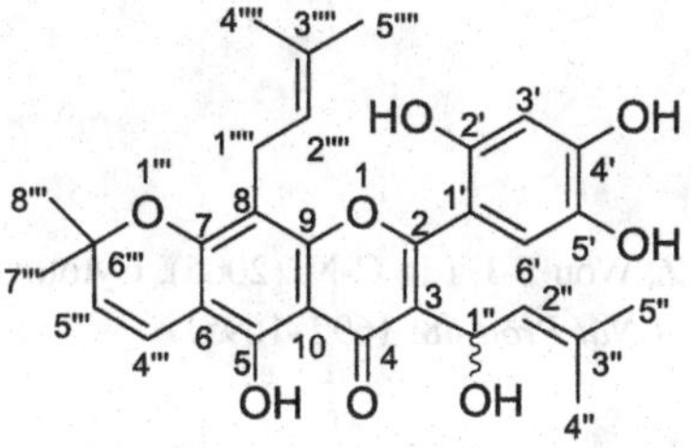

Source: *Artocarpus elasticus* (Family: Moraceae); root bark

Molecular formula: $C_{30}H_{32}O_8$

Molecular weight: 520

State: Orange gum

UV: λ_{max} (MeOH) (log ε): 210 (4.49), 277 (4.40), 290 (sh, 3.95), 315 (3.61) nm

IR (KBr): v_{max} 3395 (OH), 1652 (α,β-unsaturated carbonyl), 1620 cm^{-1}

^{1}H-NMR (CDCl$_3$, 400 MHz): δ 6.45 (1H, s, H-3′), 7.25 (1H, s, H-6′), 6.18 (1H, d, $J = 9.2$ Hz, H-1″), 5.44 (1H, d, $J = 9.2$ Hz, H-2″), 1.67 (3H, s, H-4″), 1.92 (3H, s, H-5″), 6.69 (1H, d, $J = 10.0$ Hz, H-4‴), 5.59 (1H, d, $J = 10.0$ Hz, H-5‴), 1.45 (3H, s, H-7‴), 1.44 (3H, s, H-7‴), 3.45 (2H, m, H-1⁗), 5.22 (1H, d, $J = 6.4$ Hz, H-2⁗), 1.66 (3H, s, H-4⁗), 1.81 (3H, s, H-5⁗), 12.80 (1H, s, 5-OH).

^{13}C-NMR (CDCl$_3$, 100 MHz): δ 151.5 (C-2), 109.7 (C-3), 178.8 (C-4), 154.4 (C-5), 105.2 (C-6), 156.5 (C-7), 107.8 (C-8), 153.6 (C-9), 105.3 (C-10), 107.9 (C-1′), 155.4 (C-2′), 104.8 (C-3′), 149.7 (C-4′), 138.9 (C-5′), 109.3 (C-6′), 69.3 (C-1″), 121.0 (C-2″), 139.1 (C-3″), 25.9 (C-4″), 18.6 (C-5″), 115.9 (C-4‴), 127.9 (C-5‴), 77.7 (C-6‴), 28.1 (C-7‴), 28.2 (C-8‴), 21.5 (C-1⁗), 122.1 (C-2⁗), 131.7 (C-3⁗), 25.7 (C-4⁗), 18.1 (C-5⁗).

EIMS: m/z (%rel.) 518 ([M − 2]$^+$, 9), 502 ([M − H$_2$O]$^+$, 48), 447 ([M − H$_2$O − C$_4$H$_7$]$^+$, 100), 391 ([M − H$_2$O − (C$_4$H$_7$)$_2$ − H]$^+$, 18), 261 (21), 216 (7), 205 (29), 153 (17).

HR-EIMS: m/z 502.1994 ([M − H$_2$O]$^+$, Calcd. for C$_{30}$H$_{30}$O$_7$, 502.1994).

Reference

Ko H-H, Lu Y-H, Yang S-Z, Won S-J, Lin C-N. (2005). Cytotoxic prenylflavonoids from *Artocarpus elasticus. J Nat Prod* **68**: 1692–1695.

Asphodelin A 4'-*O*-β-D-glucoside

IUPAC name: 4,7-Dihydroxy-3-(2-hydroxy-4-(((2*S*,3*R*,4*S*,5*S*,6*R*)-3,4,5-trihydroxy-6-(hydroxymethyl)tetrahydro-2*H*-pyran-2-yl)oxy)phenyl)-2*H*-chromen-2-one

Sub-class: 3-Aryl-substituted coumarin glycoside

Chemical structure

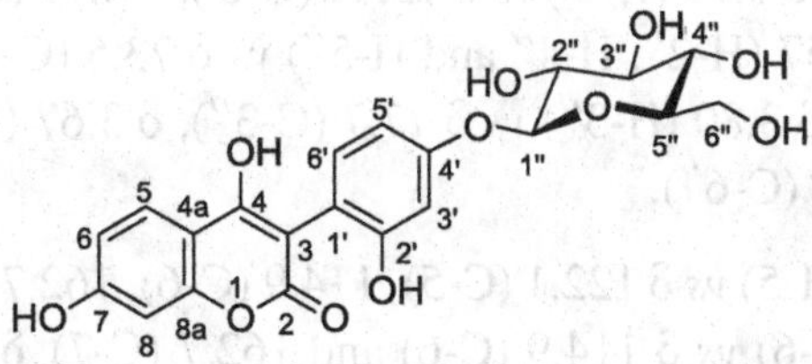

Source: *Asphodelus microcarpus* (Family: Asphodelaceae/Liliaceae); bulbs and roots

Molecular formula: $C_{21}H_{20}O_{11}$

Molecular weight: 448

State: Yellowish white solid

Melting point: 143–146 °C

Bioactivity studied: Antimicrobial

Specific rotation: $[\alpha]_D^{22}$ +5.7° (MeOH, *c* 0.05)

UV (MeOH): λ_{max} (logε) 208 (4.39), 242 (3.58), 325 (3.71) nm

IR (KBr): v_{max} 3425 (OH), 1696 (C=O), 1635 (C=C), 1616, 1540 (aromatic unsaturaion) cm^{-1}

^{1}H-NMR (Acetone-d_6, 400 MHz): δ 7.83 (1H, d, J = 8.3 Hz, H-5), 6.89 (1H, dd, J = 8.3, 2.2 Hz, H-6), 6.85 (1H, d, J = 2.2 Hz, H-8), 7.14 (1H, d, J = 2.2 Hz, H-3'), 7.01 (1H, dd, J = 8.3, 2.2 Hz, H-5'), 7.77 (1H, d, J = 8.3 Hz, H-6'), 4.99 (1H, d, J = 7.3 Hz, H-1''), 3.46-3.57 (3H, m, H-2'', H-4'' and H-5''), 3.80 (1H, m, H-3''), 3.67 (1H, dd, J = 11.0, 2.8, 2.5 Hz, H-6''a), 3.80 (1H, dd, J = 11.0, 5.5, 5.3 Hz, H-6''b).

^{13}C-NMR (Acetone-d_6, 100 MHz): δ 160.5 (C-2), 103.8 (C-3), 161.8 (C-4), 106.1 (C-4a), 122.1 (C-5), 114.9 (C-6), 162.7 (C-7), 104.2 (C-8), 156.4 (C-8a), 118.4 (C-1'), 158.0 (C-2'), 100.4 (C-3'), 157.5 (C-4'), 115.6 (C-5'), 125.4 (C-6'), 102.3 (C-1''), 73.5 (C-2''), 77.7 (C-3''), 70.7 (C-4''), 77.5 (C-5''), 61.4 (C-6'').

HMQC: δ 7.83 (H-5) *vs* δ 122.1 (C-5), δ 6.89 (H-6) *vs* δ 114.9 (C-6), δ 6.85 (H-8) *vs* δ 104.2 (C-8), δ 7.14 (H-3'') *vs* δ 100.4 (C-3''), δ 7.01 (H-5') *vs* δ 115.6 (C-5'), δ 7.77 (H-6') *vs* δ 125.4 (C-6'), δ 4.99 (H-1'') *vs* δ 102.3 (C-1''), δ 3.46-3.57 (H-2'', H-4'' and H-5'') *vs* δ 73.5 (C-2''), 70.7 (C-4'') and 77.5 (C-5''), δ 3.80 (H-3'') *vs* δ 77.7 (C-3''), δ 3.67 (H-6''a) and 3.80 (H-6''b) *vs* δ 61.4 (C-6'').

HMBC: δ 7.83 (H-5) *vs* δ 122.1 (C-5), 114.9 (C-6), 162.7 (C-7) and 106.1 (C-4a), δ 6.89 (H-6) *vs* δ 114.9 (C-6) and 162.7 (C-7), δ 6.85 (H-8) *vs* δ 162.7 (C-7) and 156.4 (C-8a), δ 7.01 (H-5') *vs* δ 157.5 (C-4') and 125.4 (C-6'), δ 4.99 (H-1'') *vs* δ 157.5 (C-4') (selected HMBC correlations were shown).

HREIMS: m/z 448.1071 ([M]$^+$, Calcd. for $C_{21}H_{20}O_{11}$, 448.1066).

Reference

El-Seedi HR. (2007). Antimicrobial arylcoumarins from *Asphodelus microcarpus*. *J Nat Prod* **70**: 118–120.

Bismurrangatin

IUPAC name: 8,8′-(Oxybis(2-hydroxy-3-methylbut-3-ene-1,1-diyl)) bis(7-methoxy-2H-chromen-2-one)

Sub-class: Biscoumarin

Chemical structure

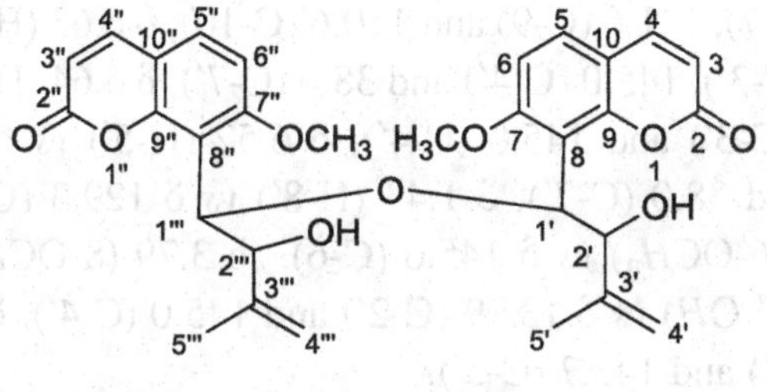

Source: *Murraya exotica* (Family: Rutaceae); branches

Molecular formula: $C_{30}H_{30}O_9$

Molecular weight: 534

State: Colorless oil

Specific rotation: $[\alpha]_D$ +2.5° (MeOH, c .014)

UV (MeOH): λ_{max} 204, 264, 316 nm

FT-IR (CHCl$_3$): ν_{max} 3510 (OH), 1730 (lactone carbonyl), 1608 (aromatic unsaturation) cm^{-1}

^{1}H-NMR (CDCl$_3$, 400 MHz): δ 5.95 (1H, d, J = 9.5 Hz, H-3), 7.23 (1H, d, J = 9.5 Hz, H-4), 6.95 (1H, d, J = 8.4 Hz, H-5), 6.57 (1H, d, J = 8.4 Hz,

H-6), 3.91 (3H, s, 7-OCH_3), 5.44 (1H, d, J = 8.1 Hz, H-1′), 5.00 (1H, d, J = 8.1 Hz, H-2′), 4.72 (2H, s, H-4′), 1.72 (3H, s, H-5′), 6.01 (1H, d, J = 9.5 Hz, H-3″), 7.33 (1H, d, J = 9.5 Hz, H-4″), 7.00 (1H, d, J = 8.8 Hz, H-5″), 6.54 (1H, d, J = 8.8 Hz, H-6″), 3.88 (3H, s, 7″-OCH_3), 5.10 (1H, d, J = 8.8 Hz, H-1‴), 5.02 (1H, d, J = 8.8 Hz, H-2‴), 4.63 (1H, s, H-4‴a), 4.60 (1H, s, H-4‴b), 1.69 (3H, s, H-5‴).

^{13}C-NMR (CDCl$_3$, 100 MHz): δ 160.4 (C-2), 112.4 (C-3), 143.0 (C-4), 128.1 (C-5), 106.9 (C-6), 160.9 (C-7), 114.4 (C-8), 153.5 (C-9), 111.9 (C-10), 78.7 (C-1′), 77.3 (C-2′), 143.8 (C-3′), 113.9 (C-4′), 17.5 (C-5′), 56.0 (7-OCH_3), 160.1 (C-2″), 112.3 (C-3″), 143.1 (C-4″), 128.2 (C-5″), 107.6 (C-6″), 161.5 (C-7″), 114.9 (C-8″), 152.2 (C-9″), 111.5 (C-10″), 56.2 (7″-OCH_3), 76.9 (C-1‴), 76.8 (C-2‴), 143.4 (C-3‴), 114.0 (C-4‴), 17.4 (C-5‴).

HMBC: δ 7.66 (H-4) vs δ 160.6 (C-2), 129.3 (C-3), 104.3 (C-5), 141.6 (C-9), 110.6 (C-10) and 38.0 (C-7′), δ 7.02 (H-5) vs δ 138.5 (C-4), 145.6 (C-6), 143.0 (C-7), 141.6 (C-9) and 110.6 (C-10), δ 6.63 (H-2′) vs δ 118.1 (C-1′), 114.9 (C-3′), 145.0 (C-4′) and 38.0 (C-7′), δ 6.64 (H-5′) vs δ 118.1 (C-1′), 114.9 (C-3′) and 145.0 (C-4′), δ 6.52 (H-5′) vs δ 118.1 (C-1′), 145.0 (C-4′) and 38.0 (C-7′), δ 1.43 (H-8′) vs δ 129.3 (C-3) and 118.1 (C-1′), δ 3.79 (6-OCH_3) vs δ 145.6 (C-6), δ 3.79 (8-OCH_3) vs δ 134.5 (C-8), δ 8.75 (3′-OH) vs δ 135.9 (C-2′) and 145.0 (C-4′), δ 8.72 (4′-OH) vs δ 114.9 (C-3′) and 143.7 (C-5′).

EI-MS: m/z (rel%) 463 (15), 259 (71), 242 (58), 231 (82), 205 (100, base peak), 203 (22), 189 (71), 131 (25).

HR-FAB-MS: m/z 557.1780 ([M + Na])$^+$, Calcd. for C$_{30}$H$_{30}$O$_9$Na, 557.1788).

Reference

Negi N, Ochi A, Kurosawa M, Ushijima K, Kitaguchi Y, Kusakabe E, Okasho F, Kimachi T, Teshima N, Ju-Ichi M, Abou-Douh A M, Ito C, Furukawa H. (2005). Two new dimeric coumarins isolated from *Murraya exotica*. *Chem Pharm Bull* **53**: 1180–1182.

Brosimacutin K
[(2S)-3′,4′-dihydroxy-7,8-(2,2-dimethyl-3-hydroxy-2,3-dihydro-4H-pyrano)-flavan]

IUPAC name: 4-((2S)-9-Hydroxy-8,8-dimethyl-2,3,4,8,9,10-hexahydropyrano[2,3-f]chromen-2-yl)benzene-1,2-diol

Sub-class: Dihydropyrano-fused flavan derivative

Chemical structure

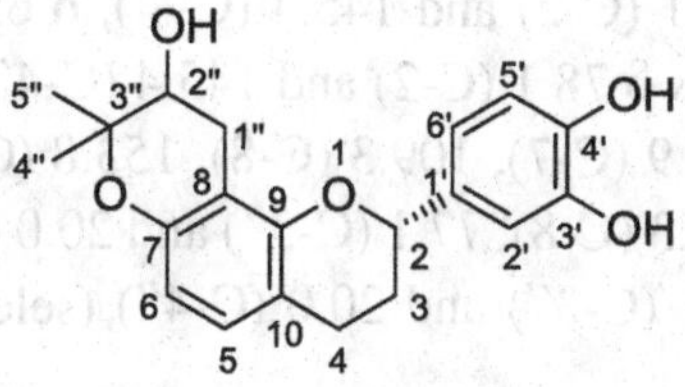

Source: *Brosimum acutifolium* Huber (Family: Moraceae); barks

Molecular formula: $C_{20}H_{22}O_5$

Molecular weight: 342

State: Colorless amorphous solid

Specific rotation: $[\alpha]^{22}_D$ −252.5° (MeOH, c 0.12)

Bioactivity studied: Cytotoxic against murine leukemia P388 cells

UV: λ_{max} (MeOH) (log ε): 211 (4.70), 283 (3.93) nm

IR (KBr): ν_{max} 3422 (OH), 2926, 1614, 1595, 1520, 1487, 1444, 1383, 1321 (aromatic unsaturaion), 1265, 1213, 1159, 1086, 1022 1083, 962 cm^{-1}

^{1}H-NMR (Acetone-d_6, 500 MHz): δ 4.96 (1H, dd, J = 10.0, 2.3 Hz, H-2β), 1.91 (1H, m, H-3a), 2.16 (1H, m, H-3b), 2.65 (1H, ddd, J = 4.1, 15.8 Hz, H-4a), 2.90 (1H, m, H-4b), 6.79 (1H, d, J = 8.3 Hz, H-5), 6.27 (1H, d, J = 8.3 Hz, H-6), 6.97 (1H, d, J = 1.7 Hz, H-2′), 6.83 (1H, d, J = 8.0 Hz, H-5′), 6.79 (1H, dd, J = 8.0, 1.7 Hz, H-6′), 2.47 (1H, dd, J = 17.1, 8.2 Hz, H-1″a), 2.92 (1H, m, H-1″b), 3.75 (1H, dd, J = 8.2, 5.6 Hz, H-2″), 1.19 (3H, s, H-4″), 1.33 (3H, s, H-5″).

^{13}C-NMR Acetone-d_6, 125 MHz): δ 78.1 (C-2), 30.9 (C-3), 25.2 (C-4), 128.1 (C-5), 109.5 (C-6), 152.9 (C-7), 109.3 (C-8), 153.8 (C-9), 113.5 (C-10), 134.9 (C-1′), 114.0 (C-2′), 145.9 (C-3′), 145.4 (C-4′), 115.9 (C-5′), 118.2 (C-6′), 27.4 (C-1″), 70.0 (C-2″), 77.1 (C-3″), 20.0 (C-4″), 26.1 (C-5″).

HMBC: δ 4.96 (H-2β) *vs* δ 134.9 (C-1′), 114.0 (C-2′) and 118.2 (C-6′), δ 2.65 (H-4a) and 2.90 (H-4b) *vs* δ 128.1 (C-5), 153.8 (C-9) and 113.5 (C-10), δ 6.79 (H-5) *vs* δ 25.2 (C-4), 152.9 (C-7), 153.8 (C-9) and 113.5 (C-10), δ 6.27 (H-6) *vs* δ 152.9 (C-7), 109.3 (C-8) and 113.5 (C-10), δ 6.97 (H-2′) *vs* δ 78.1 (C-2) and 145.4 (C-4′), δ 6.83 (H-5′) *vs* δ 134.9 (C-1′), δ 6.79 (H-6′) *vs* δ 78.1 (C-2) and 145.4 (C-4′), δ 2.47 (H-1″a) and 2.92 (H-1″b) *vs* δ 152.9 (C-7), 109.3 (C-8), 153.8 (C-9) and 77.1 (C-3″), δ 3.75 (H-2″) *vs* δ 109.3 (C-8), 77.1 (C-3″) and 20.0 (C-4″), δ 1.33 (H-5″) *vs* δ 70.0 (C-2″), 77.1 (C-3″) and 20.0 (C-4″) (selected 2D-correlations were shown).

HR-FABMS: *m/z* 343.1537 ([M + H]$^+$, Calcd. for $C_{20}H_{23}O_5$, 343.1546).

Reference

Takashima J, Komiyama K, Ishiyama H, Kobayashi J, Ohsaka A. (2005). Brosimacutins J-M, four new flavonoids from *Brosimum acutifolium* and their cytotoxic activity. *Planta Med* **71**: 654–658.

Celtiside A [Isoswertisin 8-C-[α-L-rhamnopyranosyl-(1→6)-β-D-glucopyranoside]

IUPAC name: 5-Hydroxy-2-(4-hydroxyphenyl)-7-methoxy-8-((2*S*,3*R*,5*S*,6*R*)-3,4,5-trihydroxy-6-(((((2*S*,3*R*,5*S*,6*R*)-3,4,5-trihydroxy-6-methyltetrahydro-2*H*-pyran-2-yl)oxy)methyl)tetrahydro-2*H*-pyran-2-yl)-4*H*-chromen-4-one

Sub-class: Flavone gylcoside

Chemical structure

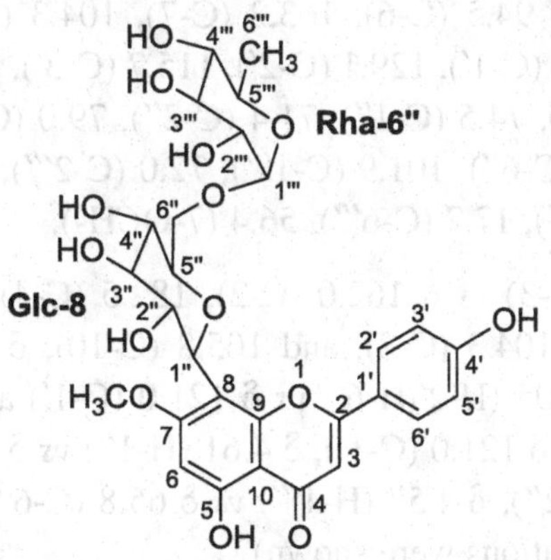

Source: *Celtis africana* Brum.f. (Family:Ulmaceae); aerial parts

Molecular formula: $C_{28}H_{32}O_{14}$

Molecular weight: 592

State: Yellow amorphous powder

Melting point: 208–210 °C

Specific rotation: $[\alpha]^{20}_{D}$ −74.0° (MeOH, c 0.04)

Bioactivity studied: Antioxidant (DPPH radical scavenging activity) and urease inhibitory activity

UV (MeOH): λ_{max} (log ε) 274 (4.30), 330 (4.35) nm

IR (KBr): v_{max} 3370 (OH), 1680 (α,β-unsaturated carbonyl), 1545, 1490 (aromatic unsaturaion) cm^{-1}

^{1}H-NMR (DMSO-d_6, 400 MHz): δ 6.87 (1H, s, H-3), 6.50 (1H, s, H-6), 8.05 (2H, d, J = 9.0 Hz, H-2′ and H-6′), 6.90 (2H, d, J = 9.0 Hz, H-3′ and H-5′), 13.30 (1H, s, 5-OH), 3.85 (3H, s, 7-OCH_3), 4.61 (1H d, J = 9.1 Hz, H-1″), 3.42 (1H, m, H-2″), 3.36 (1H, m, H-3″), 3.18 (1H, m, H-4″), 3.48 (1H, m, H-5″), 3.78 (1H, dd, J = 12.2, 4.6 Hz, H-6″a), 3.95 (1H, m, H-6″b), 4.57 (1H, br s, H-1‴), 4.25 (1H, m, H-2‴), 3.41 (1H, m, H-3‴), 3.21 (1H, m, H-4‴), 3.97 (1H, m, H-5‴), 0.49 (3H, d, J = 6.0 Hz, H-6‴).

^{13}C-NMR (DMSO-d_6, 100 MHz): δ 162.0 (C-2), 102.6 (C-3), 182.5 (C-4), 160.7 (C-5), 94.5 (C-6), 163.2 (C-7), 104.3 (C-8), 155.0 (C-9), 105.8 (C-10), 121.0 (C-1′), 129.1 (C-2′), 115.7 (C-3′), 161.3 (C-4′), 115.7 (C-5′), 129.1 (C-6′), 74.5 (C-1″), 71.4 (C-2″), 79.0 (C-3″), 71.9 (C-4″), 73.5 (C-5″), 65.8 (C-6″), 101.9 (C-1‴), 72.0 (C-2‴), 71.2 (C-3‴), 71.6 (C-4‴), 70.1 (C-5‴), 17.7 (C-6‴), 56.4 (7-OCH_3).

HMBC: δ 6.87 (H-3) *vs* δ 162.0 (C-2), 182.5 (C-4) and 121.0 (C-1′), δ 6.50 (H-6) *vs* δ 104.3 (C-8), and 105.8 (C-10), δ 3.85 (7-OCH_3) *vs* δ 163.2 (C-7), δ 8.05 (H-2′/H-6′) *vs* δ 121.0 (C-1′) and 161.3 (C-4′), δ 6.90 (H-3′/H-5′) *vs* δ 121.0 (C-1′), δ 4.61 (H-1″) *vs* δ 104.3 (C-8), 155.0 (C-9) and 71.4 (C-2″), δ 4.57 (H-1‴) *vs* δ 65.8 (C-6″) and 72.0 (C-2‴) (selected 2D-correlations were shown).

NOSEY: δ 3.85 (7-OCH_3) *vs* δ 6.50 (H-6).

EIMS: *m/z* (%rel.) 297 (32), 180 (20), 118 (100).

HR-FAB-MS (−ve mode): *m/z* 591.1695 ([M − H]⁻, Calcd. for $C_{28}H_{31}O_{14}$, 591.17123).

Reference

Azza S P, El-Shafae M, Al-Taweel A, Fawzy G A, Malik A, Afza N, Latif M, Iqbal L. (2011). Antioxidant and urease inhibitory C-glycosylflavonoids from *Celtis africana*. *J Asian Nat Prod Res* **13**: 799–804.

Celtiside B [7-Methoxyluteolin 8-*C*-[α-L-rhamnopyranosyl-(1→2)-β-D-glucopyranoside]

IUPAC name: 8-((2*S*,3*R*,5*S*,6*R*)-4,5-dihydroxy-6-(hydroxymethyl)-3-(((2*R*,3*R*,5*S*,6*R*)-3,4,5-trihydroxy-6-methyltetrahydro-2*H*-pyran-2-yl)oxy)tetrahydro-2*H*-pyran-2-yl)-2-(3,4-dihydroxyphenyl)-5-hydroxy-7-methoxy-4*H*-chromen-4-one

Sub-class: Flavone gylcoside

Chemical structure

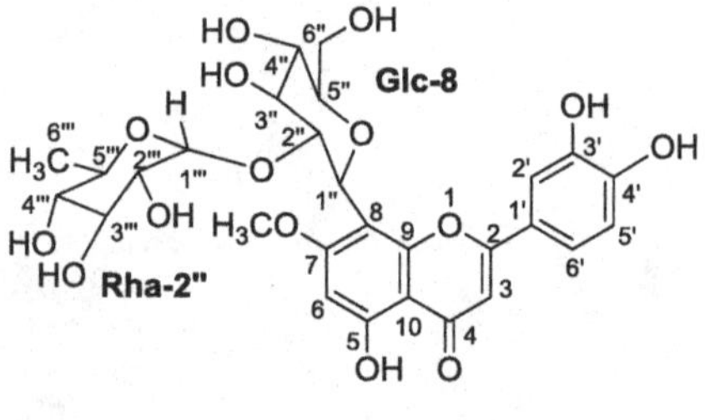

Source: *Celtis africana* Brum.f. (Family:Ulmaceae); aerial parts

Molecular formula: $C_{28}H_{32}O_{15}$

Molecular weight: 608

State: Yellow amorphous powder

Melting point: 219–221 °C

Specific rotation: $[\alpha]^{20}_D$ −35.0° (MeOH, c 0.03)

Bioactivity studied: Antioxidant (DPPH radical scavenging activity) and urease inhibitory activity

UV (MeOH): λ_{max} (log ε) 275 (4.28), 328 (4.32) nm

IR (KBr): ν_{max} 3372 (OH), 1678 (α,β-unsaturated carbonyl), 1546, 1490 (aromatic unsaturaion) cm^{-1}

^{1}H-NMR (DMSO-d_6, 400 MHz): δ 6.85 (1H, s, H-3), 6.47 (1H, s, H-6), 7.45 (1H, d, J = 2.0 Hz, H-2′), 6.89 (1H, d, J = 8.5 Hz, H-5′), 7.50 (1H, dd, J = 8.5, 2.0 Hz, H-6′), 13.33 (1H, s, 5-OH), 3.83 (3H, s, 7-OCH_3), 4.78 (1H d, J = 9.3 Hz, H-1″), 4.02 (1H, t, J = 9.3 Hz, H-2″), 3.51 (1H, m, H-3″), 3.41 (1H, m, H-4″), 3.23 (1H, m, H-5″), 3.55 (1H, dd, J = 12.1, 4.8 Hz, H-6″a), 3.74 (1H, J = 12.1, 4.8 Hz, H-6″b), 5.12 (1H, br s, H-1‴), 3.76 (1H, m, H-2‴), 3.01 (1H, m, H-3‴), 2.88 (1H, m, H-4‴), 3.01 (1H, m, H-5‴), 0.46 (3H, d, J = 6.0 Hz, H-6‴).

^{13}C-NMR (DMSO-d_6, 100 MHz): δ 162.4 (C-2), 103.3 (C-3), 182.3 (C-4), 160.5 (C-5), 95.2 (C-6), 162.8 (C-7), 104.6 (C-8), 154.8 (C-9), 105.7 (C-10), 121.9 (C-1′), 113.1 (C-2′), 145.67 (C-3′), 149.5 (C-4′), 116.0 (C-5′), 118.9 (C-6′), 71.5 (C-1″), 75.1 (C-2″), 79.8 (C-3″), 70.4 (C-4″), 81.9 (C-5″), 61.0 (C-6″), 100.4 (C-1‴), 70.3 (C-2‴), 70.2 (C-3‴), 71.4 (C-4‴), 68.3 (C-5‴), 17.5 (C-6‴), 56.5 (7-OCH_3).

HMBC: δ 6.85 (H-3) vs δ 162.4 (C-2), 182.3 (C-4) and 121.9 (C-1′), δ 6.47 (H-6) vs δ 104.6 (C-8), and 105.7 (C-10), δ 8.05 (H-2′) vs δ 121.9 (C-1′) and 149.5 (C-4′), δ 6.89 (H-5′) vs δ 121.9 (C-1′) and 145.67 (C-3′), δ 7.50 (H-6′) vs δ 121.9 (C-1′), 113.1 (C-2′) and 149.5 (C-4′), δ 4.78 (H-1″) vs δ 104.6 (C-8), 154.8 (C-9) and 75.1 (C-2″), δ 5.12 (H-1‴) vs δ 75.1 (C-2″) and 70.3 (C-2‴) (selected 2D-correlations were shown).

NOSEY: δ 3.83 (7-OCH_3) vs δ 6.47 (H-6).

EIMS: m/z (%rel.) 313 (15), 180 (25), 134 (55).

HR-FAB-MS (−ve mode): m/z 607.1650 ([M − H]$^-$, Calcd. for $C_{28}H_{31}O_{15}$, 607.1663).

Reference

Azza S P, El-Shafae M, Al-Taweel A, Fawzy G A, Malik A, Afza N, Latif M, Iqbal L. (2011). Antioxidant and urease inhibitory C-glycosylflavonoids from *Celtis africana*. *J Asian Nat Prod Res* **13**: 799–804.

Citrumedin-B

IUPAC name: 2,3,3,9,9-Pentamethyl-6-(2-methylbut-3-en-2-yl)-2,3-dihydrofuro[2,3-*f*]pyrano[2,3-*h*]chromen-5(9*H*)-one

Sub-class: Pyrano-furano-fused coumarin

Chemical structure

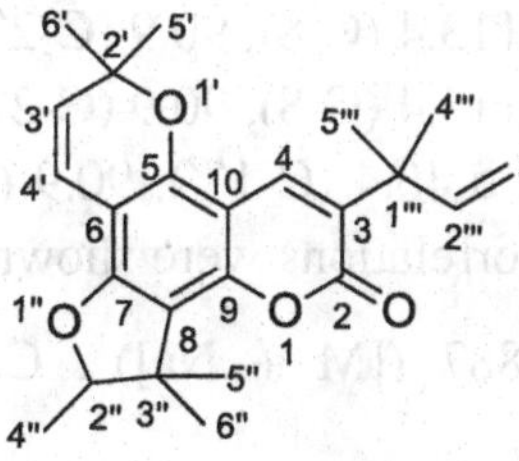

Source: *Citrus medica* L. var. *sarcodactylis* Swingle (Family: Rutaceae); root barks

Molecular formula: $C_{24}H_{28}O_4$

Molecular weight: 380

State: Pale yellow syrup

Specific rotation: $[\alpha]_D$ +14.4° (CHCl$_3$, *c* .03)

UV (MeOH): λ_{max} (log ε) 225, 273 (sh), 283, 302 (sh), 339 nm

FT-IR (KBr): ν_{max} 1724 (lactone carbonyl), 1604 (aromatic unsaturation) cm^{-1}

¹H-NMR (CDCl₃, 400 MHz): δ 7.79 (1H, s, H-4), 5.53 (1H, d, J = 9.8 Hz, H-3′), 6.45 (1H, d, J = 9.8 Hz, H-4′), 1.45 (6H, s, H-5′ and H-6′), 4.45 (1H, q, J = 6.6 Hz, H-2″), 1.38 (3H, d, J = 6.6 Hz, H-4″), 1.47 (3H, s, H-5″), 1.24 (3H, s, H-6″), 6.19 (1H, dd, J = 17.7, 11.4 Hz, H-2‴), 5.08 (1H, d, J = 17.7 Hz, H-3‴a), 5.06 (1H, d, J = 11.4 Hz, H-3‴b), 1.45 (6H, s, H-4‴ and H-5‴).

¹³C-NMR (CDCl₃, 100 MHz): δ 159.8 (C-2), 128.7 (C-3), 133.1 (C-4), 149.9 (C-5), 101.7 (C-6), 157.2 (C-7), 113.4 (C-8), 150.7 (C-9), 103.7 (C-10), 77.7 (C-2′), 127.5 (C-3′), 116.1 (C-4′), 28.1 (C-5′), 27.9 (C-6′), 90.9 (C-2″), 44.0 (C-3″), 14.2 (C-4″), 25.6 (C-5″), 21.6 (C-6″), 40.4 (C-1‴), 145.9 (C-2‴), 111.7 (C-3‴), 26.3 (C-4‴), 26.3 (C-5‴).

HMBC: δ 7.79 (H-4) *vs* δ 159.8 (C-2), 149.9 (C-5), 150.7 (C-9), and 40.4 (C-1‴), δ 5.53 (H-3′) *vs* δ 101.7 (C-6) and 77.7 (C-2′), δ 6.45 (H-4′) *vs* δ 77.7 (C-2′), δ 1.45 (H-5′) *vs* δ 77.71 (C-2′) and 127.5 (C-3′), δ 1.45 (H-6′) *vs* δ 77.7 (C-2′) and 127.5 (C-3′), δ 1.38 (H-4″) *vs* δ 90.9 (C-2″) and 44.0 (C-3″), δ 1.47 (H-5″) *vs* δ 113.4 (C-8), 90.9 (C-2″), 44.0 (C-3″) and 21.6 (C-6″), δ 1.24 (H-6″) *vs* δ 113.4 (C-8), 90.9 (C-2″), 44.0 (C-3″) and 25.6 (C-5″), δ 5.06 (H-3‴b) *vs* δ 40.4 (C-1‴), 90.9 (C-2″), 44.0 (C-3″) and 21.6 (C-6″) (selected 2D-correlations were shown).

HR-FAB-MS: *m/z* 403.1887 ([M + Na])⁺, Calcd. for $C_{24}H_{28}O_4Na$, 403.1885).

Reference

Chan Y-Y, Li C-H, Shen Y-C, Wu T-S. (2010). Anti-inflammatory principles from the stem and root barks of *Citrus medica*. *Chem Pharm Bull* **58**: 61–65.

Corbulain Ib [2-(4-Hydroxyphenyl)-3,4,9,10-tetrahydro-3,5-dihydroxy-10-(3,4-dihydroxyphenyl)-(2*R*,3*R*,10*S*)-2*H*,8*H*-benzo[1,2-b:3,4-b']dipyran-8-one]

IUPAC name: (2*R*,3*R*,10*S*)-10-(3,4-Dihydroxyphenyl)-3,5-dihydroxy-2-(4-hydroxyphenyl)-3,4,9,10-tetrahydropyrano[2,3-*f*]chromen-8(2*H*)-one

Sub-class: Coumaroflavanol

Chemical structure

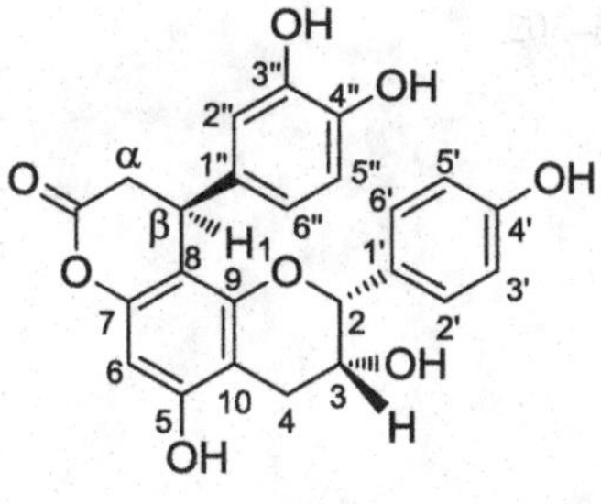

Source: *Smilax corbularia* Kunth. (Family: Smilacaceae); rhizomes

Molecular formula: $C_{24}H_{20}O_8$

Molecular weight: 436

State: Dark red amorphous powder

Specific rotation: $[\alpha]^{25}_D$ +27.9° (MeOH, c 0.3)

UV (MeOH): λ_{max} (log ε) 211 (4.72), 281 (3.88), 315 (3.16), 330 (3.88) nm

^{1}H-NMR (CD$_3$OD, 400 MHz): δ 4.86 (1H, br s, H-2β), 4.27 (1H, m, H-3β), 2.92 (1H, dd, J = 17.5, 4.0 Hz, H-4a), 2.85 (1H, dd, J = 17.5, 3.0 Hz, H-4b), 6.20 (1H, s, H-6), 7.28 (2H, d, J = 8.5 Hz, H-2′ and H-6′), 6.77 (2H, d, J = 8.5 Hz, H-3′ and H-5′), 6.52 (1H, d, J = 2.0 Hz, H-2″), 6.60 (1H, d, J = 8.0 Hz, H-5″), 6.42 (1H, dd, J = 8.0, 2.0 Hz, H-6″), 3.01 (2H, dd, J = 16.5, 7.5 Hz, H-α), 2.85 (1H, dd, J = 16.5, 1.5 Hz, H-β).

^{13}C-NMR (CD$_3$OD, 100 MHz): δ 79.8 (C-2), 66.4 (C-3), 29.7 (C-4), 157.4 (C-5), 96.3 (C-6), 152.1 (C-7), 106.1 (C-8), 163.5 (C-9), 105.3 (C-10), 131.2 (C-1′), 128.9 (C-2′), 115.9 (C-3′), 157.9 (C-4′), 115.9 (C-5′), 128.9 (C-6′), 135.5 (C-1″), 115.0 (C-2″), 146.3 (C-3″), 145.1 (C-4″), 116.5 (C-5″), 119.2 (C-6″), 38.6 (C- α), 35.5 (C-β), 170.87 (lactone carbonyl carbon).

HRFABMS: m/z 436.1187 [M]$^+$ (Calcd. for C$_{24}$H$_{20}$O$_8$, 436.1158).

Reference

Wungsintaweekul B, Umehara K, Miyase T, Noguchi H. (2011). Estrogenic and anti-estrogenic compounds from the Thai medicinal plant, *Smilax corbularia* (Smilacaceae). *Phytochemistry* **72**: 495–502.

Cyanidin 3-*O*-(2″-(5‴-(*E-p*-coumaroyl)-β-apiofuranosyl)-β-xylopyranoside)-5-*O*-β-glucopyranoside

Sub-class: Anthocyanin glycoside

Chemical structure

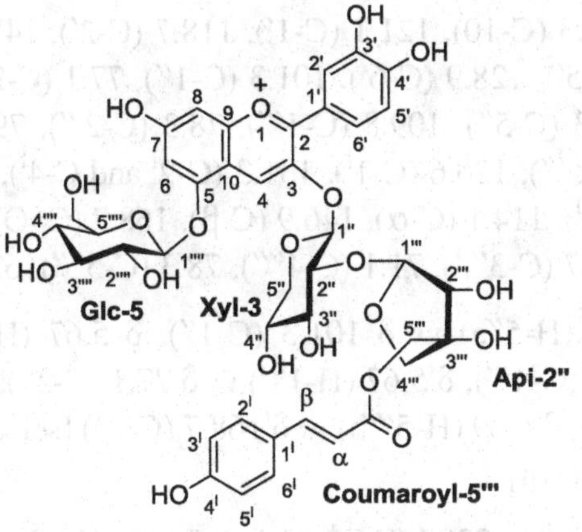

Source: *Synadenium grantii* (Family: Euphorbiaceae); leaves

Molecular formula: $C_{40}H_{43}O_{21}^{+}$

Molecular weight: 859

State: Pigment

UV: λ_{max} 550, 320, 281 nm

^{1}H-NMR [CF$_3$COOD-CD$_3$OD (5:95; v/v), 600 MHz]: δ 8.97 (1H, d, J = 0.6 Hz, H-4), 7.11 (1H, d, J = 1.9 Hz, H-6), 6.97 (1H, dd, J = 1.9, 0.6 Hz, H-8), 8.05 (1H, d, J = 2.4 Hz, H-2'), 7.09 (1H, d, J = 8.7 Hz, H-5'), 8.28 (1H, dd, J = 8.7, 2.4 Hz, H-6'), 5.68 (1H, d, J = 7.2 Hz, H-1''), 4.08 (1H, dd, J = 8.9, 7.3 Hz, H-2''), 3.81 (1H, dd, J = 13.2, 8.6 Hz, H-3''), 3.75 (1H, ddd, J = 13.2, 8.2, 4.4 Hz, H-4''), 4.09 (1H, dd, J = 11.2, 8.2 Hz, H-5''a), 3.58 (1H, dd, J = 11.2, 4.4 Hz, H-5''b), 5.67 (1H, d, J = 1.2 Hz, H-1'''), 3.96 (1H, d, J = 1.2 Hz, H-2'''), 3.86 (1H, d, J = 9.7 Hz, H-4'''a), 3.61 (1H, d, J = 9.7 Hz, H-4'''b), 4.31 (1H, d, J = 11.6 Hz, H-5'''a), 4.09 (1H, d, J = 11.6 Hz, H-5'''b), 7.36 (1H, d, J = 8.7 Hz, H-2^l), 6.87 (2H, d, J = 8.7 Hz, H-3^l and H-5^l), 7.36 (1H, d, J = 8.7 Hz, H-6^l), 6.09 (1H, d, J = 15.9 Hz, H-α), 7.27 (1H, d, J = 15.9 Hz, H-β), 5.30 (1H, d, J = 7.8 Hz, H-1''''), 3.76 (1H, dd, J = 9.1, 7.8 Hz, H-2''''), 3.67 (1H, dd, J = 9.1, 8.3 Hz, H-3''''), 3.56 (1H, dd, J = 9.2, 8.3 Hz, H-4''''), 3.69 (1H, m, H-5''''), 4.04 (1H, dd, J = 12.3, 2.5 Hz, H-6''''a), 3.83 (1H, dd, J = 12.3, 6.1 Hz, H-6''''b).

^{13}C-NMR [CF$_3$COOD-CD$_3$OD (5:95; v/v), 150 MHz]: δ 164.6 (C-2), 145.7 (C-3), 133.3 (C-4), 157.1 (C-5), 105.7 (C-6), 169.3 (C-7), 97.4 (C-8), 156.6 (C-9), 113.3 (C-10), 121.1 (C-1'), 118.7 (C-2'), 147.5 (C-3'), 156.5 (C-4'), 117.4 (C-5'), 128.9 (C-6'), 101.3 (C-1''), 77.1 (C-2''), 78.2 (C-3''), 70.6 (C-4''), 67.3 (C-5''), 109.8 (C-1'''), 78.3 (C-2'''), 79.0 (C-3'''), 74.7 (C-4'''), 67.8 (C-5'''), 126.6 (C-1^l), 131.3 (C-2^l and C-4^l), 116.9 (C-3^l and C-5^l), 161.2 (C-4^l), 114.1 (C-α), 146.9 (C-β), 168.7 (C=O), 102.5 (C-1''''), 74.6 (C-2''''), 77.7 (C-3''''), 71.1 (C-4''''), 78.3 (C-5''''), 62.4 (C-6'''').

HMBC: δ 4.09 (H-5''a) *vs* δ 101.3 (C-1''), δ 5.67 (H-1''') *vs* δ 79.0 (C-3''') and 74.7 (C-4'''), δ 5.67 (H-1''') *vs* δ 77.1 (C-2''), δ 4.31 (H-5'''a) *vs* δ 168.7 (C=O), δ 4.09 (H-5'''b) *vs* δ 168.7 (C=O) [selected HMBC correlations were shown].

HR-ESI-MS: *m/z* 859.2297 ([M]$^+$, Calcd. for $C_{40}H_{43}O_{21}{}^+$, 859.2312.

Reference

Andersen Ø M, Jordheim M, Byamukama R, Mbabazi A, Ogweng G, Skaar I, Kiremire B. (2010). Anthocyanins with unusual furanose sugar (apiose) from leaves of *Synadenium grantii* (Euphorbiaceae). *Phytochemistry* **71**: 1558–1563.

7-De-*O*-methylaciculatin [8-*C*-β-D-digitoxopyranosylapigenin]

IUPAC name: 8-((2*R*,4*S*,5*S*,6*R*)-4,5-dihydroxy-6-methyltetrahydro-2*H*-pyran-2-yl)-5,7-dihydroxy-2-(4-hydroxyphenyl)-4*H*-chromen-4-one

Sub-class: Flavone *C*-glycoside

Chemical structure

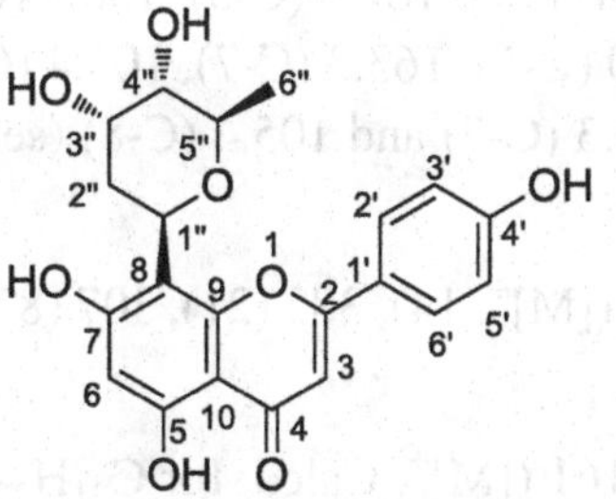

Source: *Chrysopogon aciculatis* (Family: Poaceae); whole grass

Molecular formula: $C_{21}H_{20}O_8$

Molecular weight: 400

State: Light yellow powder

Melting point: 188–190 °C

Bioactivity studied: Cytotoxic [IC_{50} values of 6.35 and 4.42 µM, respectively against MCF-7 and CEM cancer cell lines]

Specific rotation: $[\alpha]_D$ +79° (MeOH, c 0.54)

UV (MeOH): λ_{max} (log ε) 213 (4.55), 271 (4.31), 331 (4.31) nm

IR (KBr): ν_{max} 3500 (OH), 1659 (γ-pyrone carbonyl), 1611, 1576, 1544, 1449 (aromatic unsaturaion), 1354, 1243, 1176 cm^{-1}

^{1}H-NMR (Acetone-d_6, 600 MHz): δ 6.67 (1H, s, H-3), 6.17 (1H, s, H-6), 8.07 (2H, d, J = 9.0 Hz, H-2′ and H-6′), 7.01 (2H, d, J = 9.0 Hz, H-3′ and H-5′), 5.77 (1H, dd, J = 11.4, 2.4 Hz, H-1″), 2.20 (1H, ddd, J = 14.4, 3.6, 2.4 Hz, H-2″a), 2.11(1H, ddd, J = 14.4, 11.4, 2.4 Hz, H-2″b), 4.18 (1H, br q, H-3″), 3.49 (1H, dd, J = 9.6, 3.0 Hz, H-4″), 4.01 (1H, dq, J = 9.6, 6.6 Hz, H-5″), 1.38 (3H, d, J = 6.6 Hz, H-6″), 12.95 (1H, s, 5-OH).

^{13}C-NMR (Acetone-d_6, 150 MHz): δ 164.7 (C-2), 103.7 (C-3), 183.2 (C-4), 162.0 (C-5), 100.4 (C-6), 163.3 (C-7), 105.4 (C-8), 154.3 (C-9), 105.3 (C-10), 123.0 (C-1′), 129.5 (C-2′ and C-6′), 116.9 (C-3′ and C-5′), 162.1 (C-4′), 70.3 (C-1″), 38.9 (C-2″), 68.0 (C-3″), 73.6 (C-4″), 74.8 (C-5″), 18.7 (C-6″).

HMBC: δ 12.95 (C$_5$-OH) vs δ 162.0 (C-5), 100.4 (C-6) and 105.4 (C-10), δ 6.17 (H-6) vs δ 162.0 (C-5), 163.3 (C-7), 105.4 (C-8) and 105.4 (C-10), δ 5.77 (H-1″) vs δ 163.3 (C-7) and 105.4 (C-8) (selected HMBC correlations were shown).

EIMS: m/z (%rel) 400 ([M]$^+$, 14), 382 (21), 307 (87), 284 (31), 270 (100), 256 (60), 189 (36).

HR-EIMS: m/z 400.1161 ([M]$^+$, Calcd. for $C_{21}H_{20}O_8$, 400.1164).

Reference

Shen C-C, Cheng J-J, Lay H-L, Wu S-Y, Ni C-L, Teng C-M, Chen C-C. (2012). Cytotoxic apigenin derivatives from *Chrysopogon aciculatis*. *J Nat Prod* **75**: 198–201.

Delphinidin 3-O-β-galactopyranoside-3′, 5′-di-O-β-glucopyranoside

Sub-class: Anthocyanin glycoside

Chemical structure

Source: *Cornus alba* (Siberian dogwood; Family: Cornaceae); berries

Molecular formula: $C_{33}H_{41}O_{22}{}^{+}$

Molecular weight: 789

State: Pigment

UV-vis: λ_{max} 511 nm

¹H-NMR [CF₃COOD-CD₃OD (5:95; v/v), 600 MHz]: δ 9.14 (1H, d, $J = 0.9$ Hz, H-4), 6.74 (1H, d, $J = 2.0$ Hz, H-6), 7.13 (1H, dd, $J = 2.0$, 0.9 Hz, H-8), 8.35 (1H, d, $J = 2.4$ Hz, H-2′), 8.35 (1H, s, H-6′), 5.35 (1H, d,

J = 7.7 Hz, H-1″), 4.11 (1H, dd, J = 9.6, 7.7 Hz, H-2″), 3.76 (1H, dd, J = 9.6, 3.4 Hz, H-3″), 4.04 (1H, dd, J = 3.4, 0.8 Hz, H-4″), 3.93 (1H, m, H-5″), 3.86 (2H, m, H-6″), 5.19 (1H, d, J = 7.7 Hz, H-1‴), 3.3.67 (1H, m, H-2‴), 3.66 (1H, m, H-3‴), 3.47 (1H, dd J = 9.8, 8.9 Hz, H-4‴), 3.69 (1H, ddd, J = 9.8, 6.6, 2.2 Hz, H-5‴), 4.06 (1H, dd, J = 12.3, 2.2 Hz, H-6‴a), 3.78 (1H, dd, J = 12.3, 6.6 Hz, H-6‴b), 5.19 (1H, d, J = 7.7 Hz, H-1⁗), 3.67 (1H, m, H-2⁗), 3.66 (1H, m, H-3⁗), 3.47 (1H, dd, J = 9.8, 8.9 Hz, H-4⁗), 3.69 (1H, ddd, J = 9.8, 6.6, 2.2 Hz, H-5⁗), 4.06 (1H, dd, J = 12.3, 2.2 Hz, H-6⁗a), 3.78 (1H, dd, J = 12.3, 6.6 Hz, H-6⁗b).

^{13}C-NMR [CF$_3$COOD-CD$_3$OD (5:95; v/v), 150 MHz]: δ 163.16 (C-2), 145.8 (C-3), 137.02 (C-4), 158.80 (C-5), 103.57 (C-6), 171.03 (C-7), 95.57 (C-8), 158.20 (C-9), 114.24 (C-10), 120.39 (C-1′), 116.68 (C-2′ and C-6′), 147.69 (C-3′ and C-5′), 146.67 (C-4′), 103.84 (C-1″), 72.19 (C-2″), 74.98 (C-3″), 70.6 (C-4″), 77.81 (C-5″), 62.34 (C-6″), 103.84 (C-1‴), 74.98 (C-2‴), 77.49 (C-3‴), 71.52 (C-4‴), 78.86 (C-5‴), 62.80 (C-6′), 103.84 (C-1⁗), 74.98 (C-2⁗), 77.49 (C-3⁗), 71.52 (C-4⁗), 78.86 (C-5⁗), 62.80 (C-6⁗).

ESI-TOFMS: m/z 789.2095 ([M]$^+$, Calcd. for C$_{33}$H$_{41}$O$_{22}^+$, 789.2090.

Reference

Bjorøy Ø, Fossen T, Andersen Ø M. (2007). Anthocyanin 3-galactosides from *Cornus alba* 'Sibirica' with glucosidation of the B-ring. *Phytochemistry* **68**: 640–645.

Denticulatain C

IUPAC name: 2-(3,4-Dihydroxyphenyl)-3,5,7-trihydroxy-6-((*E*)-3-methyl-5-((1*S*,4a*S*,8a*S*)-5,5,8a-trimethyl-2-methylenedecahydronaphthalen-1-yl)pent-2-en-1-yl)-4*H*-chromen-4-one

Sub-class: Flavonol-diterpene heterodimer

Chemical structure

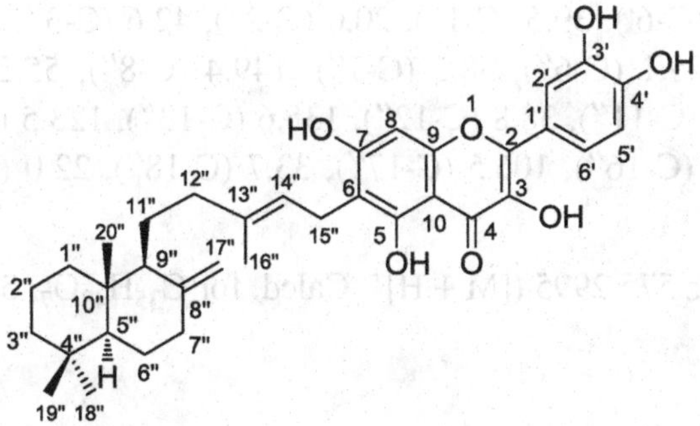

Source: *Macaranga denticulate* (Family: Euphorbiaceae); fronds

Molecular formula: $C_{35}H_{42}O_7$

Molecular weight: 574

State: Yellow oil

Specific rotation: $[\alpha]^{25}_{D}$ +1.4° (MeOH, *c* 0.14)

UV (MeOH): λ_{max} (log ε) 208 (4.64), 257 (4.36), 370 (4.38) nm

IR (KBr): v_{max} 3423 (OH), 2925, 2845, 1647 (C=O), 1626, 1602, 1564, 1483, 1443, 1367, 1319, 1268, 1196, 1157, 1089, 1035, 959, 886, 812 cm^{-1}

^{1}H-NMR (Acetone-d_6, 400 MHz): δ 6.60 (1H, s, H-8), 7.79 (1H, s, H-2′), 6.97 (1H, d, J = 8.4 Hz, H-5′), 7.66 (1H, d, J = 8.4 Hz, H-6′), 1.65 (1H, m, H-1″a), 0.80 (1H, m, H-1″b), 1.49 (1H, m, H-2″a), 1.35 (1H, m, H-2″b), 1.29 (1H, m, H-3″a), 1.05 (1H, m, H-3″b), 0.85 (1H, m, H-5″), 1.61 (1H, m, H-6″a), 1.18 (1H, m, H-6″b), 2.23 (1H, m, H-7″a), 1.88 (1H, m, H-7″b), 1.54 (1H, m, H-9″), 1.52 (1H, m, H-11″a), 1.38 (1H, m, H-11″b), 2.04 (1H, m, H-12″a), 1.75 (1H, m, H-12″b), 5.25 (1H, t, J = 7.2 Hz, H-14″), 3.42 (1H, m, H-15″a), 3.29 (1H, m, H-15″b), 1.78 (3H, m, H-16″), 4.74 (1H, s, H-17″a), 4.47 (1H, s, H-17″b), 0.73 (3H, s, H-18″), 0.71 (3H, s, H-19″), 0.61 (3H, s, H-20″), 12.44 (1H, s, 5-O*H*).

^{13}C-NMR (Acetone-d_6, 100 MHz): δ 148.2 (C-2), 136.7 (C-3), 176.5 (C-4), 158.8 (C-5), 112.0 (C-6), 162.6 (C-7), 93.8 (C-8), 155.5 (C-9), 104.0 (C-10), 123.8 (C-1′), 115.6 (C-2′), 145.7 (C-3′), 146.6 (C-4′), 116.1 (C-5′), 121.3 (C-6′), 39.5 (C-1″), 20.0 (C-2″), 42.6 (C-3″), 33.8 (C-4″), 56.1 (C-5″), 25.1 (C-6″), 38.5 (C-7″), 149.4 (C-8″), 55.2(C-9″), 39.8 (C-10″), 21.9 (C-11″), 38.8 (C-12″), 135.6 (C-13″), 123.5 (C-14″), 21.9 (C-15″), 16.2 (C-16″), 106.5 (C-17″), 33.7 (C-18″), 22.0 (C-19″), 14.9 (C-20″).

HREIMS: m/z 575.2995 ([M + H]$^+$, Calcd. for $C_{35}H_{43}O_7$, 575.3008).

Reference

Yang D-S, Li Z-L, Peng W-B, Yang Y-P, Wang X, Liu K-C, Li X-L, Xiao W-L. (2015). Three new prenylated flavonoids from *Macaranga denticulata* and their anticancer effects. *Fitoterapia* **103**: 165–170.

Denticulatain D

IUPAC name: (*E*)-3,5,7-trihydroxy-6-(6-hydroxy-3,7-dimethylocta-2,7-dien-1-yl)-2-(4-hydroxyphenyl)-4*H*-chromen-4-one

Sub-class: Modified geranylated flavonoid

Chemical structure

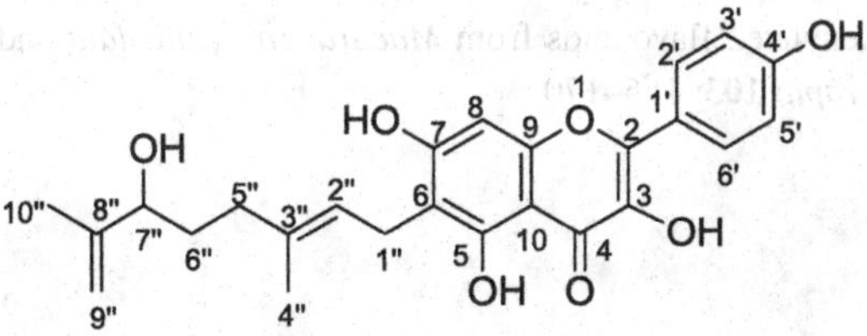

Source: *Macaranga denticulate* (Family: Euphorbiaceae); fronds

Molecular formula: $C_{25}H_{26}O_7$

Molecular weight: 438

State: Yellow amorphous powder

Specific rotation: $[\alpha]^{18}_D$ −2.5° (MeOH, *c* 0.31)

UV (MeOH): λ_{max} (log ε) 202 (4.48), 222 (4.79), 272 (4.51), 348 (4.02), 428 (4.48) nm

IR (KBr): ν_{max} 3377 (OH), 3377, 2920, 1651 (C=O), 1622, 1607, 1564, 1483, 1367, 1316, 1267, 1228, 1182, 1086, 1024, 894, 839, 805 cm^{-1}

¹H-NMR (Acetone-*d₆*, 400 MHz): δ 6.59 (1H, s, H-8), 8.12 (1H, d, *J* = 8.6 Hz, H-2′), 6.99 (1H, d, *J* = 8.6 Hz, H-3′), 6.99 (1H, d, *J* = 8.6 Hz,

H-5′), 8.12 (1H, d, J = 8.6 Hz, H-6′), 3.36 (2H, d, J = 6.8 Hz, H-1″), 5.30 (1H, t, J = 6.8 Hz, H-2″), 1.78 (3H, s, H-4″), 1.95 (2H, m, H-5″), 1.56 (2H, m, H-6″), 3.95 (1H, t, J = 6.2 Hz, H-7″), 4.84 (1H, s, H-9″a), 4.69 (1H, s, H-9″b), 1.65 (3H, s, H-10″), 12.41 (1H, s, 5-OH).

^{13}**C-NMR (Acetone-d_6, 100 MHz):** δ 146.7 (C-2), 135.5 (C-3), 176.5 (C-4), 158.9 (C-5), 111.7 (C-6), 162.7 (C-7), 93.8 (C-8), 155.6 (C-9), 104.0 (C-10), 123.4 (C-1′), 130.4 (C-2′), 116.3 (C-3′), 160.0 (C-4′), 116.3 (C-5′), 130.4 (C-6′), 21.9 (C-1″), 122.9 (C-2″), 136.6 (C-3″), 16.3 (C-4″), 36.4 (C-5″), 34.5 (C-6″), 75.2 (C-7″), 149.3 (C-8″), 110.3 (C-9″), 17.8 (C-10″).

HREIMS: m/z 437.1597 ([M − H]⁻, Calcd. for $C_{25}H_{25}O_7$, 437.1600).

Reference

Yang D-S, Li Z-L, Peng W-B, Yang Y-P, Wang X, Liu K-C, Li X-L, Xiao W-L. (2015). Three new prenylated flavonoids from *Macaranga denticulata* and their anticancer effects. *Fitoterapia* 103: 165–170.

Devenyol

IUPAC name: 8-[(2*S*),3-Dihydroxy-3-methylbutyl]-7-hydroxychromen-2-one

Sub-class: Coumarin

Chemical structure

Source: *Seseli devenyense* Simonkai (Family: Apiaceae); fruits

Molecular formula: $C_{14}H_{16}O_5$

Molecular weight: 264

State: Amorphous solid

Specific rotation: $[\alpha]_D^{25}$ −44.6° (MeOH, *c* 0.1)

UV (MeOH): λ_{max} (log ε) 256 (sh), 326 (3.55) nm

IR (MeOH): ν_{max} 1711 (C=O), 1607, 1250 cm^{-1}

^{1}H-NMR (CD$_3$OD, 400 MHz): δ 6.19 (1H, d, *J* = 9.3 Hz, H-3), 7.86 (1H, d, *J* = 9.3 Hz, H-4), 7.35 (1H, d, *J* = 8.5 Hz, H-5), 6.84 (1H, d, *J* = 8.5 Hz, H-6), 2.93 (1H, dd, *J* = 13.8, 10.1 Hz, H-1′a), 3.17 (1H, dd, *J* = 13.8, 1.5

Hz, H-1'b), 3.68 (1H, dd, J = 10.1, 1.5 Hz, H-2'), 1.29 (3H, s, H-4'), 1.30 (3H, s, H-5').

^{13}C-NMR (CD$_3$OD, 100 MHz): δ 163.8 (C-2), 111.9 (C-3), 146.7 (C-4), 128.3 (C-5), 114.2 (C-6), 161.3 (C-7), 115.5 (C-8), 113.4 (C-9), 155.1 (C-10), 26.4 (C-1'), 79.5 (C-2'), 74.0 (C-3'), 25.4 (C-4'), 25.6 (C-5').

CIMS: m/z 265 ([M + H]$^+$

HR-FABMS: m/z 265.1069 ([M + H]$^+$, Calcd. for $C_{14}H_{17}O_5$, 265.1076).

Reference

Widelski J, Melliou E, Fokialakis N, Magiatis P, Glowniak K, Chinou I. (2005). Coumarins from the fruits of *Seseli devenyense*. *J Nat Prod* **68**: 1637–1641.

(2*R*,3*S*)-6,8-Di-*C*-methyldihydrokaempferol

IUPAC name: (2*R*,3*S*)-3,5,7-Trihydroxy-2-(4-hydroxyphenyl)-6,8-dimethylchroman-4-one

Sub-class: Flavanone

Chemical structure

Source: *Diplomorpha canescens* (Meisn.) C. A. Meyer (Family: Thymelaeaceae) (Synonym: *Wikstroemia canescens* Meisn.); aerial parts

Molecular formula: $C_{17}H_{16}O_6$

Molecular weight: 316

State: Pale yellow amorphous powder

Specific rotation: $[\alpha]^{20}_D$ –80.5° (MeOH, *c* 0.52)

CD (MeOH, *c* = 0.022): $\Delta\varepsilon$ (nm) –27.1 (297), + 6.2 (347)

^{1}H-NMR (CD$_3$OD, 500 MHz): δ 4.20 (1H, d, *J* = 2.7 Hz, H-2β), 5.32 (1H, d, *J* = 2.7 Hz, H-3β), 7.35 (2H, d, *J* = 8.2 Hz, H-2′ and H-6′), 6.79

(2H, d, J = 8.2 Hz, H-3′ and H-5′), 2.00 (3H, s, 6-CH_3), 2.04 (3H, s, 8-CH_3).

^{13}C-NMR (CD$_3$OD, 125 MHz): δ 73.0 (C-2), 82.5 (C-3), 197.0 (C-4), 160.7 (C-5), 105.0 (C-6), 164.4 (C-7), 104.2 (C-8), 158.7 (C-9), 101.7 (C, C-10), 128.6 (C-1′), 129.6 (C-2′, C-6′), 115.9 (C-3′, C-5′), 158.4 (C-4′), 7.4 (6-CH_3), 8.1 (8-CH_3).

HMQC: δ 4.2 (H-2) vs δ 73.0 (C-2), δ 5.32 (H-3) vs δ 82.5 (C-3), δ 7.35 (H-2′/H-6′) vs δ 129.6 (C-2′/C-6′), δ 6.79 (H-3′/5′) vs δ 115.9 (C-3′/C-5′), δ 2.00 (6-CH_3) vs δ 7.4 (6-CH_3), δ 2.04 (8-CH_3) vs δ 8.1 (8-CH_3).

HMBC: δ 4.2 (H-2) vs δ 82.5 (C-3), 128.6 (C-1′) and 129.6 (C-2′), δ 5.32 (H-3) vs δ 73.0 (C-2) and 197.0 (C-4), δ 2.00 (6-CH_3) vs δ 160.7 (C-5), 105.0 (C-6) and 164.4 (C-7), δ 2.04 (8-CH_3) vs δ 164.4 (C-7), 104.2 (C-8), 158.7 (C-9).

HR-FAB-MS: m/z 317.0997 ([M + H]$^+$, Calcd. for C$_{17}$H$_{17}$O$_6$, 317.1025).

Reference

Devkota H P, Watanabe M, Watanabe T, Yahara S. (2010). Flavonoids from the aerial parts of *Diplomorpha canescens*. *Chem Pharm Bull* **58**: 859–861.

(3*R*)-2′,7-Dihydroxy-3′-(3-methylbut-2-enyl)-2′′′, 2′′′-dimethylpyrano[5′′′,6′′′:4′,5′] isoflavan

IUPAC name: (*R*)-6-(7-Hydroxychroman-3-yl)-2,2-dimethyl-8-(3-methylbut-2-en-1-yl)-2*H*-chromen-7-ol

Sub-class: Prenylated pyrano-fused isoflavan

Chemical structure

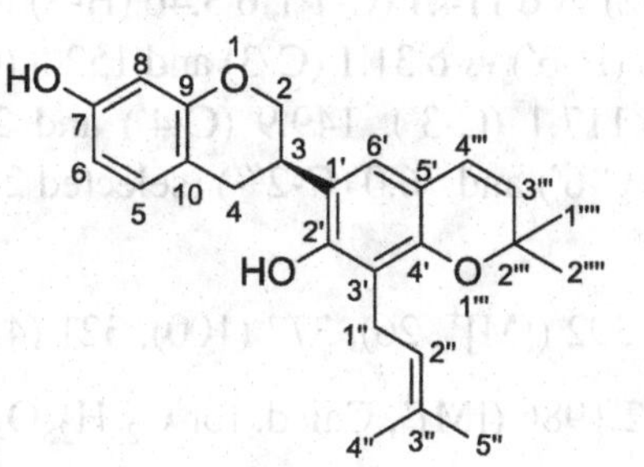

Source: *Erythrina mildbraedii* (Family: Leguminosae); root bark

Molecular formula: $C_{25}H_{28}O_4$

Molecular weight: 392

State: Amorphous gummy substance

Bioactivity studied: *In vitro* protein tyrosine phosphatase 1B (PTP1B) inhibitory activity (IC$_{50}$ = 5.5 ± 0.3 µM)

Specific rotation: [α]^{25}$_D$ −28.8° (MeOH, *c* 0.49)

UV: λ$_{max}$ (MeOH) (log ε): 280 (4.25), 312 (3.92) nm

^{1}H-NMR (CDCl$_3$, 400 MHz): δ 4.02 (1H, t-like, *J* = 10.8 Hz, H-2$_{ax}$), 4.37 (1H, ddd, *J* = 10.8, 3.6, 2.0 Hz, H-2$_{eq}$), 3.46 (1H, m, H-3), 2.87 (1H, ddd, *J* = 16.0, 4.4, 2.0 Hz, H-4$_{ax}$), 2.94 (1H, dd, *J* = 16.0, 10.8 Hz, H-4$_{eq}$), 6.98 (1H, d, *J* = 8.8 Hz, H-5), 6.38 (1H, dd, *J* = 8.8, 2.4 Hz, H-6), 6.30 (1H, d, *J* = 2.4 Hz, H-8), 6.56 (1H, s, H-6′), 3.31 (2H, br d, *J* = 7.6 Hz, H-1″), 5.24 (1H, m, H-2″), 1.79 (3H, br s, H-4″), 1.67 (3H, br s, H-5″), 5.48 (1H, d, *J* = 10.0 Hz, H-3‴), 6.24 (1H, d, *J* = 10.0 Hz, H-4‴), 1.41 (6H, br s, H-1‴′ and H-2‴′).

^{13}C-NMR (CDCl$_3$, 100 MHz): δ 70.1 (C-2), 31.1 (C-3), 31.9 (C-4), 128.6 (C-5), 108.1 (C-6), 155.3 (C-7), 103.3 (C-8), 154.6 (C-9), 120.5 (C-10), 114.1 (C-1′), 152.8 (C-2′), 117.1 (C-3′), 149.9 (C-4′), 114.8 (C-5′), 124.4 (C-6′), 22.3 (C-1″), 123.2 (C-2″), 130.9 (C-3″), 26.1 (C-4″), 18.1 (C-5″), 76.0 (C-2‴), 128.4 (C-3‴), 122.6 (C-4‴), 28.1 (C-1‴′ and C-2‴′).

HMBC: δ 4.02 (H-2) *vs* δ 114.1 (C-1′), δ 3.46 (H-3) *vs* δ 152.8 (C-2′) and 124.4 (C-6′), δ 6.56 (H-6′) *vs* δ 31.1 (C-3) and 152.8 (C-2′), δ 3.31 (H-1″) *vs* δ 152.8 (C-2′), 117.1 (C-3′), 149.9 (C-4′) and 26.1 (C-4″), δ 6.24 (H-4‴) *vs* δ 124.4 (C-6′) and 76.0 (C-2‴) (selected 2D-correlations were shown).

EIMS: *m/z* (%rel.) 392 ([M]$^+$, 26), 377 (100), 321 (4).

HR-EIMS: *m/z* 392.1986 ([M]$^+$, Calcd. for C$_{25}$H$_{28}$O$_4$, 392.1988).

Reference

Jang J, Na M, Thuong P T, Njamen D, Mbafor J T, Fomum Z T, Woo E-R, Oh W K. (2008). Prenylated flavonoids with PTP1B inhibitory activity from the root bark of *Erythrina mildbraedii*. *Chem Pharm Bull* **56**: 85–88.

(3*R*)-5,4′-Dihydroxy-2′-methoxy-3′-(3-methylbut-2-enyl)-(6″, 6″-dimethylpyrano)-(7,6:2″,3″)-isoflavanone

IUPAC name: (*R*)-5-Hydroxy-3-(4-hydroxy-2-methoxy-3-(3-methylbut-2-en-1-yl)phenyl)-8,8-dimethyl-2,3-dihydropyrano[3,2-*g*]chromen-4(8*H*)-one

Sub-class: Prenylated pyrano-fused isoflavanone

Chemical structure

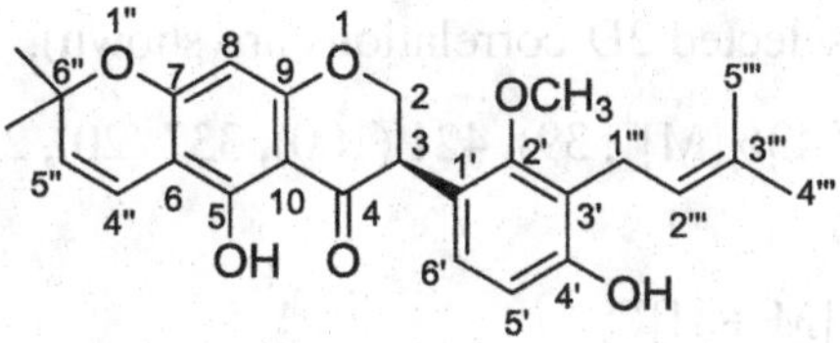

Source: *Campylotropis hirtella* (Franch.) Schindl. (Family: Leguminosae); roots

Molecular formula: $C_{26}H_{28}O_6$

Molecular weight: 436

State: Yellow oil

Bioactivity studied: *In vitro* immunosuppressive activity

Specific rotation: $[\alpha]^{25}_D$ +37.6° (MeOH, c 1.21)

UV: λ_{max} (MeOH): 227, 272, 298 nm

IR (KBr): ν_{max} 3447 (OH), 2972, 2924, 2854, 1645, 1570, 1462, 1140, 1427, 1296, 1281, 1157, 1119, 1097, 1053 cm^{-1}

^{1}H-NMR (CDCl$_3$, 400 MHz): δ 4.45 (2H, m, H-2), 4.34 (1H, dd, J = 11.2, 5.6 Hz, H-3$_{ax}$), 5.94 (1H, s, H-8), 6.62 (1H, d, J = 8.4 Hz, H-5′), 6.87 (1H, d, J = 8.4 Hz, H-6′), 6.62 (1H, d, J = 10.0 Hz, H-4″), 5.47 (1H, d, J = 10.0 Hz, H-5″), 1.45 (6H, s, 2 × 6″-CH_3), 3.73 (3H, s, 2′-OCH_3), 3.43 (2H, d, J = 6.4 Hz, H-1‴), 5.25 (1H, br s, H-2‴), 1.80 (3H, s, H-4‴), 1.76 (3H, s, H-5‴).

^{13}C-NMR (CDCl$_3$, 100 MHz): δ 71.4 (C-2), 45.5 (C-3), 198.6 (C-4), 158.9 (C-5), 102.8 (C-6), 161.9 (C-7), 94.5 (C-8), 163.2 (C-9), 103.4 (C-10), 119.6 (C-1′), 158.3 (C-2′), 121.9 (C-3′), 156.4 (C-4′), 111.5 (C-5′), 127.5 (C-6′), 115.1 (C-4″), 126.7 (C-5″), 78.4 (C-6″), 22.8 (C-1‴), 123.6 (C-2‴), 130.7 (C-3‴), 17.4 (C-4‴), 25.2 (C-5‴), 27.9 (2 × 6″-CH$_3$), 61.7 (2′-OCH$_3$).

HMBC: δ 4.34 (H-3) *vs* δ 71.4 (C-2), 158.3 (C-2′) and 127.5 (C-6′), δ 3.73 (2′-OCH_3) *vs* δ 158.3 (C-2′), δ 6.62 (H-4″) *vs* δ 158.9 (C-5), 163.2 (C-9), 103.4 (C-10) and 78.4 (C-6″), δ 5.47 (H-5″) *vs* δ 103.4 (C-10), 78.4 (C-6″) and 27.9 (6″-CH$_3$), δ 3.43 (H-1‴) *vs* δ 158.3 (C-2′), 121.9 (C-3′) and 156.4 (C-4′) (selected 2D-correlations are shown).

EIMS: m/z (%rel.) 436 [M]$^+$, 39), 421 (100), 337 (20), 217 (38), 203 (34), 165 (12), 149 (12).

ESI-MS: m/z 437 [M + H]$^+$

HR-EIMS: m/z 436.1882 ([M$^+$], Calcd. for C$_{26}$H$_{28}$O$_6$, 436.1886).

Reference

Shou Q, Tan Q, Shen Z. (2010). Isoflavonoids from the roots of *Campylotropis hirtella*. *Planta Med* **76**: 803–808.

5,2'-Dihydroxy-3,6, 7-trimethoxyflavone-5-*O*-β-D-xylopyranosyl-(1→4)-*O*-β-D-glucopyranoside

IUPAC name: 5-((((2S,3R,4R,5S,6R)-3,4-Dihydroxy-6-(hydroxymethyl)-5-((((2S,3R,4S,5R)-3,4,5-trihydroxytetrahydro-2*H*-pyran-2-yl)oxy)tetrahydro-2*H*-pyran-2-yl)oxy)-2-(2-hydroxyphenyl)-3,6,7-trimethoxy-4*H*-chromen-4-one

Sub-class: Flavone glycoside

Chemical structure

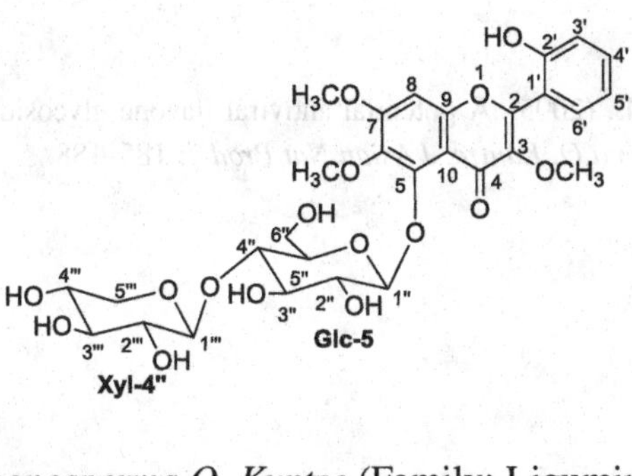

Source: *Betea monosperma O. Kuntze* (Family: Liguminosae); seeds

Molecular formula: $C_{29}H_{34}O_{16}$

Molecular weight: 638

State: Light brownish needles

Melting point: 268–269 °C

FT-IR (KBr): n_{max} 3244 (OH), 2907, 1652 (α,β-unsaturated carbonyl), 1599 (aromatic moiety) cm^{-1}

^{1}H-NMR (CDCl$_3$, 300 MHz): δ 6.52 (1H, s, H-8), 7.12 (1H, br d, J = 8.2 Hz, H-3′), 7.46 (1H, br t, J = 8.2 Hz, H-4′), 7.08 (1H, br t, J = 8.2 Hz, H-5′), 7.67 (1H, dd, J = 8.3, 2.2 Hz, H-6′), 3.88 (3H, s, 3-OCH_3), 3.95 (3H, s, 6-OCH_3), 3.92 (3H, s, 7-OCH_3), 4.42 (1H, d, J = 7.9 Hz, H-1″), 2.94–3.10 (4H, m, H-2″, H-3″, H-4″, H-5″), 3.94 (2H, dd, J = 4.5, 2.1 Hz, H-6″), 5.44 (1H, d, J = 8.5 Hz, H-1‴), 3.78–3.95 (5H, m, H-2‴, H-3‴, H-4‴, H-5‴).

^{13}C-NMR (DMSO-d_6, 90 MHz): δ 155.3 (C-2), 137.3 (C-3), 175.3 (C-4), 152.7 (C-5), 132.6 (C-6), 159.2 (C-7), 90.4 (C-8), 153.1 (C-9), 106.5 (C-10), 118.1 (C-1′), 155.4 (C-2′), 120.7 (C-3′), 133.6 (C-4′), 119.7 (C-5′), 129.6 (C-6′), 104.7 (C-1″), 74.9 (C-2″), 74.7 (C-3″), 71.2 (C-4″), 78.1 (C-5″), 62.5 (C-6″), 106.2 (C-1‴), 72.6 (C-2‴), 76.2 (C-3‴), 71.1 (C-4‴), 68.7 (C-5‴) [^{13}C-NMR data for the methoxyl groups were not indicated]

EIMS: m/z 638 ([M]$^+$).

Reference

Yadava R N, Tiwari L. (2005). A potential antiviral flavone glycoside from the seeds of *Butea monosperma O. Kuntze. J Asian Nat Prod* **7**: 185–188.

5,4'-Dihydroxy-8-(3''-methylbut-2''-enyl)-2'''-(4'''-hydroxy-4'''-methylethyl)furano-[4''',5''';6,7]isoflavone

IUPAC name: 4-Hydroxy-6-(4-hydroxyphenyl)-2-(2-hydroxypropan-2-yl)-9-(3-methylbut-2-en-1-yl)-5*H*-furo[3,2-*g*]chromen-5-one

Sub-class: Furano-prenylated isoflavone

Chemical structure

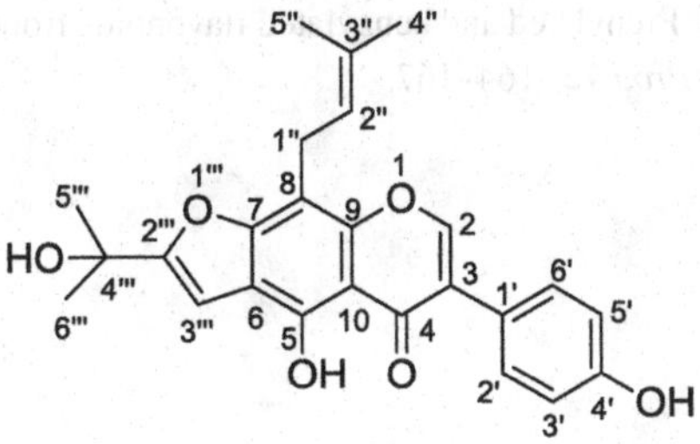

Source: *Cudrania tricuspidata* (carr.) Bur. (Family: Moraceae); fruits

Molecular formula: $C_{25}H_{24}O_6$

Molecular weight: 420

State: Yellow powder

Melting point: 160–162 °C

UV (MeOH): l_{max} (log e) 268.9 (4.6) nm

IR (dried film): n_{max} 3435 (OH), 1631 (α,β-unsaturated carbonyl), 604 cm^{-1}

^{1}H-NMR (CDCl$_3$, 500 MHz): δ 8.01 (1H, s, H-2), 7.44 (2H, d, J = 7.3 Hz, H-2′ and H-6′), 6.91 (2H, d, J = 7.3 Hz, H-3′ and H-5′), 3.69 (2H, d, J = 7.1 Hz, H-1″), 5.33 (1H, t, J = 7.1 Hz, H-2″), 1.70 (3H, s, H-4″), 1.87 (3H, s, H-5″), 6.81 (1H, s, H-3‴), 1.26 (3H, s, H-5‴), 1.70 (3H, s, H-6‴), 13.40 (1H, s, 5-OH).

^{13}C-NMR (CDCl$_3$, 125 MHz): δ 153.4 (C-2), 122.4 (C-3), 182.9 (C-4), 153.4 (C-5), 113.4 (C-6), 157.4 (C-7), 104.0 (C-8), 151.1 (C-9), 106.8 (C-10), 122.4 (C-1′), 123.3 (C-2′), 115.6 (C-3′), 130.4 (C-4′), 115.6 (C-5′), 123.3 (C-6′), 22.2 (C-1″), 121.1 (C-2″), 132.8 (C-3″), 28.6 (C-4″), 17.9 (C-5″), 162.9 (C-2‴), 98.4 (C-3‴), 69.2 (C-4‴), 25.8 (C-5‴), 29.7 (C-6‴).

HR-FABMS: m/z 421.1653 ([M + H]$^+$, Calcd. for C$_{25}$H$_{25}$O$_6$, 421.1651).

Reference

Han X H, Hong S S, Jin Q, Li D, Kim H-K, Lee J, Kwon S H, Lee D, Lee C-K, Lee M K, Hwang B Y. (2009). Prenylated and benzylated flavonoids from the fruits of *Cudrania tricuspidata*. *J Nat Prod* **72:** 164–167.

5,6-Dihydroxy-7,8,4'-trimethoxyflavone

IUPAC name: 5,6-Dihydroxy-7,8-dimethoxy-2-(4-methoxyphenyl)-4H-chromen-4-one

Sub-class: Flavone

Chemical structure

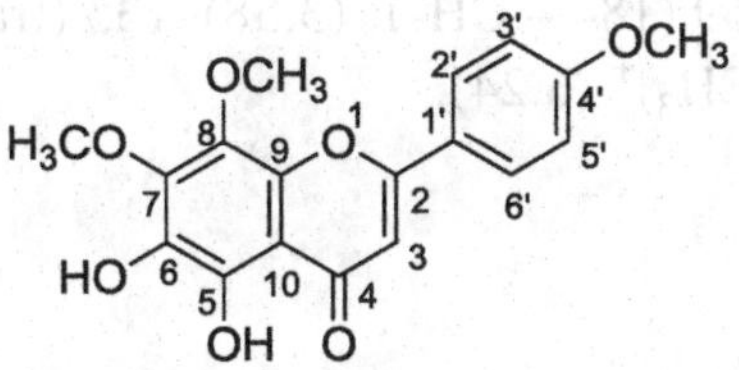

Source: *Limnophila indica* Linn. (Druce) (Family: Scrophulariaceae); aerial parts and roots

Molecular formula: $C_{18}H_{16}O_7$

Molecular weight: 344

State: Yellow crystalline solid

Melting point: 184–186 °C

UV (MeOH): λ_{max} 282, 329 nm

FT-IR (KBr): ν_{max} 3411 (OH), 2939, 2842, 1661 (α,β-unsaturated carbonyl), 1590, 1508, 1388 (aromatic unsaturation), 1266, 1025 cm^{-1}

^{1}H-NMR (CDCl$_3$, 300 MHz): δ 6.58 (1H, s, H-3), 7.89 (2H, d, J = 9.0 Hz, H-2′ and H-6′), 7.04 (2H, d, J = 9.0 Hz, H-3′ and H-5′), 4.04 (3H, s, 7-OCH_3), 4.02 (3H, s, 8-OCH_3), 3.90 (3H, s, 4′-OCH_3), 12.78 (1H, s, 5-OH), 6.44 (1H, s, 6-OH).

^{13}C-NMR (CDCl$_3$, 75 MHz): δ 164.2 (C-2), 104.1 (C-3), 183.4 (C-4), 146.2 (C-5), 131.1 (C-6), 149.2 (C-7), 127.8 (C-8), 148.8 (C-9), 104.9 (C-10), 123.9 (C-1′), 128.4 (C-2′), 115.0 (C-3′), 163.1 (C-4′), 115.0 (C-5′), 128.4 (C-6′), 62.2 (7-OCH$_3$), 61.4 (8-OCH$_3$), 55.9 (4′-OCH$_3$).

HMQC: δ 6.58 (H-3) *vs* δ 104.1 (C-3), δ 7.89 (H-2′/H-6′) *vs* δ 128.4 (C-2′/C-6′), δ 7.04 (H-3′/H-5′) *vs* δ 115.0 (C-3′/C-5′), δ 4.04 (7-OCH_3) *vs* δ 62.2 (7-OCH_3), δ 4.02 (8-OCH_3) *vs* δ 61.4 (8-OCH_3), δ 3.90 (4′-OCH_3) *vs* δ 55.9 (4′-OCH_3).

EIMS (70 eV): *m/z* (% rel.) 344 (M$^+$, base peak, 100), 329 (M$^+$ – CH$_3$, 22.51), 316 (M$^+$ – CO, 4.32), 315 (M$^+$ – CO – H, 8.45), 301 (M$^+$ – CO – CH$_3$, 5.24), 212 (15.23) and 132 (9.8) (retro-Diels-Alder fragmented ion-peaks of parent molecule), 184 (212–CO)$^+$ (4.16), 183 (184 – H)$^+$ (4.98), 169 (184 – CH$_3$)$^+$ (3.58), 135 (fragmented ion-peak, 13.21), 107 (135 – CH$_3$)$^+$ (6.24).

Reference

Brahmachari G, Jash S K, Gangopadhyay A, Sarkar S, Laskar S, Gorai D. (2008). Chemical constituents of *Limnophila indica*. *Indian J Chem* **47B**: 1898–1902.

5,7-Dihydroxy-6-(2″-hydroxy-3″-methylbut-3″-enyl)-4′-methoxylisoflavone

IUPAC name: 5,7-Dihydroxy-6-(2-hydroxy-3-methylbut-3-en-1-yl)-3-(4-methoxyphenyl)-4H-chromen-4-one

Sub-class: Prenylated isoflavone

Chemical structure

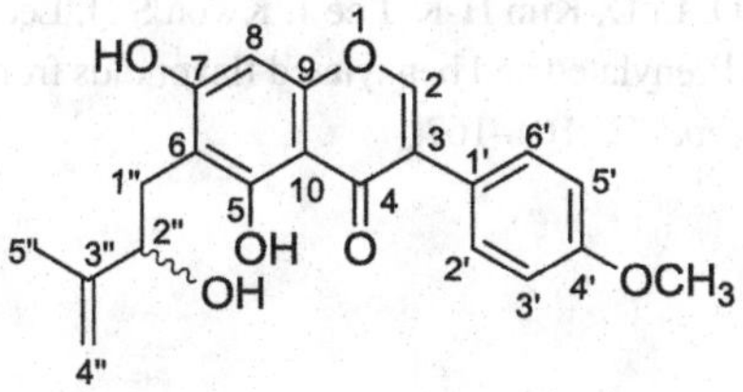

Source: *Cudrania tricuspidata* (carr.) Bur. (Family: Moraceae); fruits

Molecular formula: $C_{21}H_{20}O_6$

Molecular weight: 368

State: Pale yellow needles

Melting point: 208–210 °C

Specific rotation: $[\alpha]^{25}_{D}$ +4.44° (MeOH, *c* 0.19)

UV (MeOH): λ_{max} (log ε) 264.4 (4.5) nm

IR (dried film): ν_{max} 3433 (OH), 2081, 1639 (α,β-unsaturated carbonyl), 1247, 582 cm^{-1}

^{1}H-NMR (CDCl$_3$, 500 MHz): δ 7.83 (1H, s, H-2), 6.46 (1H, s, H-8), 7.44 (2H, d, J = 8.7 Hz, H-2' and H-6'), 6.97 (2H, d, J = 8.7 Hz, H-3' and H-5'), 3.17 (1H, d, J = 14.9, 7.8 Hz, H-1''a), 2.92 (1H, d, J = 14.9, 7.8 Hz, H-1''b), 4.41 (1H, d, J = 7.8 Hz, H-2''), 4.99 (1H, s, H-4''a), 4.88 (1H, s, H-4''b), 1.86 (3H, s, H-5''), 3.84 (3H, s, 4'-OCH_3), 13.25 (1H, s, 5-OH).

^{13}C-NMR (CDCl$_3$, 125 MHz): δ 152.6 (C-2), 123.3 (C-3), 180.9 (C-4), 160.3 (C-5), 109.2 (C-6), 163.1 (C-7), 95.2 (C-8), 156.7 (C-9), 105.6 (C-10), 123.4 (C-1'), 130.2 (C-2'), 114.1 (C-3'), 159.7 (C-4'), 114.1 (C-5'), 130.2 (C-6'), 55.4 (4'-OCH$_3$), 28.2 (C-1''), 77.5 (C-2''), 136.6 (C-3''), 110.5 (C-4''), 18.6 (C-5'').

EIMS: m/z 368 [M]$^+$

HR-FABMS: m/z 369.1333 ([M + H]$^+$, Calcd. for C$_{21}$H$_{21}$O$_6$, 369.1338).

Reference

Han X H, Hong S S, Jin Q, Li D, Kim H-K, Lee J, Kwon S H, Lee D, Lee C-K, Lee M K, Hwang B Y. (2009). Prenylated and benzylated flavonoids from the fruits of *Cudrania tricuspidata*. *J Nat Prod* **72**: 164–167.

5,7-Dihydroxy-6,8, 4'-trimethoxyflavone

IUPAC name: 5,7-Dihydroxy-6,8-dimethoxy-2-(4-methoxyphenyl)-4H-chromen-4-one

Sub-class: Flavone

Chemical structure

Source: *Limnophila heterophylla* Benth. (Family: Scrophulariaceae); aerial parts and roots

Molecular formula: $C_{18}H_{16}O_7$

Molecular weight: 344

State: Golden yellow needles

Melting point: 188–191 °C

Bioactivity studied: Antibacterial and antifungal

UV (EtOH): λ_{max} 280, 355 nm

FT-IR (KBr): v_{max} 3407 (OH), 3100, 2936, 2840, 1663 (α,β-unsaturated carbonyl), 1591, 1508 (aromatic unsaturation), 1060, 1025 cm^{-1}

^{1}H-NMR (CDCl$_3$, 300 MHz): δ 6.59 (1H, s, H-3), 7.89 (2H, dd, J = 11.7, 2.7 Hz, H-2′ and H-6′), 7.05 (2H, dd, J = 11.7, 3.0 Hz, H-3′ and H-5′), 4.04 (3H, s, 6-OCH$_3$), 4.02 (3H, s, 8-OCH$_3$), 3.90 (3H, s, 4′-OCH$_3$), 12.78 (1H, s, 5-OH).

^{13}C-NMR (CDCl$_3$, 75 MHz): δ 164.2 (C-2), 104.2 (C-3), 183.4 (C-4), 148.8 (C-5), 131.5 (C-6), 149.2 (C-7), 128.5 (C-8), 146.2 (C-9), 105.0 (C-10), 124.0 (C-1′), 127.8 (C-2′), 115.0 (C-3′), 163.1 (C-4′), 115.0 (C-5′), 127.8 (C-6′), 62.3 (6-OCH$_3$), 61.4 (8-OCH$_3$), 56.0 (4′-OCH$_3$).

HMQC: δ 6.59 (H-3) *vs* δ 104.2 (C-3), δ 7.89 (H-2′/H-6′) *vs* δ 127.8 (C-2′/C-6′), δ 7.05 (H-3′/H-5′) *vs* δ 115.0 (C-3′/C-5′), δ 4.04 (6-OCH$_3$) *vs* δ 62.3 (6-OCH$_3$), δ 4.02 (8-OCH$_3$) *vs* δ 61.4 (8-OCH$_3$), δ 3.90 (4′-OCH$_3$) *vs* δ 56.0 (4′-OCH$_3$).

EIMS (70 eV): *m/z* 344 (M$^+$), 329 (M$^+$ − CH$_3$, base peak), 316 (M$^+$ − CO), 315 (M$^+$ − CO − H), 314 (M$^+$ − 2 × CH$_3$), 312 (M$^+$ − 2 × CH$_3$ − 2H), 301 (M$^+$− CO − CH$_3$), 212 and 132 (retro-Diels–Alder ion peaks of parent molecule), 197 and 132 (retro-Diels–Alder ion peaks of mass fragment 329), 169 (197–CO)$^+$, 168 (169–H)$^+$, 153 (169 − CH$_3$)$^+$, 141 (169 − CO)$^+$, 135 (fragmented ion peak), 126 (141 − CH$_3$)$^+$.

Reference

Brahmachari G, Mandal N C, Jash S K, Roy R, Mandal L C, Mukhopadhyay A, Behera B, Majhi S, Mondal A, Gangopadhyay A. (2011). Evaluation of the antimicrobial potential of two flavonoids isolated from *Limnophila* plants. *Chem Bioderv* **8**: 1139–1151.

5,7-Dihydroxy-8-(3-methylbut-2-enyl)-4-phenyl-2*H*-chromen-2-one

Sub-class: Coumarin

Chemical structure

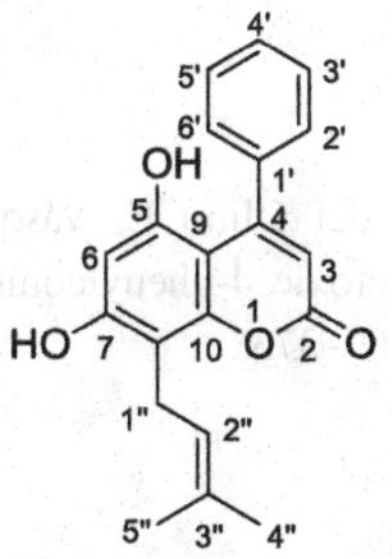

Source: *Marila pluricostata* (Family: Clusiaceae/Guttiferae); leaves

Molecular formula: $C_{20}H_{18}O_4$

Molecular weight: 322

State: White amorphous solid

Melting point: 130–132 °C

Bioactivity studied: Cytotoxic [GI_{50} (µg/ml): 3.4, 4.4 and 4.8, respectively, against MCF-7, H-460 and SF-268]

IR (CHCl₃): v_{max} 3308 (OH), 3271, 2959, 1696 (C=O), 1559, 1449, 1439, 1367 (aromatic unsaturaion), 1080 cm^{-1}

^{1}H-NMR (CDCl$_3$, 200 MHz): δ 5.96 (1H, s, H-3), 6.25 (1H, s, H-6), 6.45 (2H, br s, C$_5$-O*H* and C$_7$-O*H*), 7.3 (2H, m, H-2′ and H-6′), 7.54 (3H, m, H-3′, H-4′and H-5′), 3.56 (2H, t, *J* = 6.8 Hz, H-1″), 5.29 (1H, t, *J* = 6.8 Hz, H-2″), 1.85 and 1.75 (3H each, s, 2 × 3″-C*H*$_3$).

^{13}C-NMR ((CDCl$_3$, 100 MHz): δ 160.9 (C-2), 112.0 (C-3), 154.0 (C-4), 153.4 (C-5), 100.8 (C-6), 159.4 (C-7), 108.0 (C-8), 100.8 (C-9), 153.4 (C-10), 137.0 (C-1′), 127.5 (C-2′ and C-6′), 129.7 (C-3′ and C-5′), 129.9 (C-4′), 22.1 (C-1″), 121.1 (C-2″), 135.5 (C-3″), 18.1 (C-4″), 25.9 (C-5″).

EIMS: *m/z* (%rel) 322 ([M]$^+$, 34), 307 (20), 305 (4), 279 (17), 268 (17), 267 (100), 251 (70), 238 (5), 226 (4), 210 (2), 197 (2), 181 (3), 171 (4), 165 (9), 152 (7), 139 (7), 128 (5), 115 (14), 105 (7), 91 (5), 77 (10), 69 (15), 55 (5).

HR-FABMS: *m/z* 322.1203 ([M]$^+$, Calcd. for C$_{20}$H$_{18}$O$_4$, 322.1205).

Reference

López-Pérez J L, Olmedo D A, del Olmo E, Vásquez Y, Solís P N, Gupta M P, San Feliciano A. (2005). Cytotoxic 4-phenylcoumarins from the leaves of *Marila pluricostata*. *J Nat Prod* **68**: 369–373.

5,7-Dihydroxy-8,3', 5'-trimethoxyflavone

IUPAC name: 2-(3,5-Dimethoxyphenyl)-5,7-dihydroxy-8-methoxy-4*H*-chromen-4-one

Sub-class: Flavone

Chemical structure

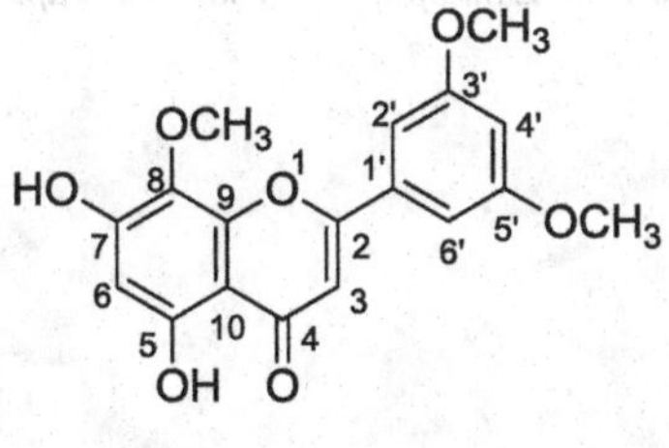

Source: *Limnophila rugosa* (Roth) Merill (Family: Scrophulariaceae); aerial parts and roots

Molecular formula: $C_{18}H_{16}O_7$

Molecular weight: 344

State: Yellow needles

Melting point: 170–172 °C

UV (MeOH): λ_{max} 221 (sh), 279, 325 nm

FT-IR (KBr): ν_{max} 3410 (OH), 2843, 1662 (α,β-unsaturated carbonyl), 1593, 1507, 1391 (aromatic unsaturation), 1265, 1029, 883, 667 cm^{-1}

¹H-NMR (CDCl₃, 400 MHz): δ 6.59 (1H, s, H-3), 6.52 (1H, s, H-6), 7.88 (2H, d, J = 2.5 Hz, H-2′ and H-6′), 7.03 (1H, d, J = 2.5 Hz, H-5′), 4.02 (3H, s, 8-OCH₃), 3.90 (6H, s, 3′-OCH₃ and 5′-OCH₃), 12.78 (1H, s, 5-OH), 8.59 (1H, s, 7-OH).

¹³C-NMR (CDCl₃, 100 MHz): δ 163.8 (C-2), 103.7 (C-3), 182.9 (C-4), 162.7 (C-5), 98.2 (C-6), 157.1 (C-7), 127.3 (C-8), 148.7 (C-9), 104.5 (C-10), 120.1 (C-1′), 128.0 (C-2′), 146.2 (C-3′), 123.5 (C-4′), 146.2 (C-5′), 128.0 (C-6′), 61.0 (8-OCH₃), 55.5 (3′-OCH₃ and 5′-OCH₃).

EIMS (70 eV): m/z 344 (M⁺, base peak), 329 (M⁺ − CH₃), 316 (M⁺ − CO), 315 (M⁺ − CO − H), 314 (M⁺ − 2 × CH₃), 301(M⁺− CO − CH₃), 182 and 162 (retro-Diels–Alder ion peaks), 154 (182–CO)⁺, 153 (154–H)⁺, 139 (154 − CH₃)⁺, 165, 135 (165–2 × CH₃)⁺.

Reference

Mukherjee K S, Gorai D, Sohel S M A, Chatterjee D, Mistri B, Mukherjee B, Brahmachari G. (2003). A new flavonoid from *Limnophila rugosa. Fitoterapia* **74**: 188–190.

7,4'-Dihydroxy-6,8-dimethoxy-4-phenylcoumarin

IUPAC name: 7-Hydroxy-4-(4-hydroxyphenyl)-6,8-dimethoxy-2*H*-chromen-2-one

Sub-class: Coumarin

Chemical structure

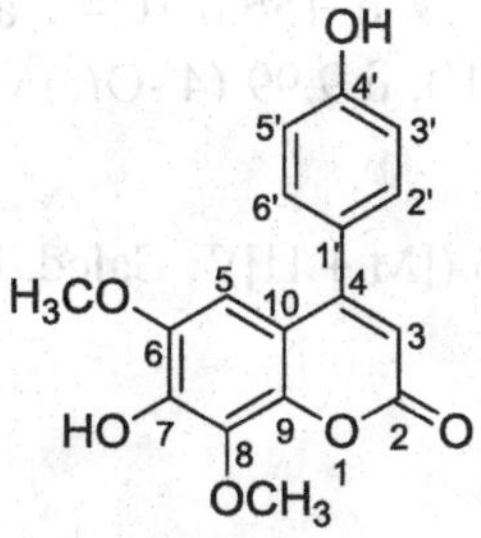

Source: *Calophyllum polyanthum* (Family: Guttiferae); seeds

Molecular formula: $C_{17}H_{14}O_6$

Molecular weight: 314

State: White amorphous powder

UV (MeOH): λ_{max} (log ε) 315 (3.78) nm

FT-IR (KBr): ν_{max} 3215 (OH), 1693 (lactone carbonyl), 1608, 1502 (aromatic unsaturation) cm^{-1}

^{1}H-NMR (DMSO-d_6, 600 MHz): δ 6.12 (1H, s, H-3), 6.75 (1H, s, H-5), 7.42 (2H, d, J = 8.4 Hz, H-2′ and H-4′), 6.94 (2H, d, J = 8.4 Hz, H-3′ and H-5′), 3.71 (3H, s, 6-OCH_3), 3.86 (3H, s, 8-OCH_3), 10.00 (1H, s, 7-OH), 9.99 (1H, s, 4′-OH).

^{13}C-NMR (DMSO-d_6, 150 MHz): δ 160.5 (C-2), 110.5 (C-3), 156.0 (C-4), 103.4 (C-5), 145.7 (C-6), 144.5 (C-7), 135.6 (C-8), 143.8 (C-9), 110.1 (C-10), 126.1 (C-1′), 130.6 (C-2′), 116.1 (C-3′), 159.1 (C-4′), 116.1 (C-5′), 130.6 (C-6′), 56.4 (6-OCH$_3$), 61.2 (8-OCH$_3$).

HSQC: δ 6.12 (H-3) *vs* δ 110.5 (C-3), δ 6.75 (H-5) *vs* δ 103.4 (C-5), δ 7.42 (H-2′/H-4′) *vs* δ 130.6 (C-2′/C-6′), δ 6.94 (H-3′/H-5′) *vs* δ 130.6 (C-3′/C-5′), δ 3.71 (6-OCH$_3$) *vs* δ 56.4 (6-OCH$_3$), δ 3.86 (8-OCH$_3$) *vs* δ 61.2 (8-OCH$_3$).

HMBC: δ 6.12 (H-3) *vs* δ 160.5 (C-2), 110.1 (C-10) and 126.1 (C-1′), δ 6.75 (H-5) *vs* δ 156.0 (C-4), 145.7 (C-6), 144.5 (C-7), 143.8 (C-9) and 110.1 (C-10), δ 3.71 (6-OCH$_3$) *vs* δ 145.7 (C-6), δ 10.00 (7-OH) *vs* δ 145.7 (C-6), 144.5 (C-7) and 135.6 (C-8), δ 3.86 (8-OCH$_3$) *vs* δ 135.6 (C-8), δ 7.42 (H-2′/H-4′) *vs* δ 156.0 (C-4) and 159.1 (C-4′), δ 6.94 (H-3′/H-5′) *vs* δ 126.1 (C-1′), δ 9.99 (4′-OH) *vs* δ 130.6 (C-3′/C-5′) and 159.1 (C-4′).

HR-ESIMS: *m/z* 315.0896 ([M + H])$^+$, Calcd. for $C_{17}H_{15}O_6$, 315.0569).

Reference

Zhong H, Ruan J-L, Yao Q-Q. (2010). Two new 4-arylcoumarins from the seeds of *Calophyllum polyanthum*. *J Asian Nat Prod Res* **12**: 562–568.

7,8-Dihydroxy-3-(3-hydroxy-4-oxo-4*H*-pyran-2-yl)-2*H*-chromen-2-one

Sub-class: Pyrone-coumarin

Chemical structure

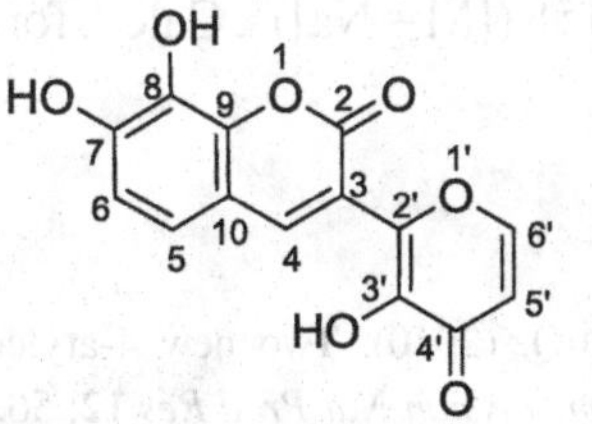

Source: *Bambusa pervariabilis* McClure (Family: Poaceae); leaves

Molecular formula: $C_{14}H_8O_7$

Molecular weight: 288

State: Yellow amorphous powder

Melting point: 278.6–279.8 °C

UV (MeOH): λ_{max} 230, 277, 311 nm

FT-IR (KBr): ν_{max} 3456 (OH), 1744 (pyrone oxo), 1650 (lactone carbonyl), 1575, 1541, 1319 (aromatic unsaturation), 1281, 843, 825 cm^{-1}

^{1}H-NMR (DMSO-d_6, 600 MHz): δ 7.23 (1H, s, H-4), 6.70 (1H, d, J = 8.4 Hz, H-5), 7.02 (1H, d, J = 8.4 Hz, H-6), 6.62 (1H, d, J = 5.4 Hz, H-5′), 8.27 (1H, d, J = 8.4 Hz, H-6′), 10.30 (1H, br s, 7-OH), 9.68 (1H, br s, 8-OH), 12.80 (1H, br s, 3′-OH).

^{13}C-NMR (DMSO-d_6, 150 MHz): δ 162.3 (C-2), 123.9 (C-3), 124.7 (C-4), 122.7 (C-5), 117.7 (C-6), 147.4 (C-7), 144.8 (C-8), 146.4 (C-9), 116.3 (C-10), 155.8 (C-2′), 144.5 (C-3′), 173.3 (C-4′), 116.3 (C-5′), 156.8 (C-6′).

^{1}H-^{1}H COSY: δ 6.70 (H-5) vs δ 7.02 (H-6) and $vice$ $versa$, δ 6.62 (H-5′) vs δ 8.27 (H-6′) and $vice$ $versa$.

HMBC: δ 7.23 (H-4) vs δ 162.3 (C-2), 122.7 (C-5), 146.4 (C-9) and 155.8 (C-2′), δ 6.70 (H-5) vs δ 147.4 (C-7), 146.4 (C-9) and 116.3 (C-10), δ 7.02 (H-6) vs δ 144.8 (C-8) and 116.3 (C-10), δ 6.62 (H-5′) vs δ 144.5 (C-3′), δ 8.27 (H-6′) vs δ 155.8 (C-2′) and 173.3 (C-4′), δ 10.30 (7-OH) vs δ 147.4 (C-7), δ 9.68 (9-OH) vs δ 144.8 (C-8) and 146.4 (C-9).

HR-ESIMS: m/z 311.0158 ([M + Na])$^+$, Calcd. for $C_{14}H_8O_7Na$, 311.0168).

Reference

Zhong H, Ruan J-L, Yao Q-Q. (2010). Two new 4-arylcoumarins from the seeds of *Calophyllum polyanthum*. *J Asian Nat Prod Res* **12**: 562–568.

8,4″-Dihydroxy-3″,4″-dihydrocapnolactone-2′,3′-diol

IUPAC name: 7-(2,3-Dihydroxy-4-(4-hydroxy-4-methyl-5-oxotetrahydrofuran-2-yl)-3-methylbutoxy)-8-hydroxy-2H-chromen-2-one

Sub-class: Coumarin

Chemical structure

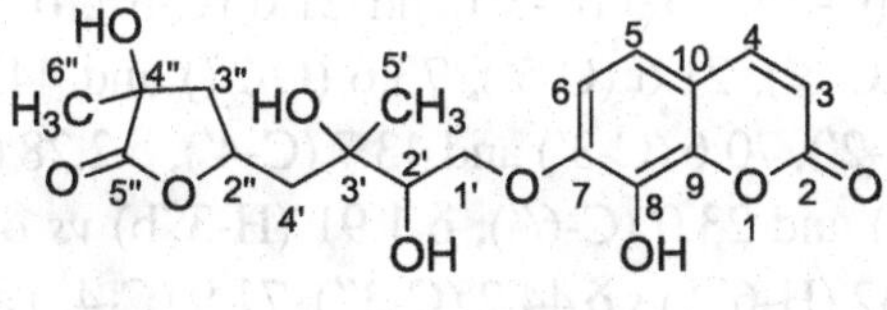

Source: *Micromelum minutum* (G. Forst) Wight and Arn (Family: Rutaceae); leaves

Molecular formula: $C_{19}H_{22}O_9$

Molecular weight: 394

State: Amorphous crystals

Melting point: 236–238 °C

UV (DMSO): λ_{max} (log ε) 289.2 (3.80), 298.0 (sh), 319.8 (4.23) nm

FT-IR (KBr): v_{max} 3400 (OH), 3339 (OH), 2972, 1772 (lactone carbonyl), 1721 (lactone carbonyl), 1613, 1572, 1457, 1347 (aromatic unsaturation), 1282, 835 cm^{-1}

^{1}H-NMR (DMSO-d_6, 500 MHz): δ 6.31 (1H, d, J = 9.5 Hz, H-3), 7.97 (1H, d, J = 9.5 Hz, H-4), 7.19 (1H, d, J = 8.6 Hz, H-5), 6.91 (1H, d, J = 8.6

89

Hz, H-6), 4.62 (1H, dd, J = 11.5, 1.9 Hz, H-1′a), 4.11 (1H, dd, J = 11.5, 8.9 Hz, H-1′b), 3.96 (1H, dd, J = 8.9, 1.9 Hz, H-3′), 2.05 (1H, dd, J = 14.9, 7.8 Hz, H-4′a), 1.86 (1H, dd, J = 14.9, 3.4 Hz, H-4′b), 1.30 (3H, s, H-5′), 4.88 (1H, m, H-2″), 2.28 (1H, dd, J = 13.3, 5.4 Hz, H-3″a), 1.91 (1H, dd, J = 13.2, 10.0 Hz, H-3″b), 1.32 (3H, s, H-6″), 5.19 (1H, s, 3′-OH), 5.58 (1H, s, 4″-OH).

^{13}C-NMR (DMSO-d_6, 125 MHz): δ 159.7 (C-2), 112.8 (C-3), 144.7 (C-4), 119.9 (C-5), 113.2 (C-6), 146.2 (C-7), 131.2 (C-8), 143.2 (C-9), 113.1 (C-10), 64.8 (C-1′), 78.6 (C-2′), 70.6 (C-3′), 43.7 (C-4′), 21.6 (C-5′), 73.6 (C-2″), 44.2 (C-3″), 71.9 (C-4″), 177.3 (C-5″), 23.0 (C-6″).

HMBC: δ 6.31 (H-3) vs δ 159.2 (C-2) and 113.1 (C-10), δ 7.97 (H-4) vs δ 159.7 (C-2), 112.8 (C-3), 119.9 (C-5), 143.2 (C-9) and 113.1 (C-10), δ 7.19 (H-5) vs δ 144.7 (C-4), 113.2 (C-6), 146.2 (C-7), 143.2 (C-9) and 113.1 (C-10), δ 6.91 (H-6) vs δ 146.2 (C-7), 131.2 (C-8) and 113.1 (C-10), δ 4.62 (H-1′a) vs δ 78.6 (C-2′), 4.11 (H-1′b) vs δ 78.6 (C-2′) and 146.2 (C-7), δ 3.96 (H-3′) vs δ 64.8 (C-1′), 70.6 (C-3′), 43.7 (C-4′), 21.6 (C-5′), δ 2.05 (H-4′a) vs δ 78.6 (C-2′), 70.6 (C-3′) and 21.6 (C-5′), δ 1.86 (H-4′b) vs δ 78.6 (C-2′), 70.6 (C-3′), 21.6 (C-5′), 73.6 (C-2″) and 44.2 (C-3″), δ 1.30 (H-5′) vs δ 78.6 (C-2′), 70.6 (C-3′) and 43.7 (C-4′), δ 2.28 (H-3″a) vs δ 43.7 (C-4′), 73.6 (C-2″) and 23.0 (C-6″), δ 1.91 (H-3″b) vs δ 71.9 (C-4″) and 177.3 (C-5″), δ 1.32 (H-6″) vs δ 44.2 (C-3″), 71.9 (C-4″) and 177.3 (C-5″).

NOESY: δ 7.97 (H-4) vs δ 7.19 (H-5) and $vice$ $versa$, δ 4.11 (H-1′b) vs δ 1.30 (H-5′), δ 3.96 (H-2′) vs δ 2.05 (H-5′), δ 4.11 (H-4′a) vs δ 1.30 (H-5′), δ 1.30 (H-5′) vs δ 4.11 (H-1′b), 3.96 (H-2′), 4.11 (H-4′a), 4.88 (H-2″) and 5.19 (3′-OH), δ 4.88 (H-2″) vs δ 1.30 (H-5′), δ 1.91 (H-3″b) vs δ 1.32 (H-6″), δ 1.32 (H-6″) vs δ 1.91 (H-3″b) and 5.58 (4″-OH), 5.19 (3′-OH) vs δ 1.30 (H-5′), δ 5.58 (4″-OH) vs δ 1.32 (H-6″).

EIMS: m/z (rel%) 394 (M$^+$, 0.16), 377 (21.07), 332 (6.91), 271 (7.29), 256 (4.23), 246 (6.86), 205 (13.11), 204 (39.68), 189 (69.63), 175 (57.28), 158 (10.02), 146 (9.95), 121 (12.61), 109 (18.42), 94 (8.09).

Reference

Susidarti R A, Rahmani M, Ismail H B M, Sukari M A, Hin T-Y Y, Lian G E C, Ali A M, Kulip J, Waterman P G. (2007). A new coumarin and triterpenes from Malaysian *Micromelum minutum*. *Nat Prod Res* **20**: 145–151.

(2*R*,3*R*)-2-(2,4-dihydroxyphenyl)-3, 5-dihydroxy-8,8-dimethyl-2, 3-dihydropyrano[2,3-*f*]-chromen-4(8*H*)-one

Sub-class: Pryrano-annulated flavanol

Chemical structure

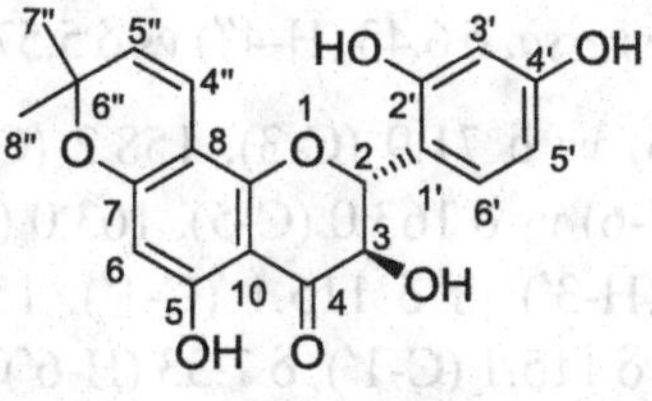

Source: *Desmodium caudatum* (Thunb.) H. Ohashi (Family: Fabaceae); roots

Molecular formula: $C_{20}H_{18}O_7$

Molecular weight: 370

State: Yellow granules

Melting point: 217–224 °C

Bioactivity studied: Anti-MRSA (methicillin-resistant *Staphylococcus aureus*) activity

Specific rotation: $[\alpha]^{20}_D$ +13.8° (MeOH, *c* 0.2)

CD (MeOH, c 2.7 × 10^{-4} M): λ_{max} ($\Delta\varepsilon$) 327 (+0.13), 293 (−0.49), 229 (+0.51) nm

^{1}H-NMR (Acetone-d_6, 400 MHz): δ 5.53 (1H, d, J = 11.6 Hz, H-2β), 4.89 (1H, d, J = 11.6 Hz, H-3α), 5.90 (1H, s, H-6), 6.47 (1H, d, J = 2.0 Hz, H-3′), 6.43 (1H, dd, J = 8.4, 2.0 Hz, H-5′), 7.33 (1H, d, J = 8.4 Hz, H-6′), 6.43 (1H, d, J = 10.0 Hz, H-4″), 5.57 (1H, d, J = 10.0 Hz, H-5″), 1.42 (3H, s, H-7″), 1.39 (3H, s, H-8″).

^{13}C-NMR (Acetone-d_6, 100 MHz): δ 79.4 (C-2), 71.9 (C-3), 199.0 (C-4), 163.0 (C-5), 97.6 (C-6), 163.0 (C-7), 102.6 (C-8), 158.2 (C-9), 101.9 (C-10), 115.1 (C-1′), 158.1 (C-2′), 103.6 (C-3′), 159.8 (C-4′), 103.6 (C-5′), 130.6 (C-6′), 115.9 (C-4″), 127.4 (C-5″), 79.0 (C-6″), 28.6 (C-7″), 28.4 (C-8″).

HMQC: δ 5.53 (H-2β) *vs* δ 79.2 (C-2), δ 4.89 (H-3α) *vs* δ 71.9 (C-3), δ 5.90 (H-6) *vs* δ 97.6 (C-6), δ 6.47 (H-3′) *vs* δ 103.6 (C-3′), δ 6.43 (H-5′) *vs* δ 103.6 (C-5′), δ 7.33 (H-6′) *vs* δ 130.6 (C-6′), δ 6.43 (H-4″) *vs* δ 115.9 (C-4″), δ 5.57 (H-5″) *vs* δ 127.4 (C-5″), δ 1.42 (H-7″) *vs* δ 28.6 (C-7″), δ 1.39 (H-8″) *vs* δ 28.4 (C-8″).

^{1}H-^{1}H COSY: δ 5.53 (H-2β) *vs* δ 4.89 (H-3α) and *vice versa*, δ 6.43 (H-5′) *vs* δ 7.33 (H-6′) and *vice versa*, δ 6.43 (H-4″) *vs* δ 5.57 (H-5″) and *vice versa*.

HMBC: δ 5.53 (H-2β) *vs* δ 71.9 (C-3), 158.2 (C-9), 115.1 (C-1′) and 158.1 (C-2′), δ 5.90 (H-6) *vs* δ 163.0 (C-5), 163.0 (C-7), 102.6 (C-8) and 101.9 (C-10), δ 6.47 (H-3′) *vs* δ 115.1 (C-1′), 158.1 (C-2′) and 159.8 (C-4′), δ 6.43 (H-5′) *vs* δ 115.1 (C-1′), δ 7.33 (H-6′) *vs* δ 79.4 (C-2), 158.1 (C-2′) and 159.8 (C-4′), δ 6.43 (H-4″) *vs* δ 163.0 (C-7), 102.6 (C-8), 158.2 (C-9) and 79.0 (C-6″), δ 5.57 (H-5″) *vs* δ 102.6 (C-8), 79.0 (C-6″) and 28.6 (C-7″), δ 1.42 (H-7″) *vs* δ 127.4 (C-5″), 79.0 (C-6″) and 28.4 (C-8″) (selected 2D correlations were shown).

HR-ESI-MS: m/z 393.0946 ([M + Na]$^+$, Calcd. for $C_{20}H_{18}O_7Na$, 393.0950).

Reference

Sasaki H, Kashiwada Y, Shibata H, Takaishi Y. (2012). Prenylated flavonoids from *Desmodium caudatum* and evaluation of their anti-MRSA activity. *Phytochemistry* **82**: 136–142.

(2*R*,3*R*)-2-(2,4-Dihydroxyphenyl)-3,5,7-trihydroxy-8-prenyl-chroman-4-one

IUPAC name: (2*R*,3*R*)-2-(2,4-Dihydroxyphenyl)-3,5,7-trihydroxy-8-(3-methylbut-2-en-1-yl)chroman-4-one

Sub-class: Prenylated flavanol

Chemical structure

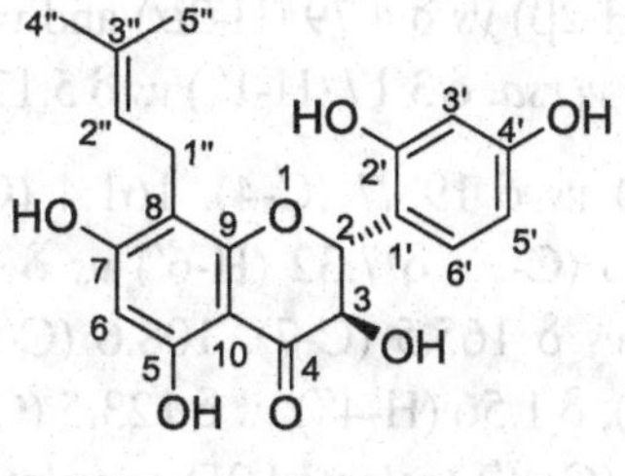

Source: *Desmodium caudatum* (Thunb.) H. Ohashi (Family: Fabaceae); roots

Molecular formula: $C_{20}H_{20}O_7$

Molecular weight: 372

State: Yellow granules

Melting point: 185–192 °C

Bioactivity studied: Anti-MRSA (methicillin-resistant *Staphylococcus aureus*) activity

Specific rotation: $[\alpha]^{20}_D$ +54.2° (MeOH, c 0.2)

CD (MeOH, c 2.7 × 10⁻⁴ M): λ_{max} ($\Delta\varepsilon$) 324 (+0.43), 291 (–5.48), 224 (+1.47) nm

¹H-NMR (Acetone-d_6, 400 MHz): δ 5.46 (1H, d, J = 11.4 Hz, H-2β), 4.79 (1H, d, J = 11.4 Hz, H-3α), 6.06 (1H, s, H-6), 6.46 (1H, d, J = 2.4 Hz, H-3'), 6.42 (1H, dd, J = 8.4, 2.4 Hz, H-5'), 7.32 (1H, d, J = 8.4 Hz, H-6'), 3.17 (2H, d, J = 7.2 Hz, H-1''), 5.17 (1H, br t, J = 7.2 Hz, H-2''), 1.56 (3H, s, H-4''), 1.57 (3H, s, H-5'').

¹³C-NMR (Acetone-d_6, 100 MHz): δ 79.3 (C-2), 72.3 (C-3), 198.7 (C-4), 162.7 (C-5), 96.6 (C-6), 165.5 (C-7), 108.6 (C-8), 161.4 (C-9), 101.6 (C-10), 115.7 (C-1'), 158.0 (C-2'), 103.8 (C-3'), 159.7 (C-4'), 107.8 (C-5'), 130.4 (C-6'), 22.1 (C-1''), 123.5 (C-2''), 131.3 (C-3''), 26.9 (C-4''), 17.7 (C-5'').

HMQC: δ 5.46 (H-2β) vs δ 79.3 (C-2), δ 4.79 (H-3α) vs δ 72.3 (C-3), δ 6.06 (H-6) vs δ 96.6 (C-6), δ 6.46 (H-3') vs δ 103.8 (C-3'), δ 6.42 (H-5') vs δ 107.8 (C-5'), δ 7.32 (H-6') vs δ 130.4 (C-6'), δ 3.17 (H-1'') vs δ 22.1 (C-1''), δ 5.17 (H-2'') vs δ 123.5 (C-2''), δ 1.56 (H-4'') vs δ 26.9 (C-4''), δ 1.57 (H-5'') vs δ 17.7 (C-5'').

¹H-¹H COSY: δ 5.46 (H-2β) vs δ 4.79 (H-3α) and $vice versa$, δ 6.42 (H-5') vs δ 7.32 (H-6') and $vice versa$, δ 3.17 (H-1'') vs δ 5.17 (H-2'') and $vice versa$.

HMBC: δ 5.46 (H-2β) vs δ 198.7 (C-4), 161.4 (C-9) and 115.7 (C-1'), δ 4.79 (H-3α) vs δ 79.3 (C-2), δ 7.32 (H-6') vs δ 108.6 (C-8) and 101.6 (C-10), δ 3.17 (H-1'') vs δ 165.5 (C-7), 108.6 (C-8), 161.4 (C-9), 123.5 (C-2'') and 131.3 (C-3''), δ 1.56 (H-4'') vs δ 123.5 (C-2'') and 131.3 (C-3''), δ 1.57 (H-5'') vs δ 26.9 (C-4'') (selected 2D correlations were shown).

HR-ESI-MS: m/z 395.1106 ([M + Na]⁺, Calcd. for $C_{20}H_{20}O_7Na$, 395.1107).

Reference

Sasaki H, Kashiwada Y, Shibata H, Takaishi Y. (2012). Prenylated flavonoids from *Desmodium caudatum* and evaluation of their anti-MRSA activity. *Phytochemistry* **82**: 136–142.

3′,5′-Dimethoxy-[2″,3″: 7,8]-furanoflavone

IUPAC name: 2-(3,5-Dimethoxyphenyl)-4H-furo[2,3-h]chromen-4-one

Sub-class: Furanoflavone

Chemical structure

Source: *Millettia erythrocalyx* Gagnep. (Family: Ligumonisae); leaves

Molecular formula: $C_{19}H_{14}O_5$

Molecular weight: 322

State: Colorless needles

Melting point: 178-180 °C

UV (MeOH): λ_{max} (log ε): 240 (3.5), 263 (3.4), 303 (3.5) nm

IR (film): ν_{max} 1630 (C=O), 1583 cm^{-1}.

^{1}H-NMR (CDCl$_3$, 400 MHz): δ 6.84 (1H, s, H-3), 8.15 (1H, d, J = 8.7 Hz, H-5), 7.55 (1H, d, J = 8.7 Hz, H-6), 7.07 (1H, d, J = 2.1 Hz, H-2′),

6.63 (1H, d, J = 2.1 Hz, H-4′), 7.07 (1H, d, J = 2.1 Hz, H-6′), 7.18 (1H, d, J = 2.1 Hz, H-4″), 7.76 (1H, d, J = 2.1 Hz, H-5″), 3.88 (6H, s, 3′-OCH_3 and 5′-OCH_3).

^{13}C-NMR (CDCl$_3$, 100 MHz): δ 162.5 (C-2), 108.5 (C-3), 178.2 (C-4), 121.8 (C-5), 110.2 (C-6), 158.4 (C-7), 117.2 (C-8), 150.8 (C-9), 119.4 (C-10), 133.7 (C-1′), 104.5 (C-2′), 161.3 (C-3′), 103.3 (C-4′), 161.3 (C-5′), 104.5 (C-6′), 55.6 (3′-OCH_3 and 5′-OCH_3), 104.2 (C-4″), 145.8 (C-5″).

HRESMS: m/z 323.0916 ([M + H])$^+$, Calcd. for $C_{19}H_{15}O_5$, 323.0919).

Reference

Likhitwitayawuid K, Sritularak B, Benchanak K, Lipipun V, Mathew J, Schinazi RF. (2005). Phenolics with antiviral activity from *Millettia Erythrocalyx* and *Artocarpus Lakoocha*. *Nat Prod Res* **19**: 177–182.

7,4′-Dimethylapigenin-6-C-β-glucopyranosyl-2″-O-α-L-arabinopyranoside

IUPAC name: 6-((2S,3R,5S,6R)-4,5-dihydroxy-6-(hydroxymethyl)-3-((((2S,3R,4S,5S)-3,4,5-trihydroxytetrahydro-2H-pyran-2-yl)oxy)tetrahydro-2H-pyran-2-yl)-5-hydroxy-7-methoxy-2-(4-methoxyphenyl)-4H-chromen-4-one

Sub-class: Flavone glycoside

Chemical structure

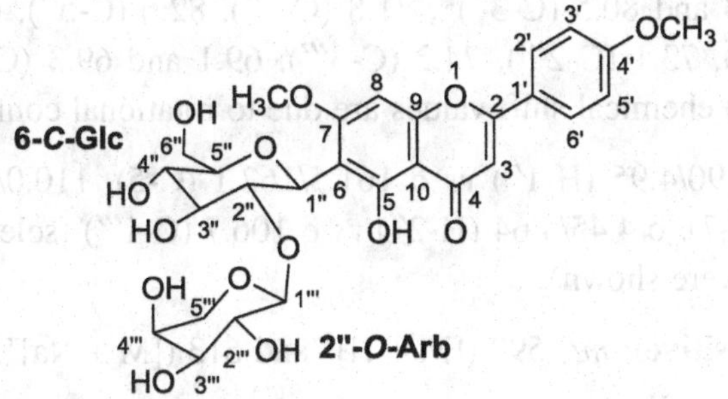

Source: *Solanum verbascifolium* (Family: Solanaceae); leaves

Molecular formula: $C_{28}H_{32}O_{14}$

Molecular weight: 592

State: Yellow amorphous solid

Bioactivity studied: TRAIL-resistance-overcoming activity

Specific rotation: $[\alpha]^{20}_{D}$ −18.0° (MeOH, c 1.0)

UV (MeOH): λ_{max} (ε) 274 (28,400) and 329 (29,300) nm

IR (ATR): ν_{max} 3350 (OH), 2920, 1650 (γ-pyrone carbonyl), 1600 (aromatic unsaturaion) cm^{-1}

^{1}H-NMR (DMSO-d_6, 600 MHz): δ 6.64 and 6.66 (1H each, s, H-3)*, 6.68 (1H, s, H-8), 7.92 (2H, d, J = 8.5 Hz, H-2′ and H-6′), 7.06 (2H, d, J = 8.5 Hz, H-3′ and H-5′), 3.90 (3H, s, C$_7$-OCH_3), 3.88 (3H, s, C$_{4'}$-OCH_3), 4.90 and 4.95 (1H each, d, J = 10.0 Hz, H-1″)*, 4.45 and 4.64 (1H each, t, J = 9.0 Hz, H-2″)*, 3.60 (1H, m, H-3″), 3.48 (1H, m, H-4″), 3.42 (1H, m, H-5″), 3.70 and 3.90 (2H each, m, H-6″)*, 4.36 and 4.37 (1H each, d, J = 6.6 Hz, H-1‴)*, 3.48 (1H, m, H-2‴), 3.42 (1H, m, H-3‴), 3.57 (1H, m, H-4‴), 3.04 and 3.17 (1H each, d, J = 6.6 Hz, H-5‴a)*, 3.22 and 3.36 (1H each, d, J = 6.6 Hz, H-5‴b)* [*two chemical shift values are due to rotational conformers].

^{13}C-NMR (DMSO-d_6, 150 MHz): δ 165.9 (C-2), 104.6 (C-3), 183.8 and 184.2 (C-4)*, 161.5 and 162.1 (C-5)*, 110.0 and 110.2 (C-6)*, 167.2 (C-7), 91.4 and 91.7 (C-8)*, 159.1 (C-9), 106.0 (C-10), 124.2 (C-1′), 129.3 (C-2′ and C-6′), 115.6 (C-3′ and C-5′), 164.4 (C-4′), 56.7 and 57.0 (C$_7$-OCH_3)*, 56.1 (C$_{4'}$-OCH_3), 71.8 and 72.7 (C-1″)*, 81.7 and 82.0 (C-2″)*, 80.3 and 80.5 (C-3″)*, 71.8 (C-4″), 82.6 (C-5″), 63.2 (C-6″), 106.7 (C-1‴), 73.1 (C-2‴), 74.2 (C-3‴), 69.1 and 69.3 (C-4‴)*, 66.4 (C-5‴) [*two chemical shift values are due to rotational conformers].

HMBC: δ 4.90/4.95 (H-1″) vs δ 161.5/162.1 (C-5), 110.0/110.2 (C-6) and 167.2 (C-7), δ 4.45/4.64 (H-2″) vs δ 106.7 (C-1‴) (selected HMBC correlations were shown).

FAB-MS (positive): m/z 593 ([M + H]$^+$ and 613 ([M + Na]$^+$.

HR-FAB-MS: m/z 593.1865 ([M + H]$^+$, Calcd. for C$_{28}$H$_{33}$O$_{14}$, 593.1871).

Reference

Ohtsuki T, Miyagawa T, Koyano T, Kowithayakorn T, Ishibashi M. (2010). Isolation and structure elucidation of flavonoid glycosides from *Solanum verbascifolium*. *Phytochemistry Lett* **3**: 88–92.

Eriodictyol 7-*O*-(6″-feruloyl)-β-D-glucopyranoside

IUPAC name: (*E*)-((2*R*,3*S*,5*R*,6*S*)-6-(((*R*)-2-(3,4-dihydroxyphenyl)-5-hydroxy-4-oxochroman-7-yl)oxy)-3,4,5-trihydroxytetrahydro-2*H*-pyran-2-yl)methyl 3-(4-hydroxy-3-methoxyphenyl)acrylate

Sub-class: Flavanone glycoside

Chemical structure

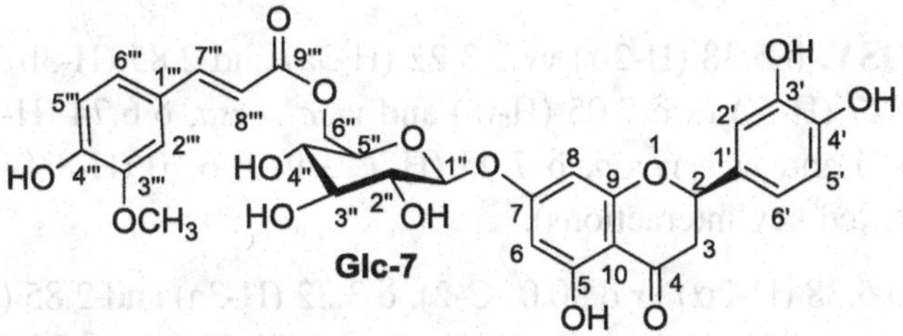

Source: *Elsholtzia bodinieri* Vaniot (Family: Labiatae); whole plants

Molecular formula: $C_{31}H_{30}O_{14}$

Molecular weight: 626

State: Yellow amorphous powder

Specific rotation: $[\alpha]^{20.9}_{D}$ −68.3° (pyridine, *c* 0.27)

UV (MeOH): λ_{max} (log ε) 203 (4.82), 285 (4.43), 327 (4.32) nm

IR (KBr): v_{max} 3420 (OH), 2926, 1690 (α,β-ester carbonyl), 1642 (α,β-unsaturated carbonyl), 1516, 1453 (aromatic ring), 1272, 1194, 1066 cm^{-1}

^{1}H-NMR (Pyridine-d_5, 400 MHz): δ 5.38 (1H, d, J = 14.6 Hz, H-2α), 3.22 (1H, dd, J = 17.1, 14.6 Hz, H-3a), 2.85 (1H, d, J = 17.1 Hz, H-3b), 6.71 (1H, d, J = 2.3 Hz, H-6), 6.44 (1H, d, J = 2.3 Hz, H-8), 7.51 (1H, br s, H-2'), 7.27 (1H, d, J = 8.0 Hz, H-5'), 7.05 (1H, br d, J = 8.0 Hz, H-6'), 5.71 (1H, d, J = 7.6 Hz, H-1''), 4.24-4.40 (4H, m, H-2'', H-3'', H-4'' and H-5''), 5.13 (1H, d, J = 12.1 Hz, H-6''a), 4.92 (1H, dd, J = 12.1, 5.8 Hz, H-6''b), 7.05 (1H, d, J = 1.2 Hz, H-2'''), 6.74 (1H, d, J = 5.6 Hz, H-5'''), 7.18 (1H, d, J = 5.6 Hz, H-6'''), 7.95 (1H, d, J = 15.7 Hz, H-7'''), 6.70 (1H, d, J = 15.7 Hz, H-8'''), 12.6 (1H, s, 5-OH), 3.89 (3H, s, 3'''-OCH_3).

^{13}C-NMR (Pyridine-d_5, 100 MHz): δ 80.0 (C-2), 43.5 (C-3), 197.5 (C-4), 164.7 (C-5), 97.6 (C-6), 166.5 (C-7), 96.8 (C-8), 163.6 (C-9), 104.5 (C-10), 130.5 (C-1'), 115.4 (C-2'), 147.6 (C-3'), 149.1 (C-4'), 116.7 (C-5'), 118.9 (C-6'), 101.6 (C-1''), 74.7 (C-2''), 78.3 (C-3''), 71.3 (C-4''), 75.8 (C-5''), 64.4 (C-6''), 126.6 (C-1'''), 111.6 (C-2'''), 151.3 (C-3'''), 148.1 (C-4'''), 116.9 (C-5'''), 123.8 (C-6'''), 146.1 (C-7'''), 115.1 (C-8'''), 167.7 (C-9'''), 56.1 (3'''-OCH_3).

^{1}H-^{1}H-COSY: δ 5.38 (H-2α) *vs* δ 3.22 (H-3a) and 2.85 (H-3b) and *vice versa*, δ 7.27 (H-5') *vs* δ 7.05 (H-6') and *vice versa*, δ 6.74 (H-5''') *vs* δ 7.18 (H-6''') and *vice versa*, δ 7.95 (H-7''') *vs* δ 6.70 (H-8''') and *vice versa* (selected key interactions).

HMQC: δ 5.38 (H-2α) *vs* δ 80.0 (C-2), δ 3.22 (H-3a) and 2.85 (H-3b) *vs* δ 43.5 (C-3), δ 6.71 (H-6) *vs* δ 97.6 (C-6), δ 6.44 (H-8) *vs* δ 96.8 (C-8), δ 7.51 (H-2') *vs* δ 115.4 (C-2'), δ 7.27 (H-5') *vs* δ 116.7 (C-5'), δ 7.05 (H-6') *vs* δ 118.9 (C-6'), δ 5.71 (H-1'') *vs* δ 101.6 (C-1''), δ 4.24-4.40 (H-3''/H-4''/H-5'') *vs* δ 78.3 (C-3'')/71.3 (C-4'')/75.8 (C-5''), δ 5.13 (H-6''a)/4.92 (H-6''b) *vs* δ 64.4 (C-6''), δ 7.05 (H-2''') *vs* δ 111.6 (C-2'''), δ 6.74 (H-5''') *vs* δ 116.9 (C-5'''), δ 7.18 (H-6''') *vs* δ 123.8 (C-6'''), δ 7.95 (H-7''') *vs* δ 146.1 (C-7'''), δ 6.70 (H-8''') *vs* δ 115.1 (C-8'''), δ 3.89 (3'''-OCH_3) *vs* δ 56.1 (3'''-OCH_3).

HMBC: δ 5.38 (H-2α) *vs* δ 115.4 (C-2') and 197.5 (C-4), δ 3.22 (H-3a)/2.85 (H-3b) *vs* δ 130.5 (C-1'), δ 12.6 (5-OH) *vs* δ 164.7 (C-5), 97.6

(C-6) and 104.5 (C-10), δ 5.71 (H-1″) *vs* δ 166.5 (C-7), δ 5.13 (H-6″a)/4.92 (H-6″b) *vs* δ 167.7 (C-9‴), δ 7.05 (H-2‴) *vs* δ 151.3 (C-3‴) and 148.1 (C-4‴), δ 6.74 (H-5‴) *vs* δ 148.1 (C-4‴) and 123.8 (C-6″), δ 7.18 (H-6‴) *vs* δ 115.1 (C-8‴), δ 7.95 (H-7‴) *vs* δ 111.6 (C-2‴), 123.8 (C-6‴) and 167.7 (C-9‴), δ 6.70 (H-8‴) *vs* δ 126.6 (C-1‴), δ 3.89 (3‴-OCH_3) *vs* δ 151.3 (C-3‴) (Selected 2D-correlations were shown).

HR-FAB-MS: *m/z* 625.1543 ([M − H]⁻, Calcd. for $C_{31}H_{29}O_{14}$, 625.1557).

Reference

Li R-T, Li J-T, Wang J-K, Han Q-B, Zhu Z-Y, and Sun H-D. (2008). Three new flavonoid glycosides isolated from *Elsholtzia bodinieri*. *Chem Pharm Bull* **56**: 592–594.

Eriodictyol 7-*O*-sophoroside [Eriodictyol-7-*O*-β-D-glucopyranosyl-(1→2)-β-D-glucopyranoside]

IUPAC name: 7-((((2*S*,3*R*,5*S*,6*R*)-4,5-Dihydroxy-6-(hydroxymethyl)-3-(((2*S*,3*R*,5*S*,6*R*)-3,4,5-trihydroxy-6-(hydroxymethyl)tetrahydro-2*H*-pyran-2-yl)oxy)tetrahydro-2*H*-pyran-2-yl)oxy)-2-(3,4-dihydroxyphenyl)-5-hydroxychroman-4-one

Sub-class: Flavanone glycoside

Chemical structure

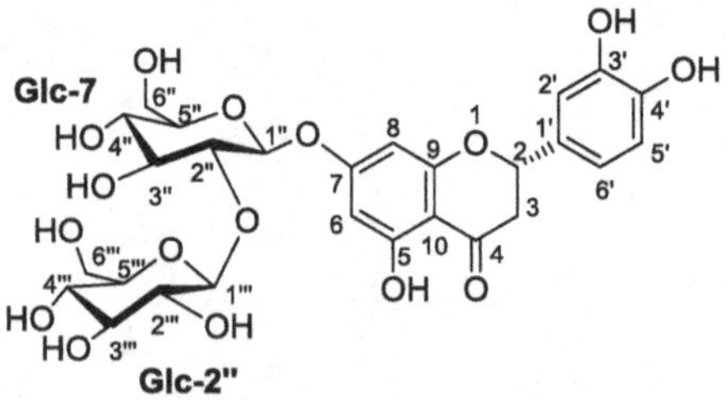

Source: *Globularia alypum* Linn. (Family: Globulariaceae); aerial parts

Molecular formula: $C_{27}H_{32}O_{16}$

Molecular weight: 612

State: Amorphous solid

Specific rotation: $[\alpha]^{20}_{D}$ −35.6° (MeOH, *c* 0.1)

Bioactivity studied: Antioxidant [showed DPPH radical scavenging activity with IC_{50} value of 12.0 µM]

UV (MeOH): λ_{max} (log ε) 289 (4.2) and 324sh (3.7) nm

IR (dried flim): ν_{max} 3420 (OH), 1650 (α,β-unsaturated carbonyl), 1610, 1540, 1520 (aromatic ring) cm^{-1}

^{1}H-NMR (DMSO-d_6, 300 MHz): δ 5.34 (1H, dd, *J* = 11.0, 5.0 Hz, H-2), 3.17 (1H, dd, *J* = 17.0, 5.0 Hz, H-3a), 2.78 (1H, dd, *J* = 17.0, 11.0 Hz, H-3b), 6.92 (1H, d, *J* = 2.0 Hz, H-6), 6.36 (1H, d, *J* = 2.0 Hz, H-8), 6.22 (1H, d, *J* = 2.5 Hz, H-2′), 6.39 (1H, d, *J* = 8.5 Hz, H-5′), 7.62 (1H, dd, *J* = 8.5, 2.5 Hz, H-6′), 5.20 (1H, d, *J* = 7.8 Hz, H-1″), 3.70 (2H, m, H-2″ and H-6″b), 3.50 (1H, m, H-3″), 3.65 (2H, m, H-4″ and H-6‴b), 3.69 (1H, m, H-5″), 3.89 (1H, dd, *J* = 11.8, 2.0 Hz, H-6″a), 4.65 (1H, d, J = 7.8 Hz, H-1‴), 3.23 (1H, m, H-2‴), 3.40 (1H, m, H-3‴), 3.53 (1H, m, H-4‴), 3.42 (1H, m, H-5‴), 3.68 (1H, m, H-6‴a).

^{13}C-NMR (DMSO-d_6, 75 MHz): δ 79.3 (C-2), 43.4 (C-3), 198.4 (C-4), 167.6 (C-5), 92.3 (C-6), 163.1 (C-7), 96.7 (C-8), 165.8 (C-9), 101.4 (C, C-10), 130.7 (C-1′), 112.9 (C-2′), 147.9 (C, C-3′), 148.1 (C-4′), 116.0 (C-5′), 120.4 (C-6′), 100.9 (C-1″), 79.0 (C-2″), 72.9 (C-3″), 69.6 (C-4″), 75.0 (C-5″), 62.8 (C-6″), 103.2 (C-1‴), 72.9 (C-2‴), 73.7 (C-3‴), 71.8 (C-4‴), 76.5 (C-5‴), 61.1 (C-6‴).

HMQC: δ 5.34 (H-2) *vs* δ 79.3 (C-2), δ 3.17 (H-3a) and 2.78 (H-3b) *vs* δ 43.4 (C-3), δ 6.92 (H-6) *vs* δ 92.3 (C-6), δ 6.36 (H-8) *vs* δ 96.7 (C-8), δ 6.22 (H-2′) *vs* δ 112.9 (C-2′), δ 6.39 (H-5′) *vs* δ 116.0 (C-5′), δ 7.62 (H-6′) *vs* δ 120.4 (C-6′), δ 5.20 (H-1″) *vs* δ 100.9 (C-1″), δ 3.70 (H-2″/ H-6″b) *vs* δ 79.0 (C-2″) and 62.8 (C-6″), δ 3.50 (H-3″) *vs* δ 72.9 (C-3″), δ 3.65 (H-4″/H-6‴b) *vs* δ 69.6 (C-4″) and 61.1 (C-6‴), δ 3.69 (H-5″) *vs* δ 75.0 (C-5″), δ 3.89 (H-6″a) *vs* δ 62.8 (C-6″), δ 4.65 (H-1‴) *vs* δ 103.2 (C-1‴), δ 3.23 (H-2‴) *vs* δ 72.9 (C-2‴), δ 3.40 (H-3‴) *vs* δ 73.7 (C-3‴), δ 3.53 (H-4‴) *vs* δ 71.8 (C-4‴), δ 3.42 (H-5‴) *vs* δ 76.5 (C-5‴), δ 3.68 (H-6‴a) *vs* δ 61.1 (C-6‴).

Negative LRESIMS: *m/z* (%rel) 611 ([M − H]$^{-}$, 100), 475 (53), 287 (28), 151 (99).

Positive LRESIMS: m/z (%rel) 613 ([M + H]$^+$, 52), 451 (86), 433(21), 331 (11), 289 (100).

HRMS: m/z 613.1786 ([M + H]$^+$, Calcd. for $C_{27}H_{33}O_{16}$, 613.1769).

Reference

Es-Safi N-E, Khlifi S, Kerhoas L, Kollmann A, El Abbouyi A, Ducrot P-H. (2005). Antioxidant constituents of the aerial parts of *Globularia alypum* growing in Morocco. *J Nat Prod* **68**: 1293–1296.

7-*O*-Feruloylorientin

IUPAC name: (*E*)-2-(3,4-dihydroxyphenyl)-5-hydroxy-4-oxo-8-((2*S*,3*R*,4*R*,5*S*,6*R*)-3,4,5-trihydroxy-6-(hydroxymethyl)tetrahydro-2*H*-pyran-2-yl)-4*H*-chromen-7-yl 3-(4-hydroxy-3-methoxyphenyl)acrylate

Sub-class: Flavone *C*-glycoside

Chemical structure

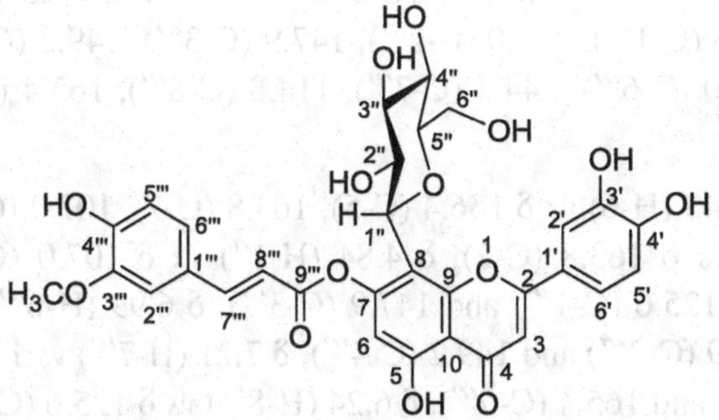

Source: *Gentiana piasezkii* (Family: Gentianaceae); whole plants

Molecular formula: $C_{31}H_{28}O_{14}$

Molecular weight: 624

State: Yellow oil

Melting point: 143–144 °C

Specific rotation: $[\alpha]^{20}_{D}$ −157° (MeOH, *c* 0.25)

Bioactivity studied: Anti-oxidant (DPPH assay)

UV (MeOH): λ_{max} (log ε) 202.6 (41.4), 329.4 (23.8) nm

FT-IR (KBr): ν_{max} 3264 (OH), 2935, 2257, 2128, 1706 ((α,β-unsaturated ester carbonyl), 1650 (α,β-unsaturated carbonyl), 1606, 1516, 1489, 1445 (aromatic moiety), 1357, 1273, 1227, 1176, 1124, 1085, 1027, 998, 823, 765, 688, 563 cm^{-1}

^{1}H-NMR (DMSO-d_6, 400 MHz): δ 6.63 (1H, s, H-3), 6.43 (1H, s, H-6), 7.36 (1H, s, H-2′), 6.86 (1H, d, J = 8.0 Hz, H-5′), 7.38 (1H, d, J = 8.0 Hz, H-6′), 13.63 (1H, s, 5-OH), 4.84 (1H, d, J = 9.6 Hz, H-1″), 7.19 (1H, br s, H-2″), 3.5-3.7 (5H, m, H-3″, H-4″, H-5″ and H-6″), 7.19 (1H, br s, H-2‴), 6.73 (1H, d, J = 8.0 Hz, H-5‴), 6.99 (1H, d, J = 8.0 Hz, H-6‴), 7.21 (1H, d, J = 16.0 Hz, H-7‴), 6.24 (1H, d, J = 16.0 Hz, H-8‴), 3.76 (3H, s, 3‴-OCH_3).

^{13}C-NMR (DMSO-d_6, 100 MHz): δ 163.8 (C-2 and C-7), 102.9 (C-3 and C-6), 181.8 (C-4), 156.4 (C-5 and C-9), 107.0 (C-8 and C-10), 121.4 (C-1′), 119.0 (C-2′), 145.7 (C-3′), 149.7 (C-4′), 116.1 (C-5′), 113.3 (C-6′), 76.4 (C-1″), 71.9 (C-2″), 80.1 (C-3″), 70.7 (C-4″), 81.8 (C-5″), 61.4 (C-6″), 125.6 (C-1‴), 111.0 (C-2‴), 147.9 (C-3‴), 149.2 (C-4‴), 115.5 (C-5‴), 123.0 (C-6‴), 144.5 (C-7‴), 114.8 (C-8‴), 165.4 (C-9‴), 55.7 (3‴-OCH_3).

HMBC: δ 6.43 (H-6) *vs* δ 156.4 (C-5), 163.8 (C-7), 107.0 (C-8/C-10), δ 7.36 (H-2′) *vs* δ 163.8 (C-2), δ 4.84 (H-1″) *vs* δ 107.0 (C-8), δ 6.73 (H-5‴) *vs* δ 125.6 (C-1‴) and 147.9 (C-3‴), δ 6.99 (H-6‴) *vs* δ 111.0 (C-2‴), 147.9 (C-3‴) and 149.2 (C-4‴), δ 7.21 (H-7‴) *vs* 111.0 (C-2‴), 123.0 (C-6‴) and 165.4 (C-9‴), δ 6.24 (H-8‴) *vs* δ 125.6 (C-1‴), δ 3.76 (3‴-OCH_3) *vs* δ 147.9 (C-3‴) (selected 2D-correlations were shown).

HR-ESI-MS: m/z 625.1544 ([M + H]$^+$, Calcd. for $C_{31}H_{29}O_{14}$, 625.1552).

Reference

Wu Q-X, Li Y, Shi Y-Y. (2006). Antioxidant phenolic glucosides from *Gentiana piasezkii*. *J Asian Nat Prod* **8**: 391–396.

Flemingichromone

IUPAC name: 1,3,8,10-Tetrahydroxy-2-methyl-4,9-bis(3-methylbut-2-en-1-yl)-11*H*-benzofuro[2,3-*b*]chromen-11-one

Sub-class: Prenylated isoflavone derivative

Chemical structure

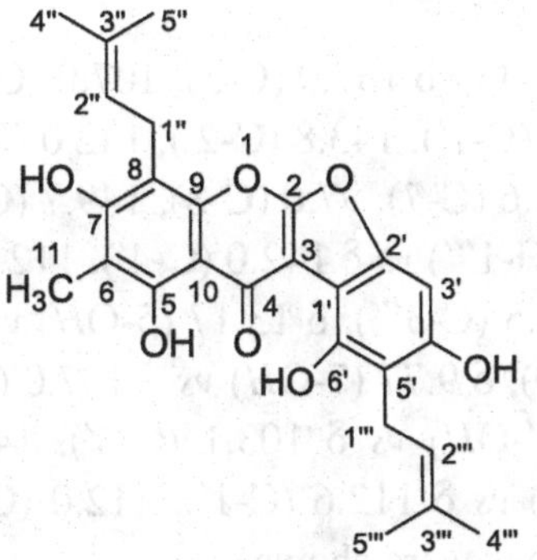

Source: *Flemingia macrophylla* (Linn.) (Family: Leguminosae); aerial parts

Molecular formula: $C_{26}H_{26}O_7$

Molecular weight: 450

State: Colorless amorphous solid

Melting point: 280–282 °C

Bioactivity studied: Neuroprotective (evaluated to possess protective activity against neuronal cell-death on fibril Aβ [f-Aβ-(25-35)]-induced neurotoxicity with EC_{50} value of 31.43 ± 3.16 μM).

UV: λ_{max} (MeOH) (log ε): 262 (4.03), 288 (3.85), 348 (3.54) nm

IR (KBr): ν_{max} 3509 (OH), 1648 (carbonyl), 1622, 1596, 1518, 1385, 1376 (aromatic unsaturaion), 1083, 962 cm^{-1}

^{1}H-NMR (DMSO-d_6, 500 MHz): δ 7.20 (1H, s, H-3′), 3.48 (2H, d, J = 7.0 Hz, H-1″), 5.16 (1H, t, J = 7.0 Hz, H-2″), 1.80 (3H, s, H-4″), 1.65 (3H, s, H-5″), 3.49 (2H, d, J = 7.0 Hz, H-1‴), 5.28 (1H, t, J = 7.0 Hz, H-2‴), 1.84 (3H, s, H-4‴), 1.66 (3H, s, H-5‴), 2.09 (3H, s, 6-CH_3), 13.17 (1H, br s, 5-OH), 9.77 (1H, br s, 7-OH), 8.69 (1H, br s, 4′-OH), 9.67 (1H, br s, 6′-OH).

^{13}C-NMR (CDCl$_3$, 125 MHz): δ 164.0 (C-2), 107.8 (C-3), 178.5 (C-4), 157.0 (C-5), 107.0 (C-6), 158.6 (C-7), 97.0 (C-8), 149.7 (C-9), 102.5 (C-10), 112.6 (C-1′), 143.8 (C-2′), 103.1 (C-3′), 142.0 (C-4′), 112.0 (C-5′), 141.7 (C-6′), 21.7 (C-1″), 121.4 (C-2″), 131.0 (C-3″), 25.4 (C-4″), 17.6 (C-5″), 22.6 (C-1‴), 121.9 (C-2‴), 131.5 (C-3‴), 25.5 (C-4‴), 17.6 (C-5‴), 8.11 (6-CH_3).

HMBC: δ 8.11 (6-CH_3) *vs* δ 157.0 (C-5), 107.0 (C-6) and 158.6 (C-7), δ 7.20 (H-3′) *vs* δ 112.6 (C-1′), 143.8 (C-2′), 142.0 (C-4′) and 112.0 (C-5′), δ 3.48 (H-1″) *vs* δ 158.6 (C-7), 97.0 (C-8), 149.7 (C-9), 121.4 (C-2″) and 131.0 (C-3″), δ 3.49 (H-1‴) *vs* δ 142.0 (C-4′), 112.0 (C-5′), 141.7 (C-6′), 121.9 (C-2‴) and 131.5 (C-3‴), δ 13.17 (5-OH) *vs* δ 157.0 (C-5), 107.0 (C-6) and 102.5 (C-10), δ 9.77 (7-OH) *vs* δ 107.0 (C-6), 158.6 (C-7) and 97.0 (C-8), δ 8.69 (4′-OH) *vs* δ 103.1 (C-3′), 142.0 (C-4′) and 112.0 (C-5′), δ 9.67 (6′-OH) *vs* δ 112.6 (C-1′), 112.0 (C-5′) and 141.7 (C-6′) (selected 2D-correlations were shown).

EIMS: m/z (%rel.) 450 ([M]$^+$, 100), 435 (15), 394 (61), 296 (30).

HR-EIMS: m/z 450.1656 ([M]$^+$, Calcd. for C$_{26}$H$_{26}$O$_7$, 450.1679).

Reference

Shiao Y-J, Wang C-N, Wang W-Y, Lin Y-L. (2005). Neuroprotective flavonoids from *Flemingia macrophylla*. *Planta Med* **71**: 835–840.

Fleminginin

IUPAC name: 5,7-Dihydroxy-3-(4-hydroxy-3-(2-methylbut-3-en-2-yl)phenyl)-6,8-bis(3-methylbut-2-en-1-yl)-4H-chromen-4-one

Sub-class: Prenylated isoflavone

Chemical structure

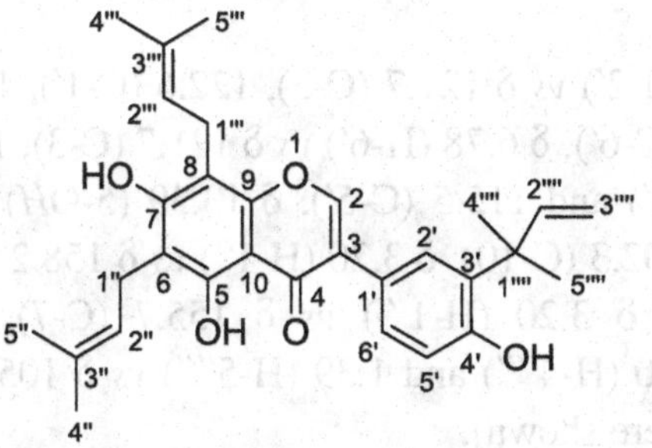

Source: *Flemingia macrophylla* (Linn.) (Family: Leguminosae); aerial parts

Molecular formula: $C_{30}H_{34}O_5$

Molecular weight: 474

State: Pale yellow needles (crystallized from ethanol)

Melting point: 199–202 °C

UV: λ_{max} (MeOH) (log ε): 277 (4.11), 342 (3.22) nm

IR (KBr): ν_{max} 3436 (OH), 1647 (carbonyl), 1615, 1578, 1516 (aromatic unsaturaion), 1079, 1017, 988 cm^{-1}

¹H-NMR (DMSO-d_6, 500 MHz): δ 8.34 (1H, s, H-2), 6.97 (1H, d, J = 2.0 Hz, H-2′), 6.77 (1H, d, J = 8.0 Hz, H-5′), 6.78 (1H, dd, J = 8.0, 2.0 Hz, H-6′), 3.20 (2H, d, J = 7.0 Hz, H-1″), 5.13 (1H, t, J = 7.0 Hz, H-2″), 1.70 (3H, s, H-4″), 1.71 (3H, s, H-5″), 3.20 (2H, d, J = 7.0 Hz, H-1‴), 5.14 (1H, t, J = 7.0 Hz, H-2‴), 1.58 (3H, s, H-4‴), 1.60 (3H, s, H-5‴), 5.91 (1H, dd, J = 18.5, 10.5 Hz, H-2⁗), 5.04 (1H, d, J = 10.5 Hz, H-3⁗a), 5.09 (1H, d, J = 5.09 Hz, H-3⁗b), 1.30 (3H, s, H-4⁗), 1.39 (3H, s, H-5⁗), 13.39 (1H, br s, 5-OH), 9.78 (1H, br s, 7-OH), 9.07 (1H, br s, 4′-OH).

¹³C-NMR (CDCl₃, 125 MHz): δ 154.0 (C-2), 121.7 (C-3), 180.8 (C-4), 158.2 (C-5), 111.0 (C-6), 155.7 (C-7), 110.6 (C-8), 149.9 (C-9), 102.3 (C-10), 122.6 (C-1′), 116.7 (C-2′), 105.2 (C-3′), 145.0 (C-4′), 115.5 (C-5′), 120.2 (C-6′), 21.0 (C-1″), 121.8 (C-2″), 131.0 (C-3″), 25.6 (C-4″), 17.8 (C-5″), 21.0 (C-1‴), 121.9 (C-2‴), 129.0 (C-3‴), 25.7 (C-4‴), 17.9 (C-5‴), 41.3 (C-1⁗), 146.8 (C-2⁗), 112.1 (C-3⁗), 28.0 (C-4⁗), 27.5 (C-5⁗).

HMBC: δ 6.97 (H-2′) vs δ 121.7 (C-3), 122.6 (C-1′), 105.2 (C-3′), 145.0 (C-4′) and 120.2 (C-6′), δ 6.78 (H-6′) vs δ 121.7 (C-3), 122.6 (C-1′), 116.7 (C-2′), 145.0 (C-4′) and 115.5 (C-5′), δ 13.39 (5-OH) vs δ 158.2 (C-5), 111.0 (C-6) and 102.3 (C-10), δ 3.20 (H-1″) vs δ 158.2 (C-5), 111.0 (C-6) and 155.7 (C-7), δ 3.20 (H-1‴) vs δ 155.7 (C-7), 110.6 (C-8) and 149.9 (C-9), δ 1.30 (H-4⁗) and 1.39 (H-5⁗) vs δ 105.2 (C-3′) (selected 2D-correlations were shown).

EIMS: m/z (%rel.) 474 ([M]⁺, 100), 459 (47), 419 (40), 363 (16).

HR-EIMS: m/z 474.2305 ([M]⁺, Calcd. for $C_{30}H_{34}O_5$, 474.2407).

Reference

Shiao Y-J, Wang C-N, Wang W-Y, Lin Y-L. (2005). Neuroprotective flavonoids from *Flemingia macrophylla*. *Planta Med* **71**: 835–840.

3'-Formyl-5,4'-dihydroxy-7-methoxyisoflavone

IUPAC name: 2-Hydroxy-5-(5-hydroxy-7-methoxy-4-oxo-4H-chromen-3-yl)benzaldehyde

Sub-class: Formylisoflavone

Chemical structure

Source: *Erythrina poeppigiana* (Walp.) O.F. Cook (Family: Fabaceae); stem barks

Molecular formula: $C_{17}H_{12}O_6$

Molecular weight: 312

Bioactivity studied: Binding affinity for estrogen receptor

State: Yellow amorphous solid

UV (MeOH): λ_{max} (log ε): 266 (4.24), 365 (3.86) nm

IR (KBr): ν_{max} 3392–3200 (OH), 1658 (C=O), 1619 cm^{-1}

^{1}H-NMR (CDCl$_3$, 600 MHz): δ 7.93 (1H, s, H-2), 6.44 (1H, d, J = 2.0 Hz, H-6), 6.41 (1H, d, J = 2.0 Hz, H-8), 3.89 (3H, s, 7-OCH_3), 12.68 (1H, s, 5-OH), 7.83 (1H, d, J = 2.0 Hz, H-2′), 7.08 (1H, d, J = 8.4 Hz, H-5′), 7.67 (1H, dd, J = 8.4, 2.0 Hz, H-6′), 9.96 (1H, s, 3′-CHO), 11.11 (1H, s, 4′-OH).

^{13}C-NMR (CDCl$_3$, 150 MHz): δ 152.6 (C-2), 122.3 (C-3), 180.2 (C-4), 162.4 (C-5), 98.1 (C-6), 165.2 (C-7), 92.4 (C-8), 157.7 (C-9), 105.5 (C-10), 122.3 (C-1′), 134.1 (C-2′), 120.4 (C-3′), 161.5 (C-4′), 117.9 (C-5′), 137.2 (C-6′), 55.8 (7-OCH_3), 196.2 (3′-CHO).

HR-ESIMS: m/z 312.2780 ([M]$^+$, Calcd. for C$_{17}$H$_{12}$O$_6$, 312.2784).

Reference

Djiogue S, Halabalaki M, Alexi X, Njamen D, Fomum Z T, Alexis M N, Skaltsounis A-L. (2009). Isoflavonoids from *Erythrina poeppigiana*: evaluation of their binding affinity for the estrogen receptor. *J Nat Prod* **72**: 1603–1607.

7-*O*-β-D-Glucopyranosylchamaejasmin

IUPAC name: (2*S*,2′*S*,3*R*,3′*R*)-5,5′,7-trihydroxy-2,2′-bis(4-hydroxyphenyl)-7′-((((2*R*,3*S*,4*R*,5*R*,6*S*)-3,4,5-trihydroxy-6-(hydroxymethyl)tetrahydro-2*H*-pyran-2-yl)oxy)-[3,3′-bichroman]-4,4′-dione

Sub-class: Biflavonoid (bis-flavanone)

Chemical structure

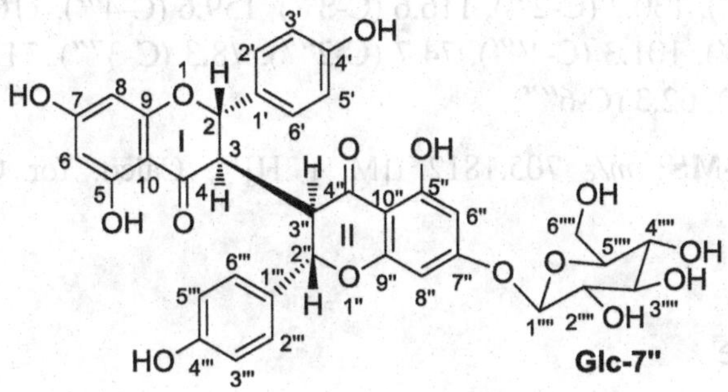

Glc-7″

Source: *Ormocarpum kirkii* S. Moore (Family: Papilionaceae); roots

Molecular formula: $C_{36}H_{32}O_{15}$

Molecular weight: 704

State: Yellow amorphous powder

Bioactivity studied: Antiplasmodial

Specific rotation: $[\alpha]_D^{20}$ +82° (MeOH, *c* 0.4)

CD (MeOH): λ_{max} ($\Delta\varepsilon$) 312 (+4.0 × 10^4), 285 (−7.1 × 10^4), 219 (+9.0 × 10^4) nm

UV (MeOH): λ_{max} 210, 215, 285, 300 nm

^{1}H-NMR (CD$_3$OD, 400 MHz): δ 5.77 (1H, d, J = 12.1 Hz, H-2β), 2.79 (1H, br d, J = 12.1 Hz, H-3α), 5.88 (1H, d, J = 2.2 Hz, H-6), 5.80 (1H, d, J = 2.2 Hz, H-8), 6.89 (2H, d, J = 8.5 Hz, H-2′ and H-6′), 6.76 (2H, d, J = 8.5 Hz, H-3′ and H-5′), 5.81 (1H, d, J = 11.5 Hz, H-2″β), 2.80 (1H, d, J = 11.5 Hz, H-3″β), 6.18 (1H, d, J = 2.2 Hz, H-6″), 6.12 (1H, d, J = 2.2 Hz, H-8″), 6.90 (2H, d, J = 8.5 Hz, H-2‴ and H-6‴), 6.76 (2H, d, J = 8.5 Hz, H-3‴ and H-5‴), 4.93 (1H, d, J = 7.4 Hz, H-1⁗), 3.44-3.36 (4H, m, H-2⁗, H-3⁗, H-4⁗ and H-6⁗), 3.85 (1H, dd, J = 12.1, 2.0 Hz, H-6⁗a), 3.65 (1H, dd, J = 12.1, 5.2 Hz, H-6⁗b).

^{13}C-NMR (CD$_3$OD, 100 MHz): δ 84.9 (C-2), 51.4 (C-3), 197.8 (C-4), 165.5 (C-5), 97.3 (C-6), 168.4 (C-7), 96.1 (C-8), 164.4 (C-9), 103.4 (C-10), 128.9 (C-1′), 130.3 (C-2′), 116.6 (C-3′), 159.6 (C-4′), 116.6 (C-5′), 130.3 (C-6′), 85.2 (C-2″), 51.5 (C-3″), 198.8 (C-4″), 165.0 (C-5″), 98.3 (C-6″), 167.1 (C-7″), 96.9 (C-8″), 164.1 (C-9″), 105.0 (C-10″), 128.8 (C-1‴), 130.3 (C-2‴), 116.6 (C-3‴), 159.6 (C-4‴), 116.6 (C-5‴), 130.3 (C-6‴), 101.3 (C-1⁗), 74.7 (C-2⁗), 78.3 (C-3⁗), 71.2 (C-4⁗), 77.8 (C-5⁗), 62.3 (C-6⁗).

HR-QTOF-MS: m/z 705.1812 ([M + H]$^+$), Calcd. for C$_{36}$H$_{33}$O$_{15}$, 705.1819).

Reference

Dhooghe L, Maregesi S, Mincheva I, Ferreira D, Marais J P J, Lemière F, Matheeussen A, Cos P, Maes L, Vlietinck A, Apers S, Pieters L. (2010). Antiplasmodial activity of (I-3,II-3)-biflavonoids and other constituents from *Ormocarpum kirkii*. *Phytochemistry* **71**: 785–791.

Griffinoid A

IUPAC name: (*S*)-4-(6,7-Dimethoxychroman-2-yl)-2-methoxyphenol

Sub-class: Flavan

Chemical structure

Source: *Combretum griffithii* Van Heurck & Müll. Arg. (Family: Combretaceae); stems

Molecular formula: $C_{18}H_{20}O_5$

Molecular weight: 316

State: White amorphous solid

Melting point: 117–119 °C

Specific rotation: $[\alpha]^{20}_D$ −37.0° (CHCl$_3$, *c* 0.20)

UV (MeOH): λ_{max} (log ε) 205 (4.57), 235 (3.87), 288 (3.71) nm

IR (KBr): ν_{max} 3408 (OH), 2927, 2852, 1617, 1513, 1453, 1266, 1225, 1194, 1123 cm^{-1}

¹H-NMR (CD₃OD, 400 MHz): δ 4.86 (1H, dd, J = 10.4, 2.4 Hz, H-2β), 2.14-1.94 (2H, m, H-3), 2.89 (1H, ddd, J = 16.0, 11.2, 6.0 Hz, H-4$_{ax}$), 2.67 (1H, ddd, J = 16.0, 5.6, 2.8 Hz, H-4$_{eq}$), 6.65 (1H, s, H-5), 6.44 (1H, s, H-8), 6.98 (1H, d, J = 1.6 Hz, H-2′), 6.77 (1H, d, J = 8.2 Hz, H-5′), 6.84 (1H, dd, J = 8.2, 1.6 Hz, H-6′), 3.75 (6H, s, 6-OCH_3 and 7-OCH_3), 3.84 (3H, s, 3′-OCH_3), 5.65 (1H, s, 4′-OH).

¹³C-NMR (CD₃OD, 100 MHz): δ 77.7 (C-2), 29.9 (C-3), 24.3 (C-4), 113.2 (C-5), 143.1 (C-6), 148.5 (C-7), 101.1 (C-8), 149.4 (C-9), 112.8 (C-10), 133.5 (C-1′), 109.6 (C-2′), 147.5 (C-3′), 145.8 (C-4′), 114.6 (C-5′), 118.6 (C-6′), 55.9 (6-OCH₃), 55.0 (7-OCH₃), 55.0 (3′-OCH₃).

HR-ESI-TOFMS: m/z 339.1203 ([M + Na]⁺, Calcd. for $C_{18}H_{20}O_5Na$, 339.1208).

Reference

Moosophon P, Kanokmedhakul S, Kanokmedhakul K, Buayairaksa M, Noichan J, Poopasit K. (2013). Antiplasmodial and cytotoxic flavans and diarylpropanes from the stems of *Combretum griffithii*. *J Nat Prod* **76**: 1298–1302.

Griffinoid D

IUPAC name: (*S*)-4-(6-hydroxy-7-methoxychroman-2-yl)benzene-1,2-diol

Sub-class: Flavan

Chemical structure

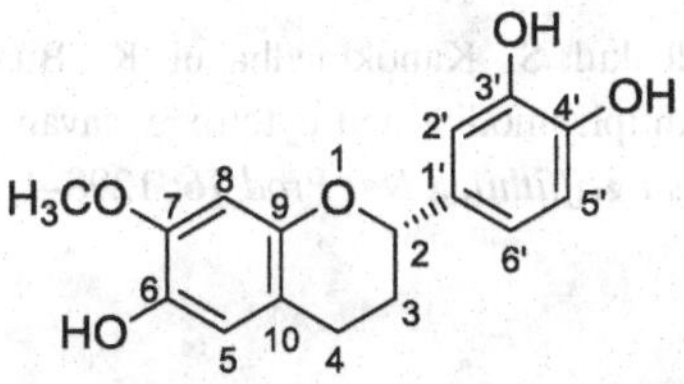

Source: *Combretum griffithii* Van Heurck & Müll. Arg. (Family: Combretaceae); stems

Molecular formula: $C_{16}H_{16}O_5$

Molecular weight: 288

State: White amorphous solid

Melting point: 133–135 °C

Specific rotation: $[\alpha]^{20}_D$ −33.2° (CHCl$_3$, *c* 0.52)

Bioactivity studied: Antimalarial (IC$_{50}$ value of 13.04 µM against *Plasmodium falciparum*), cytotoxic

UV (MeOH): λ_{max} (log ε) 204 (4.86), 289 (3.93) nm

IR (KBr): ν_{max} 3376 (OH), 2925, 2848, 1608, 1511, 1444, 1268, 1233, 1191, 1162, 1115, 1070, 894, 876, 819 cm^{-1}.

^{1}H-NMR (CD$_3$OD, 400 MHz): δ 4.76 (1H, dd, J = 10.4, 2.4 Hz, H-2β), 2.04-1.94 (2H, m, H-3), 2.79 (1H, ddd, J = 16.0, 10.4, 6.0 Hz, H-4ax), 2.60 (1H, ddd, J = 16.0, 4.8, 2.8 Hz, H-4eq), 6.54 (1H, s, H-5), 6.37 (1H, s, H-8), 6.85 (1H, d, J = 2.0 Hz, H-2'), 6.77 (1H, d, J = 8.0 Hz, H-5'), 6.74 (1H, dd, J = 8.0, 2.0 Hz, H-6'), 3.74 (3H, s, 7-OCH3).

^{13}C-NMR (CD$_3$OD, 100 MHz): δ 77.7 (C-2), 29.7 (C-3), 24.6 (C-4), 114.5 (C-5), 139.2 (C-6), 145.9 (C-7), 100.4 (C-8), 148.3 (C-9), 113.3 (C-10), 133.8 (C-1'), 113.4 (C-2'), 144.4 (C-3'), 144.1 (C-4'), 115.0 (C-5'), 118.2 (C-6'), 55.8 (7-OCH$_3$).

HR-ESI-TOFMS: m/z 311.0890 ([M + Na]$^+$, Calcd. for C$_{16}$H$_{16}$O$_5$Na, 311.0895).

Reference

Moosophon P, Kanokmedhakul S, Kanokmedhakul K, Buayairaksa M, Noichan J, Poopasit K. (2013). Antiplasmodial and cytotoxic flavans and diarylpropanes from the stems of *Combretum griffithii*. *J Nat Prod* **76**: 1298–1302.

(+)-Hexanoyllomatin

IUPAC name: (R)-8,8-Dimethyl-2-oxo-2,8,9,10-tetrahydropyrano[2,3-f]chromen-9-yl hexanoate

Sub-class: Coumarin

Chemical structure

Source: *Seseli devenyense* Simonkai (Family: Apiaceae); fruits

Molecular formula: $C_{20}H_{24}O_5$

Molecular weight: 344

State: Colorless oil

Specific rotation: $[\alpha]_D^{25}$ +30.3° (CHCl$_3$, c 0.1)

UV (MeOH): λ_{max} (log ε) 289 (sh), 326 (3.32) nm

IR (CHCl$_3$): ν_{max} 2980, 1732 (C=O), 1606, 1144 cm^{-1}

^{1}H-NMR (CDCl$_3$, 400 MHz): δ 6.24 (1H, d, J = 9.5 Hz, H-3), 7.63 (1H, d, J = 9.5 Hz, H-4), 7.26 (1H, d, J = 8.6 Hz, H-5), 6.79 (1H, d, J = 8.6 Hz, H-6), 5.13 (1H, t, J = 5.0 Hz, H-3′), 3.19 (1H, dd, J = 15.1, 5.0 Hz, H-4′a),

119

2.97 (1H, dd, J = 15.1, 5.0 Hz, H-4'b), 1.37 (3H, s, H-5'), 1.34 (3H, s, H-6'), 2.30 (2H, t, J = 7.5 Hz, H-2''), 1.59 (2H, quint, J = 7.5 Hz, H-3''), 1.24 (4H, m, H-4'' and H-5''), 0.86 (3H, t, J = 7.0 Hz, H-6'').

EIMS: m/z (%rel) 344 ([M]$^+$, 5), 228 (30), 213 (100), 187 (10), 176 (15).

HR-EIMS: m/z 344.1630 ([M]$^+$, Calcd. for $C_{20}H_{24}O_5$, 344.1624).

Reference

Widelski J, Melliou E, Fokialakis N, Magiatis P, Glowniak K, Chinou I. (2005). Coumarins from the fruits of *Seseli devenyense*. *J Nat Prod* **68**: 1637–1641.

Hildegardiol [(3*R*)-6,2′-dihydroxy-7-methoxy-4′,5′-methylenedioxyisoflavan]

IUPAC name: (*R*)-3-(6-hydroxybenzo[*d*][1,3]dioxol-5-yl)-7-methoxy-chroman-6-ol

Sub-class: Isoflavan

Chemical structure

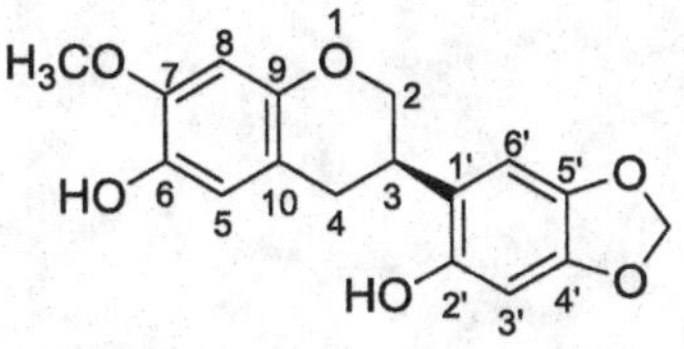

Source: *Hildegardia barteri* (Mast.) Kosterm. (Family: Sterculiaceae); roots

Molecular formula: $C_{17}H_{16}O_6$

Molecular weight: 316

State: Pale tan gum

Bioactivity studied: Antifungal

UV: λ_{max} (MeOH) (log ε): 232 (2.35), 301 (3.92) nm

IR (NaCl): ν_{max} 3401 (OH), 1633, 1510, 1167, 1036, 933 cm^{-1}

^{1}H-NMR (C_6D_6, 500 MHz): δ 3.83 (1H, dd, J = 10.2, 9.6 Hz, H-2α), 4.21 (1H, ddd, J = 10.2, 3.4, 1.8 Hz, H-2β), 3.48 (1H, m, H-3α), 2.61 (1H, ddd, J = 16.0, 10.1, 0.7 Hz, H-4α), 2.55 (1H, ddd, J = 16.0, 10.1, 1.5 Hz, H-4β), 6.74 (1H, s, H-5), 6.43 (1H, s, H-8), 5.85 (1H, s, H-3′), 6.46 (1H, s, H-6′), 5.30 and 5.29 (1H each, d, J = 1.4 Hz, -OCH_2O-), 3.10 (3H, s, 7-OCH_3), 5.09 1H, s, 6-OH), 3.66 (1H, br s, 2′-OH).

^{13}C-NMR (C_6D_6, 125 MHz): δ 69.8 (C-2), 32.3 (C-3), 146.7 (C-4), 142.1 (C-5), 140.3 (C-6), 146.0 (C-7), 100.4 (C-8), 148.3 (C-9), 114.1 (C-10), 120.2 (C-1′), 148.2 (C-2′), 98.2 (C-3′), 146.7 (C-4′), 142.1 (C-5′), 107.5 (C-6′), 101.0 (-OCH_2O-), 55.2 (6-OCH_3).

ESIMS: m/z (%rel.) 315 [M – H]⁻, 100), 151 (30).

HR-EIMS: m/z 317.0970 ([M + H]⁺, Calcd. for $C_{17}H_{17}O_6$, 317.1025).

Reference

Meragelman TL, Tucker KD, McCloud TG, Cardellina II JH, Shoemaker RH. (2005). Antifungal flavonoids from *Hildegardia barteri*. *J Nat Prod* **68**: 1790–1792.

Hirsutissimiside A [formononetin 7-*O*-β-D-(6″-ethylmalonyl)-glucopyranoside]

IUPAC name: Ethyl (((2*R*,3*S*,4*S*,5*R*,6*S*)-3,4,5-trihydroxy-6-((3-(4-methoxyphenyl)-4-oxo-4*H*-chromen-7-yl)oxy)tetrahydro-2*H*-pyran-2-yl)methyl) malonate

Sub-class: Isoflavone glycoside

Chemical structure

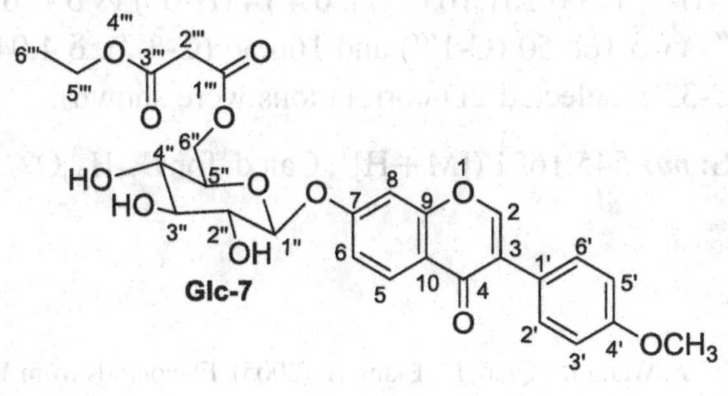

Source: *Millettia nitida* var. *hirsutissima* (Family: Liguminosae); stems

Molecular formula: $C_{27}H_{28}O_{12}$

Molecular weight: 544

State: White powder

Specific rotation: $[\alpha]^{25}_D$ −26.3° (MeOH:H$_2$O = 1:05, *c* 0.13)

UV (MeOH): λ_{max} (log ε) 206 (3.74), 258 (3.77) nm

IR (KBr): ν_{max} 3581 & 3275 (OH), 1745 & 1700 (ester carbonyl), 1621 (α,β-unsaturated carbonyl), 1511, 1444, 1349 (aromatic unsaturaion), 1249, 1184, 1078, 1035 cm^{-1}.

^{1}H-NMR (DMSO-d_6, 300 MHz): δ 8.43 (1H, s, H-2), 8.08 (1H, d, J = 9.0 Hz, H-5), 7.15 (1H, dd, J = 9.0, 2.1 Hz, H-6), 7.23 (1H, s, H-8), 7.53 (2H, d, J = 8.7 Hz, H-2′ and H-6′), 7.01 (2H, d, J = 8.7 Hz, H-3′ and H-5′), 3.79 (3H, s, 4′-OCH_3), 5.17 (1H, d, J = 6.9 Hz, H-1″), 4.14 (2H, m, H-6″) (data for other sugar proton signal were not mentioned), 3.51 (2H, s, H-2‴), 4.04 (2H, q, H-5‴), 1.11 (3H, q, H-6‴).

^{13}C-NMR (DMSO-d_6, 75 MHz): δ 153.76 (C-2), 123.99 (C-3), 174.68 (C-4), 127.06 (C-5), 115.37 (C-6), 161.16 (C-7), 103.54 (C-8), 157.01 (C-9), 118.54 (C-10), 123.37 (C-1′), 130.08 (C-2′), 113.65 (C-3′), 159.04 (C-4′), 113.65 (C-5′), 130.08 (C-6′), 55.19 (4′-OCH_3), 91.61 (C-1″), 72.99 (C-2″), 76.16 (C-3″), 69.57 (C-4″), 73.69 (C-5″), 64.19 (C-6″), 166.50 (C-1‴), 41.06 (C-2‴), 166.36 (C-3‴), 60.82 (C-5‴), 13.87 (C-6‴).

HMBC: δ 7.53 (H-2′/H-6′) vs δ 159.04 (C-4′), δ 3.79 (4′-OCH_3) vs δ 159.04 (C-4′), δ 8.08 (H-5) vs δ 161.16 (C-7), δ 7.23 (H-8) vs δ 161.16 (C-7), δ 5.17 (H-1″) vs δ 161.16 (C-7), δ 4.14 (H-6″) vs δ 166.50 (C-1‴), δ 3.51 (H-2″) vs δ 166.50 (C-1‴) and 166.36 (C-3‴), δ 4.04 (H-5‴) vs δ 1 166.36 (C-3‴) (selected 2D-correlations were shown).

HR-FAB-MS: m/z 545.1651 ([M + H]$^+$, Calcd. for $C_{27}H_{29}O_{12}$, 545.1653).

Reference

Cheng J., Zhao Y.-Y., Wang B., Qiao, L., Liang H. (2005). Flavonoids from *Millettia nitida* var. *hirsutissima*. *Chem Pharm Bull* **53**: 419–421.

Hovenin A

IUPAC name: (2*R*,3*R*)-3,5,7,5′-Tetrahydroxy-2-[(2*R*,3*R*)-3-(4-hydroxy-3,5-dimethoxyphenyl)-2-hydroxymethyl-2,3-dihydrobenzo[*b*][1,4]dioxin-6-yl]-chroman-4-one

Sub-class: Flavonolignan

Chemical structure

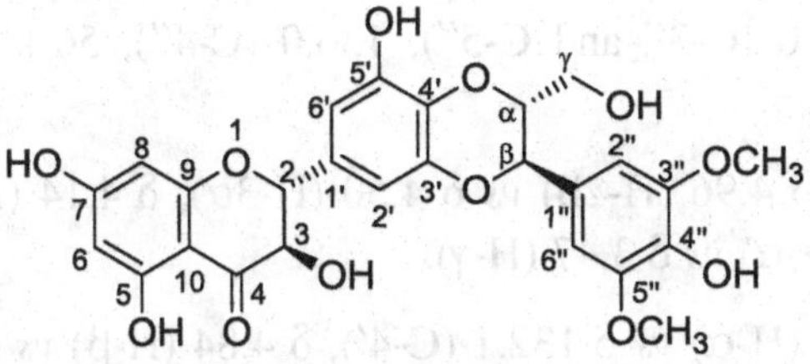

Source: *Hovenia acerba* (Family: Rhamnaceae); seeds

Molecular formula: $C_{26}H_{24}O_{12}$

Molecular weight: 528

State: White amorphous powder

Melting point: 171–173 °C

Bioactivity studied: Anti-inflammatory [exhibited inhibitory activity against production of NO and IL-6 in lipopolysaccharide (LPS)-induced murine RAW264.7 macrophases with IC_{50} values ranging from 45.11 to 63.47 µM]

Specific rotation: $[\alpha]_D^{25}$ +22.8° (MeOH, c 0.04)

CD (MeOH): λ_{max} ($\Delta\varepsilon$) 326 (+9.8), 292 (−13.0), 242 (−3.34) nm

UV (MeOH): λ_{max} (log ε) 210 (4.72), 247 (4.07), 370 (3.81) nm

IR (KBr): ν_{max} 3403 (OH), 2928, 1606, 1516, 1463, 1382, 1215, 1103, 1086, 1038 cm^{-1}

^{1}H-NMR (DMSO-d_6, 400 MHz): δ 4.96 (1H, d, J = 11.0 Hz, H-2β), 4.50 (1H, d, J = 11.0 Hz, H-3α), 5.86 (1H, d, J = 1.8 Hz, H-6), 5.82 (1H, d, J = 1.8 Hz, H-8), 4.14 (1H, m, H-α), 4.84 (1H, d, J = 7.9 Hz, H-β), 3.47 (2H, overlapped signal, H-γ), 6.55 (1H, d, J = 1.9 Hz, H-2′), 6.57 (1H, d, J = 1.9 Hz, H-6′), 6.71 (2H, s, H-2″ and H-6″), 11.91 (1H, s, 5-OH), 12.42 (1H, s, 7-OH), 3.76 (6H, s, 3″-OCH_3 and 5″-OCH_3).

^{13}C-NMR (DMSO-d_6, 100 MHz): δ 82.8 (C-2), 71.4 (C-3), 197.2 (C-4), 162.4 (C-5), 96.3 (C-6), 168.0 (C-7), 95.3 (C-8), 163.4 (C-9), 100.1 (C-10), 129.1 (C-1′), 107.3 (C-2′), 143.9 (C-3′), 132.1 (C-4′), 145.9 (C-5′), 108.2 (C-6′), 77.7 (C-α), 76.2 (C-β), 60.4 (C-γ), 126.5 (C-1″), 105.2 (C-2″ and C-6″), 148.0 (C-3″ and C-5″), 136.0 (C-4″), 56.1 (3″-OCH_3 and 5″-OCH_3).

^{1}H-^{1}H COSY: δ 4.96 (H-2β) *vs* δ 4.50 (H-3α), δ 4.14 (H-α) *vs* δ 4.84 (H-β), δ 4.14 (H-α) *vs* δ 3.47 (H-γ).

HMBC: δ 4.14 (H-α) *vs* δ 132.1 (C-4′), δ 4.84 (H-β) *vs* δ 143.9 (C-3′), 126.5 (C-1″) and 105.2 (C-2″/C-6″), 3.76 (3″-OCH_3) *vs* δ 148.0 (C-3″), and 3.76 (5″-OCH_3) *vs* δ 148.0 (C-5″) (selected HMBC correlations were shown).

HR-ESIMS: m/z 527.1182 ([M − H]$^-$, Calcd. for $C_{26}H_{23}O_{12}$, 527.1190).

Reference

Zhang X-Q, Xu F-F, Wang L, Huang M-Y, Liu Z, Zhang D-M, Wang G-C, Li Y-L, Ye W-C. (2012). Two pairs of new diastereoisomeric flavonolignans from the seeds of *Hovenia acerba*. *Phytochemistry Lett* **5**: 292–296.

Hovenin B

IUPAC name: (2*R*,3*R*)-3,5,7,5'-Tetrahydroxy-2-[(2*S*,3*S*)-3-(4-hydroxy-3,5-dimethoxyphenyl)-2-hydroxymethyl-2,3-dihydrobenzo[*b*][1,4]dioxin-6-yl]-chroman-4-one

Sub-class: Flavonolignan

Chemical structure

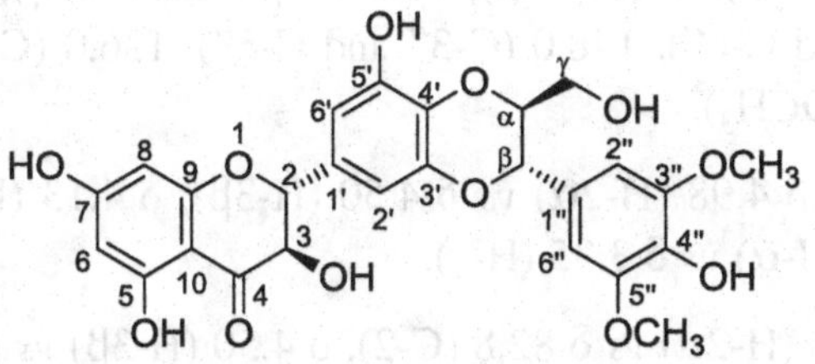

Source: *Hovenia acerba* (Family: Rhamnaceae); seeds

Molecular formula: $C_{26}H_{24}O_{12}$

Molecular weight: 528

State: White amorphous powder

Melting point: 171–172 °C

Bioactivity studied: Anti-inflammatory [exhibited inhibitory activity against production of NO and IL-6 in lipopolysaccharide (LPS)-induced murine RAW264.7 macrophases with IC_{50} values ranging from 45.11 to 63.47 µM]

127

Specific rotation: $[\alpha]_D^{25}$ −3.4° (MeOH, c 0.04)

CD (MeOH): λ_{max} ($\Delta\varepsilon$) 325 (+2.0), 291 (−9.4), 235 (+6.2) nm

UV (MeOH): λ_{max} (log ε) 210 (4.73), 247 (4.08), 371 (3.79) nm

IR (KBr): ν_{max} 3421 (OH), 2928, 1603, 1516, 1463, 1378, 1215, 1112, 1086, 1038 cm^{-1}

^{1}H-NMR (DMSO-d_6, 400 MHz): δ 4.98 (1H, d, J = 11.0 Hz, H-2α), 4.50 (1H, d, J = 11.0 Hz, H-3β), 5.91 (1H, d, J = 1.8 Hz, H-6), 5.87 (1H, d, J = 1.8 Hz, H-8), 4.13 (1H, m, H-α), 4.85 (1H, d, J = 7.9 Hz, H-β), 3.35 (2H, overlapped signal, H-γ), 6.54 (1H, d, J = 1.9 Hz, H-2′), 6.58 (1H, d, J = 1.9 Hz, H-6′), 6.72 (2H, s, H-2″ and H-6″), 11.87 (1H, s, 5-OH), 12.42 (1H, s, 7-OH), 3.77 (6H, s, 3″-OCH_3 and 5″-OCH_3).

^{13}C-NMR (DMSO-d_6, 100 MHz): δ 82.8 (C-2), 71.5 (C-3), 197.6 (C-4), 162.4 (C-5), 96.0 (C-6), 168.8 (C-7), 95.0 (C-8), 163.3 (C-9), 100.4 (C-10), 129.0 (C-1′), 107.4 (C-2′), 143.9 (C-3′), 132.1 (C-4′), 145.9 (C-5′), 108.0 (C-6′), 77.7 (C-α), 76.1 (C-β), 60.4 (C-γ), 126.5 (C-1″), 105.2 (C-2″ and C-6″), 148.0 (C-3″ and C-5″), 136.0 (C-4″), 56.1 (3″-OCH_3 and 5″-OCH_3).

^{1}H-^{1}H COSY: δ 4.98 (H-2α) *vs* δ 4.50 (H-3β), δ 4.13 (H-α) *vs* δ 4.85 (H-β), δ 4.13 (H-α) *vs* δ 3.35 (H-γ).

HMQC: δ 4.98 (H-2α) *vs* δ 82.8 (C-2), δ 4.50 (H-3β) *vs* δ 71.5 (C-3), δ 5.91 (H-6) *vs* δ 96.0 (C-6), δ 5.87 (H-8) *vs* δ 95.0 (C-8), δ 4.13 (H-α) *vs* δ 77.7 (C-α), 4.85 (H-β) *vs* δ 76.1 (C-β), δ 3.35 (H-γ) *vs* δ 60.4 (C-γ), δ 6.54 (H-2′) *vs* δ 107.4 (C-2′), δ 6.58 (H-6′) *vs* δ 108.0 (C-6′), δ 6.72 (H-2″/H-6″) *vs* δ 105.2 (C-2″/C-6″), δ 3.77 (3″-OCH_3/5″-OCH_3) *vs* δ 56.1 (3″-OCH_3/5″-OCH_3).

HR-ESIMS: m/z 527.1194 ([M − H]$^-$, Calcd. for $C_{26}H_{23}O_{12}$, 527.1190).

Reference

Zhang X-Q, Xu F-F, Wang L, Huang M-Y, Liu Z, Zhang D-M, Wang G-C, Li Y-L, Ye W-C. (2012). Two pairs of new diastereoisomeric flavonolignans from the seeds of *Hovenia acerba*. *Phytochemistry Lett* **5**: 292–296.

Hovenin C

IUPAC name: (2*R*,3*R*)-3,5,7-Trihydroxy-2-((2*R*,3*R*)-8-hydroxy-2-(4-hydroxy-3,5-dimethoxyphenyl)-3-(hydroxymethyl)-2,3-dihydrobenzo[*b*][1,4]dioxin-6-yl)chroman-4-one

Sub-class: Flavonolignan

Chemical structure

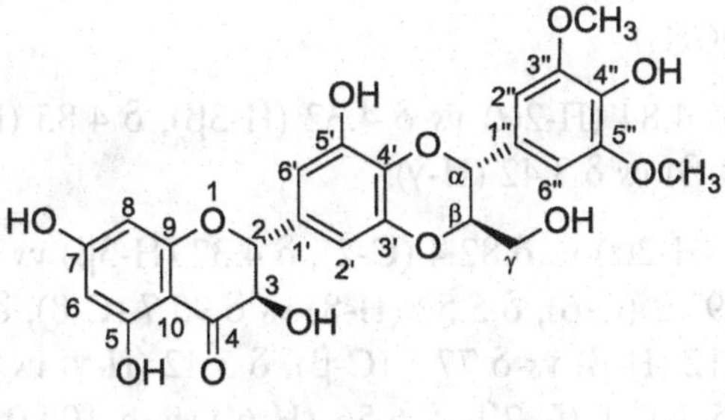

Source: *Hovenia acerba* (Family: Rhamnaceae); seeds

Molecular formula: $C_{26}H_{24}O_{12}$

Molecular weight: 528

State: White amorphous powder

Melting point: 172–174 °C

Bioactivity studied: Anti-inflammatory [exhibited inhibitory activity against production of NO and IL-6 in lipopolysaccharide (LPS)-induced murine RAW264.7 macrophases with IC_{50} values ranging from 45.11 to 63.47 μM]

129

Specific rotation: $[\alpha]_D^{25}$ −3.4° (MeOH, c 0.04)

CD (MeOH): λ_{max} ($\Delta\varepsilon$) 327 (+1.84), 292 (−9.30), 242 (−3.90) nm

UV (MeOH): λ_{max} (log ε) 210 (4.62), 248 (3.99), 371 (3.78) nm

IR (KBr): ν_{max} 3413 (OH), 2933, 1604, 1516, 1463, 1385, 1215, 1113, 1086, 1038 cm⁻1

^{1}H-NMR (DMSO-d_6, 400 MHz): δ 4.84 (1H, d, J = 10.0 Hz, H-2α), 4.32 (1H, d, J = 10.0 Hz, H-3β), 5.55 (1H, s, H-6), 5.52 (1H, s, H-8), 4.85 (1H, J = 7.5 Hz, H-α), 4.12 (1H, m, H-β), 3.42 (2H, overlapped signal, H-γ), 6.50 (1H, d, J = 1.6 Hz, H-2′), 6.56 (1H, d, J = 1.6 Hz, H-6′), 6.71 (2H, s, H-2″ and H-6″), 3.77 (6H, s, 3″-OCH_3 and 5″-OCH_3).

^{13}C-NMR (DMSO-d_6, 100 MHz): δ 82.4 (C-2), 71.2 (C-3), 195.7 (C-4), 162.1 (C-5), 97.2 (C-6), 168.0 (C-7), 95.7 (C-8), 163.7 (C-9), 100.4 (C-10), 129.6 (C-1′), 107.1 (C-2′), 143.8 (C-3′), 132.9 (C-4′), 145.8 (C-5′), 108.0 (C-6′), 76.1 (C-α), 77.7 (C-β), 60.4 (C-γ), 126.6 (C-1″), 105.3 (C-2″ and C-6″), 147.9 (C-3″ and C-5″), 135.9 (C-4″), 56.1 (3″-OCH_3 and 5″-OCH_3).

^{1}H-^{1}H COSY: δ 4.84 (H-2α) vs δ 4.32 (H-3β), δ 4.85 (H-α) vs δ 4.12 (H-β), δ 4.12 (H-β) vs δ 3.42 (H-γ).

HMQC: δ 4.84 (H-2α) vs δ 82.4 (C-2), δ 4.32 (H-3β) vs δ 71.2 (C-3), δ 5.55 (H-6) vs δ 97.2 (C-6), δ 5.52 (H-8) vs δ 95.7 (C-8), δ 4.85 (H-α) vs δ 76.1 (C-α), 4.12 (H-β) vs δ 77.7 (C-β), δ 3.42 (H-γ) vs δ 60.4 (C-γ), δ 6.50 (H-2′) vs δ 107.1 (C-2′), δ 6.56 (H-6′) vs δ 108.0 (C-6′), δ 6.71 (H-2″/H-6″) vs δ 105.3 (C-2″/C-6″), δ 3.77 (3″-OCH_3/5″-OCH_3) vs δ 56.1 (3″-OCH_3/5″-OCH_3).

HR-ESIMS: m/z 527.1196 ([M − H]⁻, Calcd. for $C_{26}H_{23}O_{12}$, 527.1190).

Reference

Zhang X-Q, Xu F-F, Wang L, Huang M-Y, Liu Z, Zhang D-M, Wang G-C, Li Y-L, Ye W-C. (2012). Two pairs of new diastereoisomeric flavonolignans from the seeds of *Hovenia acerba*. *Phytochemistry Lett* **5**: 292–296.

(2*R*)-5-Hydroxy-7,8-dimethoxy flavanone-5-*O*-β-D-glucopyranoside

IUPAC name: (*R*)-7,8-Dimethoxy-2-phenyl-5-(((2*S*,3*R*,4*S*,5*S*,6*R*)-3,4,5-trihydroxy-6-(hydroxymethyl)tetrahydro-2*H*-pyran-2-yl)oxy)chroman-4-one

Sub-class: Flavanone glycoside

Chemical structure

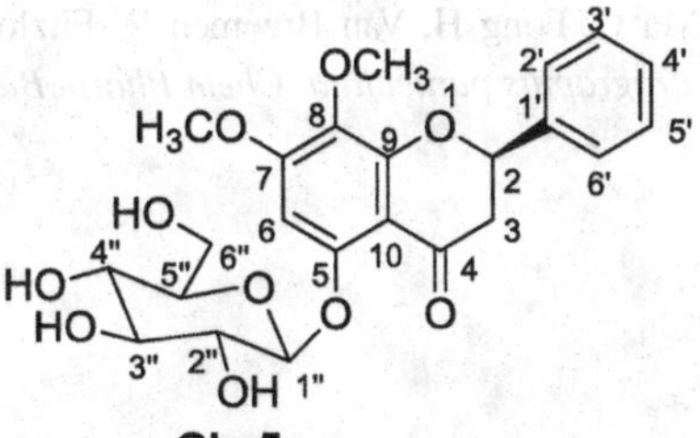

Glc-5

Source: *Andrographis paniculata* Nees (Family: Acanthaceae); whole plants

Molecular formula: $C_{23}H_{26}O_{10}$

Molecular weight: 462

State: Yellow powder

Melting point: 123–125 °C

Specific rotation: $[\alpha]^{20}_{D}$ −70° (MeOH, *c* 0.30)

UV (MeOH): λ_{max} 210.5, 235.5, 240, 287, 333, 332.5 nm

IR (KBr): v_{max} 3400 (OH), 1675 (C=O), 1602, 1594, 1205, 1145, 1075 cm^{-1}

^{1}H-NMR (DMSO-d_6, 500 MHz): δ 5.54 (1H, dd, J = 11.9, 3.1 Hz, H-2α), 3.07 (1H, dd, J = 16.3, 11.9 Hz, H-3$_{ax}$), 2.70 (1H, dd, J = 16.3, 3.1 Hz, H-3$_{eq}$), 6.60 (1H, s, H-6), 7.43 (2H, m, H-2′ and H-6′), 7.36 (3H, m, H-3′, H-4′ and H-5′), 3.77 (3H, s, 7-OCH_3), 3.57 (3H, s, 8-OCH_3), 4.75 (1H, d, J = 7.3 Hz, H-1″) (other proton signals for sugar moiety were not showed).

^{13}C-NMR (DMSO-d_6, 125 MHz): δ 79.3 (C-2), 45.5 (C-3), 192.1 (C-4), 155.7 (C-5), 96.5 (C-6), 159.1 (C-7), 132.6 (C-8), 155.7 (C-9), 107.5 (C-10), 139.8 (C-1′), 127.2 (C-2′), 129.3 (C-3′), 129.3 (C-4′), 129.3 (C-5′), 127.2 (C-6′), 55.8 (7-OCH_3), 61.8 (8-OCH_3), 103.1 (C-1″), 74.2 (C-2″), 77.2 (C-3″), 70.9 (C-4″), 78.4 (C-5″), 61.8 (C-6″).

Q-TOF-2 MS: m/z 463.1642 ([M + H])$^+$, Calcd. for $C_{23}H_{27}O_{10}$, 463.1604).

Reference

Li W, Xu X, Zhang H, Ma C, Fong H, Van Breemen R, Fitzloff J. (2007). Secondary metabolites from *Andrographis paniculata*. *Chem Pharm Bull* **55**: 455–458.

(2*S*)-3′-(2-Hydroxy-3-methylbut-3-enyl)licoflavone-4′-methyl ether

IUPAC name: (2*S*)-5,7-Dihydroxy-2-(3-(2-hydroxy-3-methylbut-3-en-1-yl)-4-methoxy-5-(3-methylbut-2-en-1-yl)phenyl)chroman-4-one

Sub-class: Flavanone

Chemical structure

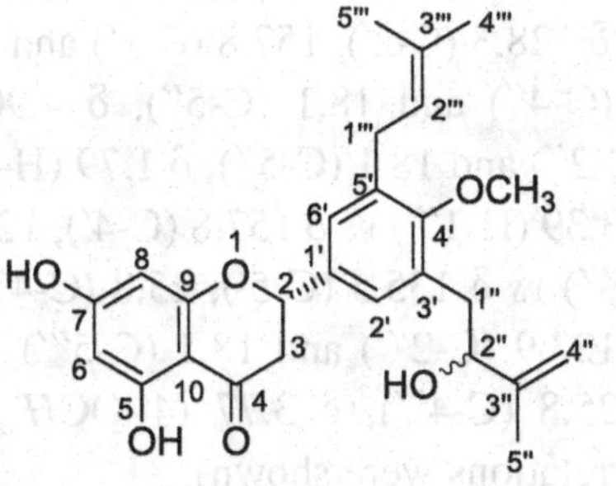

Source: *Erythrina addisoniae* (Family: Leguminosae); barks

Molecular formula: $C_{26}H_{30}O_6$

Molecular weight: 438

State: Yellowish oil

Specific rotation: $[\alpha]^{20}_{D}$ −128° (MeOH, *c* 0.15)

^{1}H-NMR (Acetone-d_6, 500 MHz): δ 5.46 (1H, dd, *J* = 13.0, 3.0 Hz, H-2β), 3.15 (1H, dd, *J* = 17.0, 13.0 Hz, H-3$_{ax}$), 2.73 (1H, dd, *J* = 17.0, 3.2

Hz, H-3$_{eq}$), 5.95 (1H, d, J = 2.2 Hz, H-6), 5.94 (1H, d, J = 2.2 Hz, H-8), 7.23 (1H, d, J = 2.1 Hz, H-2′), 7.34 (1H, d, J = 2.1 Hz, H-6′), 2.84 (1H, m, H-1″a), 2.68 (1H, dd, J = 13.23, 4.7 Hz, H-1″b), 4.23 (1H, m, H-2″), 4.90 (1H, s, H-4″a), 4.73 (1H, s, H-4″b), 1.79 (3H, s, H-5″), 3.39 (2H, d, J = 7.3 Hz, H-1‴), 5.30 (1H, t, J = 7.3 Hz, H-2‴), 1.71 (6H, s, H-4″ and H-5‴), 12.17 (1H, s, 5-OH), 9.74 (1H, s, 7-OH), 3.77 (3H, s, 4′-OCH_3).

^{13}C-NMR (Acetone-d_6, 125 MHz): δ 79.9 (C-2), 43.5 (C-3), 197.0 (C-4), 165.3 (C-5), 95.8 (C-6), 167.4 (C-7), 96.8 (C-8), 164.2 (C-9), 103.2 (C-10), 133.8 (C-1′), 128.3 (C-2′), 135.0 (C-3′), 157.8 (C-4′), 135.5 (C-5′), 127.3 (C-6′), 37.2 (C-1″), 76.1 (C-2″), 149.3 (C-3″), 110.5 (C-4″), 18.1 (C-5″), 29.0 (C-1‴), 123.9 (C-2‴), 132.9 (C-3‴), 25.8 (C-4‴), 18.1 (C-5‴), 61.3 (4′-OCH_3).

HMBC: δ 5.46 (H-2β) vs δ 128.3 (C-2′) and 127.3 (C-6′), δ 3.15 (H-3$_{ax}$) and 2.73 (H-3$_{eq}$) vs δ 197.0 (C-4), δ 5.95 (H-6) vs δ 96.8 (C-8) and 103.2 (C-10), δ 5.94 (H-8) vs δ 95.8 (C-6) and 103.2 (C-10), δ 7.23 (H-2′) vs δ 79.9 (C-2), 157.8 (C-4′), 127.3 (C-6′) and 37.2 (C-1″), δ 7.34 (H-6′) vs δ 79.9 (C-2), 128.3 (C-2′), 157.8 (C-4′) and 29.0 (C-1‴), δ 2.84 (H-1″a) and 2.68 (H-1″b) vs δ 128.3 (C-2′), 157.8 (C-4′) and 149.3 (C-3″), δ 4.23 (H-2″) vs δ 110.5 (C-4″) and 18.1 (C-5″), δ 4.90 (H-4″a) and 4.73 (H-4″b) vs δ 76.1 (C-2″) and 18.1 (C-5″), δ 1.79 (H-5″) vs δ 76.1 (C-2″) and 110.5 (C-4″), δ 3.39 (H-1‴) vs δ 157.8 (C-4′), 127.3 (C-6′) and 132.9 (C-3‴), δ 5.30 (H-2‴) vs δ 135.5 (C-5′), 25.8 (C-4‴) and 18.1 (C-5‴), δ 1.71 (H-4″) vs δ 123.9 (C-2‴) and 18.1 (C-5‴), δ 1.71 (H-5″) vs δ 123.9 (C-2‴) and 25.8 (C-4‴), δ 3.77 (4′-OCH_3) vs δ 157.8 (C-4′) (selected HMBC correlations were shown).

MS: m/z 439 [M + H]$^+$

Reference

Wätjen W, Suckow-Schnitker A K, Rohrig R, Kulawik A, Addae-Kyereme J, Wright C W, Passreiter C M. (2008). Prenylated flavonoid derivatives from the bark of *Erythrina addisoniae*. *J Nat Prod* **71**: 735–738.

2′-Hydroxy-3,4-dimethoxy-[2″,3″:4′,3′]-furanochalcone

IUPAC name: (*E*)-3-(3,4-Dimethoxyphenyl)-1-(4-hydroxybenzofuran-5-yl)prop-2-en-1-one

Sub-class: Furanoflavonol

Chemical structure

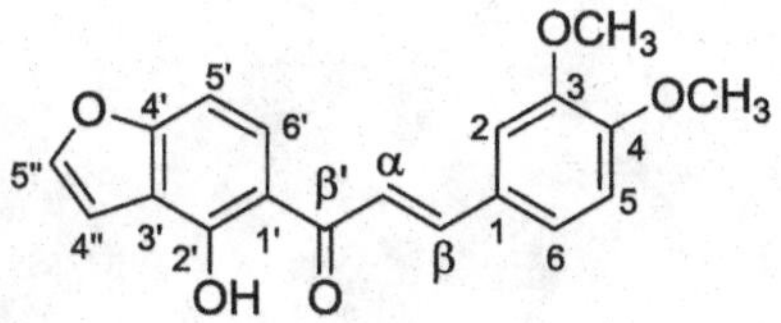

Source: *Millettia erythrocalyx* Gagnep. (Family: Leguminosae); pods

Molecular formula: $C_{19}H_{16}O_5$

Molecular weight: 324

State: Yellow powder

UV (MeOH): λ_{max} (logε) 246 (3.72), 372 (3.75) nm

IR (flim): ν_{max} 3423 (OH), 1635 (α,β-unsaturated carbonyl), 1564 cm^{-1}

^{1}H-NMR (CDCl$_3$, 500 MHz): δ 7.17 (1H, d, *J* = 2.1 Hz, H-2), 6.91 (1H, d, *J* = 2.1 Hz, H-5), 7.27 (1H, dd, *J* = 8.2, 2.1 Hz, H-6), 7.08 (1H, dd, *J* = 8.8, 0.9 Hz, H-5′), 7.85 (1H, d, *J* = 8.8 Hz, H-6′), 7.53 (1H, d, *J* = 15.5 Hz, H-α), 7.89 (1H, d, *J* = 15.5 Hz, H-β), 7.01 (1H, dd, *J* = 2.1, 0.9 Hz,

H-4″), 7.57 (1H, d, J = 2.1 Hz, H-5″), 3.96 (3H, s, 3-OCH_3), 3.93 (3H, s, 4-OCH_3), 14.07 (1H, s, 2′-OH).

^{13}C-NMR (CDCl$_3$, 125 MHz): δ 127.7 (C-1), 110.3 (C-2), 149.3 (C-3), 151.7 (C-4), 111.2 (C-5), 123.5 (C-6), 114.5 (C-1′), 160.2 (C-2′), 117.8 (C-3′), 159.7 (C-4′), 103.6 (C-5′), 125.9 (C-6′), 118.3 (C-α), 145.1 (C-β), 193.3 (C-β′), 105.1 (C-4″), 144.4 (C-5″), 56.0 (3-OCH_3 and 4-OCH_3).

HR-ESITOFMS: m/z 325.1077 ([M + H]$^+$, Calcd. for C$_{19}$H$_{17}$O$_5$, 325.1076).

Reference

Sritularak B, Likhitwitayawuid K. (2006). Flavonoids from the pods of *Millettia erythrocalyx*. *Phytochemistry* **67**: 812–817.

2′-Hydroxy-6,4′,6″,4‴-tetramethoxy-[7-*O*-7″]-bisisoflavone

IUPAC name: 3-(2-Hydroxy-4-methoxyphenyl)-6-methoxy-7-((6-methoxy-3-(4-methoxyphenyl)-4-oxo-4*H*-chromen-7-yl)oxy)-4*H*-chromen-4-one

Sub-class: Bisisoflavone

Chemical structure

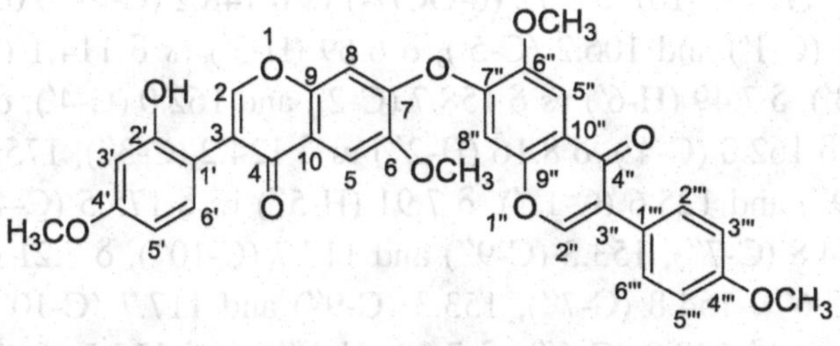

Source: *Platymiscium floribundum* (Family: Leguminosae/ Papilionoideae); heartwood

Molecular formula: $C_{34}H_{26}O_{10}$

Molecular weight: 594

State: Yellow powder

Melting point: 213–215 °C

UV (MeOH): λ_{max} (log ε) 215 (4.55), 231 (4.30), 257 (4.19), 325 (3.87) nm

IR (KBr): ν_{max} 3443 (OH), 1622 (α,β-unsaturated carbonyl), 1570, 1476, 1410 (aromatic unsaturation), 1282, 1247, 1207, 1159, 1023 cm^{-1}.

^{1}H-NMR (DMSO-d_6, 400 MHz): δ 8.33 (1H, s, H-2), 7.84 (1H, s, H-5), 7.19 (1H, s, H-8), 6.94 (1H, d, J = 2.5 Hz, H-3'), 6.69 (1H, dd, J = 8.5, 2.5 Hz, H-5'), 7.49 (1H, d, J = 8.5 Hz, H-6'), 3.72 (3H, s, 6-OCH_3), 3.68 (3H, s, 4'-OCH_3), 8.16 (1H, s, H-2''), 7.91 (1H, s, H-5''), 7.21 (1H, s, H-8''), 3.72 (3H, s, 6''-OCH_3), 7.80 (2H, d, J = 8.7 Hz, H-2''' and H-6'''), 7.07 (2H, d, J = 8.7 Hz, H-3''' and H-5'''), 3.68 (3H, s, 4'''-OCH_3).

^{13}C-NMR (DMSO-d_6, 100 MHz): δ 155.0 (C-2), 123.2 (C-3), 177.2 (C-4), 105.6 (C-5), 148.2 (C-6), 155.2 (C-7), 104.0 (C-8), 153.4 (C-9), 117.1 (C-10), 114.1 (C-1'), 158.7 (C-2'), 104.0 (C-3'), 162.0 (C-4'), 106.2 (C-5'), 132.6 (C-6'), 56.1 (6-OCH_3), 55.4 (4'-OCH_3), 152.7 (C-2''), 124.2 (C-3''), 175.5 (C-4''), 105.1 (C-5''), 148.0 (C-6''), 154.8 (C-7''), 104.2 (C-8''), 153.3 (C-9''), 117.7 (C-10''), 56.1 (6''-OCH_3), 125.6 (C-1'''), 131.0 (C-2'''), 114.4 (C-3'''), 160.1 (C-4'''), 114.5 (C-5'''), 131.0 (C-6'''), 55.4 (4'''-OCH_3).

HMBC: δ 8.33 (H-2) vs δ 123.2 (C-3), 177.2 (C-4), 153.4 (C-9) and 114.1 (C-1'), δ 7.84 (H-5) vs δ 177.2 (C-4), 148.2 (C-6), 155.2 (C-7), 153.4 (C-9) and 117.1 (C-10), δ 7.1(H-8) vs δ 148.2 (C-6), 155.2 (C-7), 153.4 (C-9) and 117.1 (C-10), δ 3.72 (6-OCH_3) vs δ 148.2 (C-6), δ 6.94 (H-3') vs δ 114.1 (C-1') and 106.2 (C-5'), δ 6.69 (H-5') vs δ 114.1 (C-1') and 104.0 (C-3'), δ 7.49 (H-6') vs δ 158.7 (C-2') and 162.0 (C-4'), δ 3.68 (4'-OCH_3) vs δ 162.0 (C-4'), δ 8.16 (H-2'') vs δ 124.2 (C-3''), 175.5 (C-4''), 153.3 (C-9'') and 125.6 (C-1'''), δ 7.91 (H-5'') vs δ 175.5 (C-4''), 148.0 (C-6''), 154.8 (C-7''), 153.3 (C-9'') and 117.7 (C-10''), δ 7.21 (H-8'') vs δ 148.0 (C-6''), 154.8 (C-7''), 153.3 (C-9'') and 117.7 (C-10''), δ 3.72 (6''-OCH_3) vs δ 148.0 (C-6''), δ 7.80 (H-2''') vs δ 175.5 (C-4''), 125.6 (C-1'''), 114.4 (C-3''') and 114.5 (C-5'''), δ 7.07 (H-3'''/H-5''') vs δ 125.6 (C-1'''), δ 7.80 (H-6''') vs δ 175.5 (C-4''), 114.4 (C-3''') and 114.5 (C-5'''), δ 3.68 (4'''-OCH_3) vs 55.4 (4'''-OCH_3).

EIMS: m/z (%rel.) 314 (49), 298 (100), 296 (37), 259 (7), 272 (2), 227 (2), 166 (21), 148 (43), 132 (7), 58 (9).

Reference

Falcão M J C, Pouliquem Y B M, Lima M A S, Gramosa N V, Costa-Lotufo LV, Militão G C G, Pessoa C, de Moraes M O, Silveira E R. (2005). Cytotoxic flavonoids from *Platymiscium floribundum*. *J Nat Prod* **68**: 423–426.

5-Hydroxy-3,7-dimethoxy-4′-prenyloxyflavone

IUPAC name: 5-Hydroxy-3,7-dimethoxy-2-(4-((3-methylbut-2-en-1-yl)oxy)phenyl)-4H-chromen-4-one

Sub-class: Flavone

Chemical structure

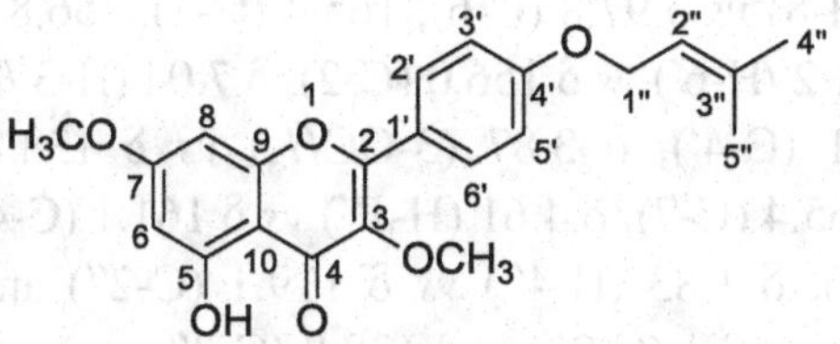

Source: *Spiranthes australis* (R. Brown) Lindl (Family: Orchidacea); whole plants

Molecular formula: $C_{22}H_{22}O_6$

Molecular weight: 382

State: Yellow needles

Melting point: 122–123 °C

UV (MeOH): λ_{max} (log ε) 208, 268 (4.35), 347 nm

IR (KBr): v_{max} 3427 (OH), 1657 (α,β-unsaturated carbonyl), 1597, 1496, 1454, 1375 (aromatic unsaturation), 1226, 1167, 1090, 947, 821 cm^{-1}

¹H-NMR (CDCl₃, 300 MHz): δ 6.36 (1H, d, J = 2.0 Hz, H-6), 6.45 (1H, d, J = 2.0 Hz, H-8), 8.08 (2H, d, J = 9.0 Hz, H-2′ and H-6′), 7.04 (2H, d, J = 9.0 Hz, H-3′ and H-5′), 3.87 (3H, s, 3-OCH_3), 3.88 (3H, s, 7-OCH_3), 4.61 (2H, d, J = 6.7 Hz, H-1″), 5.53 (1H, t, J = 6.7 Hz, H-2″), 1.83 (3H, s, H-4″), 1.79 (3H, s, H-5″).

¹³C-NMR (CDCl₃, 75 MHz): δ 156.0 (C-2), 138.8 (C-3), 178.8 (C-4), 162.1 (C-5), 97.8 (C-6), 165.4 (C-7), 92.2 (C-8), 156.8 (C-9), 106.1 (C-10), 122.7 (C-1′), 130.1 (C-2′ and C-6′), 114.7 (C-3′ and C-5′), 161.1 (C-4′), 60.1 (3-OCH_3), 55.8 (7-OCH_3), 65.0 (C-1″), 119.1 (C-2″), 138.8 (C-3″), 25.8 (C-4″), 18.2 (C-5″).

HMQC: δ 7.28 (H-6) *vs* δ 97.8 (C-6), δ 6.45 (H-8) *vs* δ 92.2 (C-8), δ 8.08 (H-2′/H-6′) *vs* δ 130.1 (C-2′/C-6′), δ 7.04 (H-3′/H-5′) *vs* δ 114.7 (C-3′/C-5′), δ 3.87 (3-OCH_3) *vs* δ 60.1 (3-OCH_3), δ 3.88 (7-OCH_3) *vs* δ 55.8 (7-OCH_3), δ 4.61 (H-1″) *vs* δ 65.0 (C-1″), δ 5.53 (H-2″) *vs* δ 119.1 (C-2″), δ 1.83 (H-4″) *vs* δ 25.8 (C-4″), δ 1.79 (H-5″) *vs* δ 18.2 (C-5″).

HMBC: δ 7.28 (H-6) *vs* δ 62.1 (C-5), 165.4 (C-7), 92.2 (C-8) and 106.1 (C-10), δ 6.45 (H-8) *vs* δ 97.8 (C-6), 165.4 (C-7), 156.8 (C-9) and 106.1 (C-10), δ 8.08 (H-2′/H-6′) *vs* δ 156.0 (C-2), δ 7.04 (H-3′/H-5′) *vs* δ 122.7 (C-1′) and 161.1 (C-4′), δ 3.87 (3-OCH_3) *vs* δ 138.8 (C-3), δ 3.88 (7-OCH_3) *vs* δ 165.4 (C-7), δ 4.61 (H-1″) *vs* δ 161.1 (C-4′), 119.1 (C-2″) and 138.8 (C-3″), δ 1.83 (H-4″) *vs* δ 119.1 (C-2″) and 138.8 (C-3″), δ 1.79 (H-5″) *vs* δ 119.1 (C-2″) and 138.8 (C-3″).

NOESY: δ 7.28 (H-6) *vs* δ 3.88 (7-OCH_3), δ 6.45 (H-8) *vs* δ 3.88 (7-OCH_3), δ 8.08 (H-2′) *vs* δ 7.04 (H-3′) and *vice versa*, δ 7.04 (H-5′) *vs* δ 8.08 (H-6′) and 4.61 (H-1″), δ 8.08 (H-6′) *vs* δ 7.04 (H-5′) and δ 3.87 (3-OCH_3), δ 4.61 (H-1″) *vs* δ 7.04 (H-5′), 5.53 (H-2″) and 1.79 (H-5″), δ 5.53 (H-2″) *vs* δ 4.61 (H-1″) and 1.83 (H-4″), δ 1.83 (H-4″) *vs* δ 5.53 (H-2″), δ 1.79 (H-5″) *vs* δ 4.61 (H-1″), δ 3.87 (3-OCH_3) *vs* δ 8.08 (H-6′) and 4.61 (H-1″), δ 3.88 (7-OCH_3) *vs* δ 7.28 (H-6) and 6.45 (H-8).

HR-FABMS: *m/z* 383.1482 ([M + H]⁺, Calcd. for $C_{22}H_{23}O_6$, 383.1494).

Reference

Dong M-L, Chen F-K, Wu L-J, Gao H-Y. (2005). A new flavonoid from the whole plant of *Spiranthes australis* (R. Brown) Lindl. *J Asian Nat Prod Res* **7**: 71–74.

5-Hydroxy-6,8-dimethoxy-3′, 4′-methylenedioxyflavone

IUPAC name: 2-(Benzo[*d*][1,3]dioxol-5-yl)-5-hydroxy-6,8-dimethoxy-4*H*-chromen-4-one

Sub-class: Flavone

Chemical structure

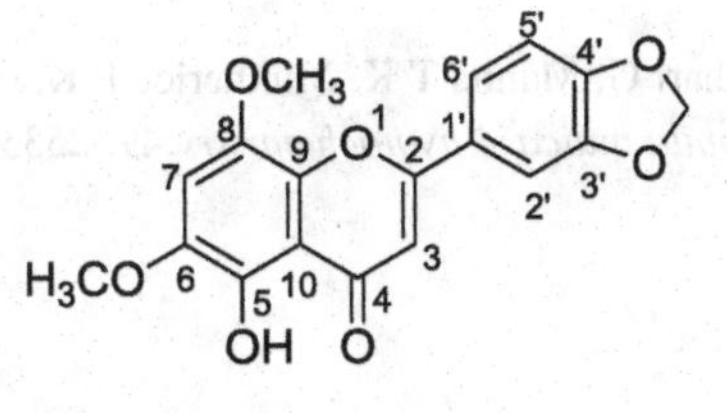

Source: *Limnophila indica* L. (Druce) (Family: Scrophulariaceae); aerial parts and roots

Molecular formula: $C_{18}H_{14}O_7$

Molecular weight: 342

State: Yellow needles

Melting point: 192–193 °C

UV (MeOH): λ_{max} (log ε): 283 (4.27), 329 (3.76) nm

FT-IR (KBr): ν_{max} 3380 (OH), 1635 (α,β-unsaturated carbonyl), 1610, 1510 (aromatic unsaturation) cm^{-1}

^{1}H-NMR (CDCl$_3$, 90 MHz): δ 6.72 (1H, s, H-3), 6.40 (1H, s, H-6), 7.30 (1H, d, J = 2.0 Hz, H-2′), 6.90 (1H, d, J = 8.0 Hz, H-5′), 7.50 (1H, dd, J = 8.0, 2.0 Hz, H-6′), 3.9 (6H, s, 6-OCH_3 and 8-OCH_3), 6.1 (2H, s, 3′,4′-OCH_2O), 13.0 (1H, s, 5-OH).

^{13}C-NMR (CDCl$_3$, 100 MHz): δ 163.8 (C-2), 104.1 (C-3), 183.6 (C-4), 149.6 (C-5), 132.0 (C-6), 118.6 (C-7), 129.1 (C-8), 152.0 (C-9), 104.1 (C-10), 122.9 (C-1′), 110.0 (C-2′), 147.1 (C-3′), 147.1 (C-4′), 112.2 (C-5′), 119.5 (C-6′), 60.8 (6-OCH_3), 61.9 (8-OCH_3), 101.0 (3′,4′-OCH_2O).

EIMS (70 eV): m/z (% rel) 342 (M$^+$, 60.2), 327 (M$^+$ – CH$_3$, 100), 314 (M$^+$ – CO, 7.9), 313 (M$^+$ – CO – H, 18.2), 299 (M$^+$ – CH$_3$CO, 20.7), 196 and 146 (retro-Diels-Alder fragmented peaks of parent molecule; 21.6, 14.7), 181 and 146 (retro-Diels-Alder fragmented peaks of mass fragment 327; 15.2, 14.7), 149 (4.3) and 132 (146 – CH$_2$, 50.6).

Reference

Mukherjee K S, Brahmachari G, Manna T K, Mukherjee P K. (1998). A methylenedioxy flavone from *Limnophila indica*. *Phyotochemistry* **49**: 2533–2534.

5-Hydroxy-7,8,2′, 4-tetramethoxyflavone

IUPAC name: 2-(2,4-Dimethoxyphenyl)-5-hydroxy-7,8-dimethoxy-4*H*-chromen-4-one

Sub-class: Flavone

Chemical structure

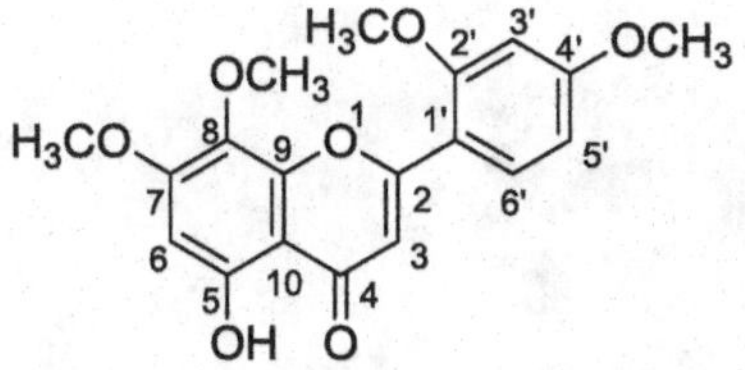

Source: *Limnophila rugosa* (Roth) Merill (Family: Scrophulariaceae); aerial parts and roots

Molecular formula: $C_{19}H_{18}O_7$

Molecular weight: 358

State: Yellow crystals

Melting point: 188–189 °C

UV (MeOH): λ_{max} (log ε) 225 (4.28), 280 (4.23), 330 (4.10) nm

IR (KBr): ν_{max} 3405 (OH), 1660 (α,β-unsaturated carbonyl), 1600, 1580 (aromatic unsaturation) cm^{-1}

^{1}H-NMR (CDCl$_3$, 90 MHz): δ 6.95 (1H, s, H-3), 6.45 (1H, s, H-6), 6.75 (1H, d, J = 2.0 Hz, H-2'), 6.90 (1H, dd, J = 9.0, 2.0 Hz, H-5'), 7.65 (1H, d, J = 9.0 Hz, H-6'), 3.95 (6H, s, 7-OCH_3 and 8-OCH_3), 3.75 (6H, s, 2'-OCH_3 and 4'-OCH_3).

EIMS: m/z 358 ([M]$^+$, Calcd. for C$_{19}$H$_{18}$O$_7$, 358).

Reference

Mukherjee K S, Chakraborty C K, Chatterjee T P. (1989). 5-Hydroxy-7,8,2',4'-tetramethoxyflavone from *Limnophila rugosa*. *Phytochmeistry* **28**: 1778–1779.

5-Hydroxy-8,8-dimethyl-4-phenyl-9,10-dihydropyrano[2,3-*f*] chromen-2(8*H*)-one

Sub-class: Coumarin

Chemical structure

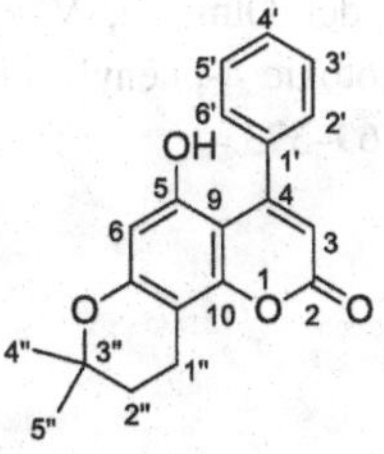

Source: *Marila pluricostata* (Family: Clusiaceae/Guttiferae); leaves

Molecular formula: $C_{20}H_{18}O_4$

Molecular weight: 322

State: Pale yellow amorphous solid

Melting point: 274–275 °C

Bioactivity studied: Cytotoxic [GI_{50} (µg/ml): 5.7, 6.7 and 7.8, respectively, against MCF-7, H-460 and SF-268]

IR (CHCl₃): ν_{max} 3503 (OH), 3271, 1712 (C=O), 1599, 1429, 1362 (aromatic unsaturaion), 1246, 1163, 1111, 1028, 754, 702 cm^{-1}

¹H-NMR (CDCl₃, 200 MHz): δ 5.94 (1H, s, H-3), 6.18 (1H, s, H-6), 5.11 (1H, br s, C₅-O*H*), 7.45 (2H, m, H-2′ and H-6′), 7.54 (3H, m, H-3′, H-4′and H-5′), 2.86 (2H, t, *J* = 6.8 Hz, H-1″), 1.85 (2H, t, *J* = 6.8 Hz, H-2″), 1.37 (6H, s, 2 × 3″-C*H₃*).

¹³C-NMR ((CDCl₃, 100 MHz): δ 161.2 (C-2), 111.3 (C-3), 155.0 (C-4), 153.4 (C-5), 100.8 (C-6), 158.3 (C-7), 102.0 (C-8), 101.0 (C-9), 153.6 (C-10), 138.0 (C-1′), 127.4 (C-2′ and C-6′), 128.7 (C-3′ and C-5′), 129.1 (C-4′), 16.4 (C-1″), 31.8 (C-2″), 75.9 (C-3″), 26.5 (C-4″ and C-5″).

EIMS: *m/z* (%rel) 322 ([M]⁺, 28), 281 (20), 267 (100), 207 (43), 153 (12), 139 (10), 114 (15), 105 (2), 84 (9), 77 (21), 68 (18), 65 (13), 63 (12), 55 (14), 50 (6).

HR-FABMS: *m/z* 322.1205 ([M]⁺, Calcd. for C₂₀H₁₈O₄, 322.1205).

Reference

López-Pérez J L, Olmedo D A, del Olmo E, Vásquez Y, Solís P N, Gupta M P, San Feliciano A. (2005). Cytotoxic 4-phenylcoumarins from the leaves of *Marila pluricostata*. *J Nat Prod* **68**: 369–373.

7-Hydroxy-4′-methoxy-3′-(3-hydroxy-3-methyl-*trans*-but-1-enyl)-5′-(3-methylbut-2-enyl) flavanone

IUPAC name: (*S,E*)-7-Hydroxy-2-(3-(3-hydroxy-3-methylbut-1-en-1-yl)-4-methoxy-5-(3-methylbut-2-en-1-yl)phenyl)chroman-4-one

Sub-class: Isoprenylated flavanone

Chemical structure

Source: *Erythrina mildbraedii* Harms (Family: Leguminosae); root barks

Molecular formula: $C_{26}H_{30}O_5$

Molecular weight: 422

State: White amorphous powder

Melting point: 95–98 °C

Specific rotation: $[\alpha]^{25}_D$ –47.7° (MeOH, c 0.2)

UV (MeOH): λ_{max} (log ε) 265 (4.00) nm

^{1}H-NMR (CDCl$_3$, 400 MHz): δ 5.39 (1H, dd, J = 13.6, 2.8 Hz, H-2), 3.05 (1H, dd, J = 16.8, 13.6 Hz, H-3$_{ax}$), 2.81 (1H, dd, J = 16.8, 2.8 Hz, H-3$_{eq}$), 7.87 (1H, d, J = 8.4 Hz, H-5), 6.55 (1H, dd, J = 8.4, 2.0 Hz, H-6), 6.48 (1H, d, J = 2.0 Hz, H-8), 7.46 (1H, d, J = 2.0 Hz, H-2′), 7.16 (1H, d, J = 2.0 Hz, H-6′), 6.91 (1H, d, J = 16.4 Hz, H-1″), 6.37 (1H, d, J = 16.4 Hz, H-2″), 1.49 (6H, br s, H-4″ and H-5″), 3.39 (2H, br d, J = 6.8 Hz, H-1‴), 5.28 (1H, m, H-2‴), 1.76 (3H, br s, H-4‴), 1.74 (3H, br s, H-5‴), 3.75 (3H, s, 4′-OCH_3).

^{13}C-NMR (CDCl$_3$, 100 MHz): δ 79.9 (C-2), 44.3 (C-3), 190.9 (C-4), 129.5 (C-5), 110.6 (C-6), 163.6 (C-7), 103.5 (C-8), 162.7 (C-9), 115.1 (C-10), 134.5 (C-1′), 122.5 (C-2′), 130.5 (C-3′), 156.2 (C-4′), 135.8 (C-5′), 127.4 (C-6′), 125.0 (C-1″), 134.8 (C-2″), 82.7 (C-3″), 24.4 (C-4″, C-5″), 28.4 (C-1‴), 122.3 (C-2‴), 133.1 (C-3‴), 25.8 (C-4‴), 17.9 (C-5‴), 61.4 (4′-OCH_3).

HMBC: δ 7.46 (H-2′) *vs* δ 79.9 (C-2) and 28.4 (C-1‴), δ 7.16 (H-6′) *vs* δ 79.9 (C-2) and 156.2 (C-4′), δ 6.91 (H-1″) *vs* δ 122.5 (C-2′), 156.2 (C-4′) and 82.7 (C-3″), δ 6.37 (H-2″) *vs* δ 130.5 (C-3′), 82.7 (C-3″) and 24.4 (C-4″/C-5″), δ 1.49 (H-4″/H-5″) *vs* δ 134.8 (C-2″) and 82.7 (C-3″), δ 3.75 (4′-OCH_3) *vs* δ 156.2 (C-4′), δ 3.39 (H-1‴) *vs* δ 156.2 (C-4′), 135.8 (C-5′), 127.4 (C-6′) and 133.1 (C-3‴) (selected 2D-correlations are shown).

HREIMS: m/z 423.2168 ([M + H]$^+$, Calcd. for $C_{26}H_{31}O_5$, 423.2165).

Reference

Na M, Jang J, Njamen D, Mbafor J T, Fomum Z T, Kim B Y, Oh W K, Ahn J S. (2006). Protein tyrosine phosphatase-1B inhibitory activity of isoprenylated flavonoids isolated from *Erythrina mildbraedii*. *J Nat Prod* **69**: 1572–1576.

7-Hydroxy-6,4′-dimethoxy-isoflavonequinone

IUPAC name: 2-(7-Hydroxy-6-methoxy-4-oxo-4H-chromen-3-yl)-5-methoxycyclohexa-2,5-diene-1,4-dione

Sub-class: Isoflavonequinone

Chemical structure

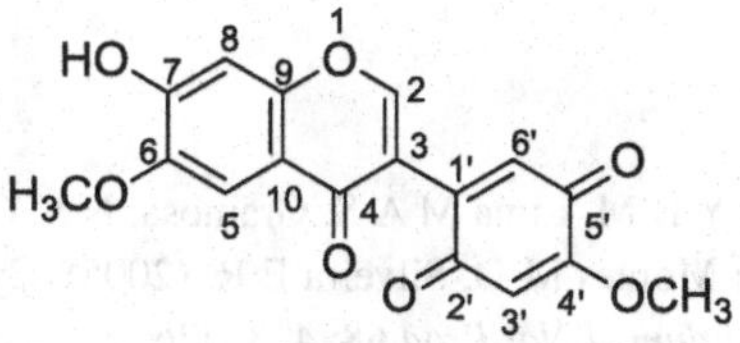

Source: *Platymiscium floribundum* (Family: Leguminosae/ Papilionoideae); heartwood

Molecular formula: $C_{17}H_{12}O_7$

Molecular weight: 328

State: Yellow powder

Melting point: >300 °C

UV (MeOH): λ_{max} (log ε) 230 (4.43), 256 (4.14), 331 (3.96) nm

IR (KBr): ν_{max} 3431 (OH), 2925, 1671 & 1625 (α,β-unsaturated carbonyl), 1514, 1477, 1442 (aromatic unsaturation), 1292, 1028 cm^{-1}

^{1}**H-NMR (DMSO-d_6, 400 MHz):** δ 8.33 (1H, s, H-2), 7.39 (1H, s, H-5), 6.99 (1H, s, H-8), 6.24 (1H, s, H-3'), 7.07 (1H, s, H-6'), 3.88 (3H, s, 6-OCH_3), 10.73 (1H, s, 7-OH), 3.83 (3H, s, 4'-OCH_3).

^{13}**C-NMR (DMSO-d_6, 100 MHz):** δ 156.2 (C-2), 116.2 (C-3), 173.1 (C-4), 104.6 (C-5), 147.3 (C-6), 153.3 (C-7), 103.0 (C-8), 151.5 (C-9), 115.8 (C-10), 139.4 (C-1'), 185.3 (C-2'), 108.0 (C-3'), 158.4 (C-4'), 181.3 (C-5'), 132.8 (C-6'), 55.9 (6-OCH_3), 56.5 (4'-OCH_3).

HMBC: δ 8.33 (H-2) vs δ 116.2 (C-3), 173.1 (C-4), 151.5 (C-9) and 139.4 (C-1'), δ 7.39 (H-5) vs δ 173.1 (C-4), 147.3 (C-6), 153.3 (C-7) and 151.5 (C-9), δ 6.99 (H-8) vs δ 147.3 (C-6) and 151.5 (C-9), δ 6.24 (H-3') vs δ 185.3 (C-2') and 158.4 (C-4'), δ 7.07 (H-6') vs δ 116.2 (C-3), 139.4 (C-1') and 158.4 (C-4').

EIMS: m/z (%rel.) 328 (62), 302 (2), 300 (3), 285 (100), 272 (2), 216 (3), 166 (5), 164 (20), 84 (5), 63 (27), 55 (15).

HR-EIMS: m/z 328.0466 ([M]$^+$, Calcd. for $C_{17}H_{12}O_7$, 328.0583).

Reference

Falcão M J C, Pouliquem Y B M, Lima M A S, Gramosa, N V, Costa-Lotufo LV, Militão G C G, Pessoa C, de Moraes M O, Silveira E R. (2005). Cytotoxic flavonoids from *Platymiscium floribundum. J Nat Prod* **68**: 423–426.

7-Hydroxy-6-methoxy-3,8-bis(3-methyl-2-butenyl)coumarin

IUPAC name: 7-Hydroxy-6-methoxy-3,8-bis(3-methylbut-2-en-1-yl)-2*H*-chromen-2-one

Sub-class: Coumarin

Chemical structure

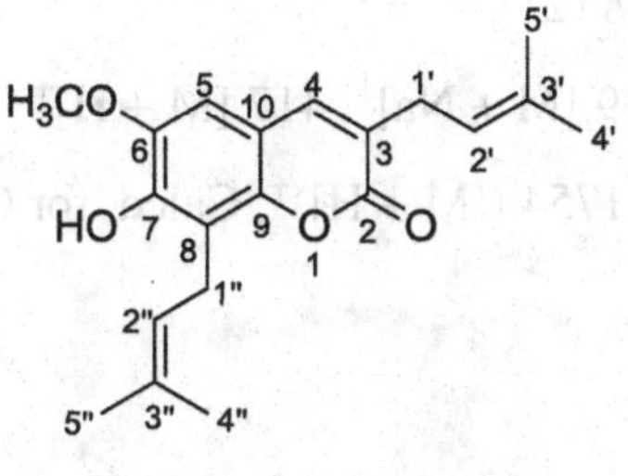

Source: *Coriaria nepalensis* (Family: Coriariaceae); leaves and stems

Molecular formula: $C_{20}H_{24}O_4$

Molecular weight: 328

State: Yellow powder

UV (MeOH): λ_{max} (log ε) 210 (4.65), 347 (4.13) nm

FT-IR (KBr): ν_{max} 3471 (OH), 2923, 2856, 1706 (lactone carbonyl), 1591, 1461 (aromatic unsaturation), 1294, 1063, 933, 913, 854, 833, 786, 762 cm^{-1}

¹H-NMR (CDCl₃, 400 MHz): δ 7.30 (1H, s, H-4), 6.67 (1H, s, H-5), 3.20 (2H, d, J = 7.2 Hz, H-1′), 5.27 (1H, m, H-2′), 1.78 (3H, s, H-4′), 1.67 (3H, s, H-5′), 3.95 (3H, s, 6-OCH_3), 6.07 (1H, s, 7-OH), 3.54 (2H, d, J = 7.2 Hz, H-1″), 5.27 (1H, m, H-2″), 1.82 (3H, s, H-4″), 1.65 (3H, s, H-5″).

¹³C-NMR (CDCl₃, 100 MHz): δ 162.5 (C-2), 125.3 (C-3), 138.5 (C-4), 104.7 (C-5), 143.6 (C-6), 146.2 (C-7), 115.8 (C-8), 147.1 (C-9), 111.9 (C-10), 28.7 (C-1′), 119.7 (C-2′), 135.2 (C-3′), 17.9 (C-4′), 25.8 (C-5′), 56.2 (6-OCH_3), 22.2 (C-1″), 120.9 (C-2″), 132.9 (C-3″), 17.8 (C-4″), 25.8 (C-5″).

HMBC: δ 6.67 (H-5) *vs* δ 138.5 (C-4), 143.6 (C-6) and 111.9 (C-10), δ 3.20 (H-1′) *vs* δ 162.5 (C-2), 125.3 (C-3), 138.5 (C-4), 119.7 (C-2′) and 135.2 (C-3′), δ 1.78 (H-4′) *vs* δ 119.7 (C-2′) and 135.2 (C-3′), δ 1.67 (H-5′) *vs* δ 119.7 (C-2′), δ 3.95 (6-OCH_3) *vs* δ 143.6 (C-6), δ 6.07 (7-OH) *vs* δ 143.6 (C-6), 146.2 (C-7) and 115.8 (C-8), δ 3.54 (H-1″) *vs* δ 146.2 (C-7), 115.8 (C-8) and 147.1 (C-9).

EI-MS (70 eV): m/z (rel%) 328 ([M]⁺, 100), 311 (8), 273 (98), 257 (29), 217 (62), 189 (28), 108 (21).

ESI-TOF-MS: m/z 439 [M + Na]⁺, 417 [M + H]⁺

HR-ESIMS: m/z 329.1754 ([M + H])⁺, Calcd. for $C_{20}H_{25}O_4$, 329.1752).

Reference

Shen Y-H, Li S-H, Han Q-B, Li R-T, Sun H-D. (2006). Coumarins from *Coriaria nepalensis*. *J Asian Nat Prod Res* **8**: 345–350.

6-Hydroxyluteolin 7-*O*-laminaribioside [6-hydroxyluteolin-7-*O*-β-D-glucopyranosyl-(1→3)-β-D-glucopyranoside]

IUPAC name: 7-((((2*S*,3*R*,5*R*,6*R*)-3,5-Dihydroxy-6-(hydroxymethyl)-4-((((2*S*,3*R*,5*S*,6*R*)-3,4,5-trihydroxy-6-(hydroxymethyl)tetrahydro-2*H*-pyran-2-yl)oxy)tetrahydro-2*H*-pyran-2-yl)oxy)-2-(3,4-dihydroxyphenyl)-5,6-dihydroxy-4*H*-chromen-4-one

Sub-class: Flavone glycoside

Chemical structure

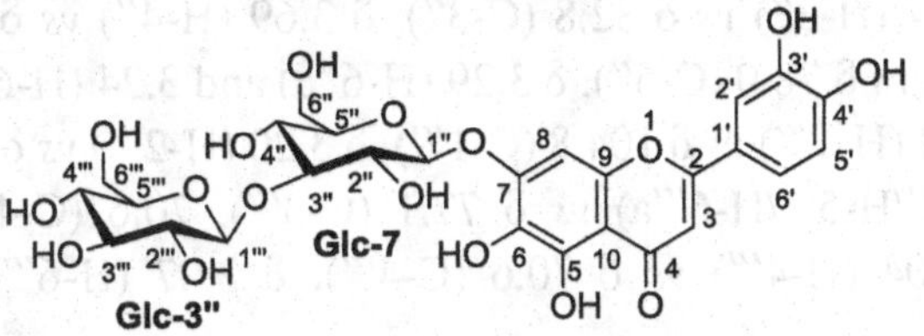

Source: *Globularia alypum* Linn. (Family: Globulariaceae); aerial parts

Molecular formula: $C_{27}H_{30}O_{17}$

Molecular weight: 626

State: Amorphous solid

Specific rotation: $[\alpha]^{20}_D$ −45.3° [MeOH-H_2O (1:1), c 0.1]

Bioactivity studied: Antioxidant [showed DPPH radical scavenging activity with IC_{50} value of 8.0 µM]

UV (MeOH): λ_{max} (log ε) 230 (3.8), 255 (3.5), 284 (4.13) and 346 (4.17) nm

IR (dried flim): v_{max} 3460 (OH), 1675 (α,β-unsaturated carbonyl), 1620, 1575, 1570, 1520, 1510 (aromatic ring) cm^{-1}

^{1}H-NMR (DMSO-d_6, 300 MHz): δ 6.71 (1H, s, H-3), 6.98 (1H, s, H-8), 7.42 (2H, m, H-2′ and H-6′), 6.91 (1H, dd, J = 8.0, 2.5 Hz, H-5′), 5.27 (1H, d, J = 8.0 Hz, H-1″), 3.57 (1H, m, H-2″), 3.54 (1H, m, H-3″), 3.69 (1H, m, H-4″), 3.30 (1H, m, H-5″), 3.29 (1H, dd, J = 12.0, 2.0 Hz, H-6″a), 3.24 (1H, dd, J = 12.0, 6.0 Hz, H-6″b), 4.64 (1H, d, J = 8.0 Hz, H-1‴), 3.20 (1H, m, H-2‴), 3.16 (3H, m, H-3‴, H-5‴ and H-6‴a), 3.04 (1H, m, H-4‴), 3.07 (1H, dd, J = 11.6, 2.0 Hz, H-6‴b).

^{13}C-NMR (DMSO-d_6, 75 MHz): δ 165.3 (C-2), 103.2 (C-3), 183.0 (C-4), 147.2 (C-5), 131.4 (C-6), 151.9 (C-7), 95.1 (C-8), 151.7 (C-9), 106.7 (C-10), 122.0 (C-1′), 114.4 (C-2′), 146.8 (C-3′), 149.9 (C-4′), 117.0 (C-5′), 120.2 (C-6′), 100.5 (C-1″), 75.3 (C-2″), 82.8 (C-3″), 70.2 (C-4″), 78.0 (C-5″), 61.5 (C-6″), 104.8 (C-1‴), 76.5 (C-2‴), 77.1 (C-3‴), 70.6 (C-4‴), 77.7 (C-5‴), 61.3 (C-6″).

HMQC: δ 6.71 (H-3) *vs* δ 165.3 (C-3), δ 6.98 (H-8) *vs* δ 103.2 (C-8), δ 7.42 (H-2′/H-6′) *vs* δ 114.4 (C-2′) and 120.2 (C-6′), δ 6.91 (H-5′) *vs* δ 117.0 (C-5′), δ 5.27 (H-1″) *vs* δ 100.5 (C-1″), δ 3.57 (H-2″) *vs* δ 75.3 (C-2″), δ 3.54 (H-3″) *vs* δ 82.8 (C-3″), δ 3.69 (H-4″) *vs* δ 70.2 (C-4″), δ 3.30 (H-5″) *vs* δ 78.0 (C-5″), δ 3.29 (H-6″a) and 3.24 (H-6″b) *vs* δ 61.5 (C-6″), δ 4.64 (H-1‴) *vs* δ 104.8 (C-1‴), δ 3.20 (H-2‴) *vs* δ 76.5 (C-2‴), δ 3.16 (H-3‴/H-5‴/H-6‴a) *vs* δ 77.1 (C-3‴), 70.6 (C-4‴) and 61.3 (C-6‴), δ 3.04 (H-4‴) *vs* δ 70.6 (C-4‴), δ 3.07 (H-6‴b) *vs* δ 61.3 (C-6‴).

Negative LRESIMS: *m/z* (%rel) 625 ([M − H]$^-$, 100), 463 (3), 445 (7), 301 (83).

Positive LRESIMS: *m/z* (%rel) 627 ([M + H]$^+$, 36), 465 (16), 303 (100).

HRMS: *m/z* 627.1575 ([M + H]$^+$, Calcd. for $C_{27}H_{31}O_{17}$, 627.1561).

Reference

Es-Safi N-E, Khlifi S, Kerhoas L, Kollmann A, El Abbouyi A, Ducrot P-H. (2005). Antioxidant constituents of the aerial parts of *Globularia alypum* growing in Morocco. *J Nat Prod* **68**: 1293–1296.

Irilone 4'-*O*-[β-D-glucopyrano-(1→ 6)-*O*-β-D-glucopyranoside]

IUPAC name: 9-Hydroxy-7-(4-(((2*R*,3*S*,4*R*,5*R*,6*S*)-3,4,5-trihydroxy-6-(((((2*S*,3*S*,4*R*,5*R*,6*S*)-3,4,5-trihydroxy-6-(hydroxymethyl)tetrahydro-2*H*-pyran-2-yl)oxy)methyl)tetrahydro-2*H*-pyran-2-yl)oxy)phenyl)-8*H*-[1,3]dioxolo[4,5-*g*]chromen-8-one

Sub-class: Isoflavone glycoside

Chemical structure

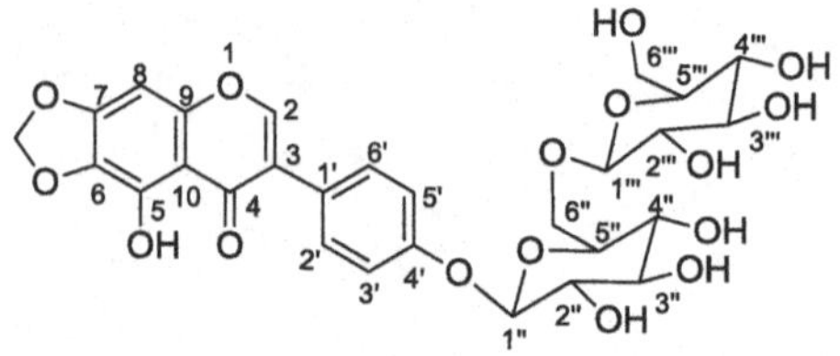

Source: *Iris pseudopumila* (Family: Iridaceae); rhizomes

Molecular formula: $C_{28}H_{30}O_{16}$

Molecular weight: 622

State: White amorphous powder

Melting point: 230–232 °C

Bioactivity studied: Antioxidant

Specific rotation: $[\alpha]^{25}_D$ −35.14° (MeOH, *c* 0.63)

UV (MeOH): λ_{max} 203, 271, 337 nm

¹H-NMR (CD₃OD, 600 MHz): δ 8.19 (1H, s, H-2), 6.63 (1H, s, H-8), 6.10 (2H, s, 6,7-O-CH_2-O-), 7.50 (2H, d, J = 8.0 Hz, H-2' and H-6'), 7.20 (2H, d, J = 8.0 Hz, H-3' and H-5'), 5.00 (1H, d, J = 7.5 Hz, H-1"), 3.25 (1H, dd, J = 9.0, 7.5 Hz, H-2"), 3.53 (1H, dd, J = 9.0 Hz, H-3"), 3.43 (1H, dd, J = 9.0 Hz, H-4"), 3.20 (1H, m, H-5"), 4.20 (1H, dd, J = 12.0, 2.0 Hz, H-6"a), 4.12 (1H, dd, J = 12.0, 4.5 Hz, H-6"b), 4.45 (1H, d, J = 7.5 Hz, H-1‴), 3.53 (1H, dd, J = 9.0, 7.5 Hz, H-2‴), 3.38 (1H, dd, J = 9.0 Hz, H-3‴), 3.31 (1H, dd, J = 9.0 Hz, H-4‴), 3.76 (1H, m, H-5‴), 3.70 (1H, dd, J = 12.0, 4.5 Hz, H-6‴a), 3.38 (1H, dd, J = 12.0, 2.0 Hz, H-6‴b).

¹³C-NMR (CD₃OD, 150 MHz): δ 153.2 (C-2), 126.3 (C-3), 177.5 (C-4), 143.0 (C-5), 137.5 (C-6), 155.1 (C-7), 94.0 (C-8), 158.7 (C-9), 114.8 (C-10), 104.0 (6,7-O-CH_2O-), 127.1 (C-1'), 131.6 (C-2'), 117.6 (C-3'), 156.4 (C-4'), 117.6 (C-5'), 131.6 (C-6'), 101.8 (C-1"), 75.2 (C-2"), 77.9 (C-3"), 71.5 (C-4"), 77.9 (C-5"), 69.6 (C-6"), 104.5 (C-1‴), 74.9 (C-2‴), 77.9 (C-3‴), 71.7 (C-4‴), 77.8 (C-5‴), 62.7 (C-6‴).

Reference

Rigano, D., Formisano, C., Grassia, A., Grassia, G., Perrone, A., Piacente, S., Vuotto, M. L., Senatore, F. (2007). Antioxidant flavonoids and isoflavonoids from rhizomes of *Iris pseudopumila*. *Planta Med* **73**: 93–96.

Kaempferol 3-*O*-β-D-apiofuranosyl-(1→6)-β-D-glucopyranoside

IUPAC name: 3-((((2*S*,4*S*,5*S*,6*R*)-6-(((((2*R*,3*R*)-3,4-Dihydroxy-4-(hydroxymethyl)tetrahydrofuran-2-yl)oxy)methyl)-3,4,5-trihydroxytet-rahydro-2*H*-pyran-2-yl)oxy)-5,7-dihydroxy-2-(4-hydroxyphenyl)-4*H*-chromen-4-one

Sub-class: Flavonol glycoside

Chemical structure

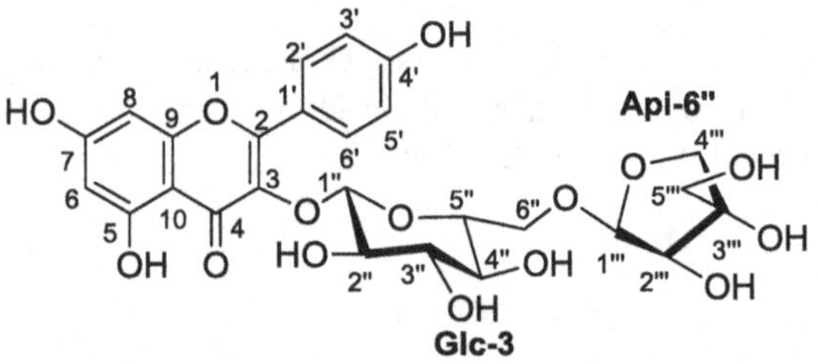

Source: *Solidago altissima* Linn. (Family: Asteraceae); leaves

Molecular formula: $C_{26}H_{28}O_{15}$

Molecular weight: 580

State: Yellow amorphous powder

Specific rotation: $[\alpha]^{25}_{D}$ −39.5° (MeOH, *c* 0.1)

UV (MeOH): λ_{max} 262, 300 sh and 350 nm

¹H-NMR (DMSO-d_6, 400 MHz): δ 6.19 (1H, d, J = 2.0 Hz, H-6), 6.41 (1H, d, J = 2.0 Hz, H-8), 8.00 (2H, d, J = 8.8 Hz, H-2′ and H-6′), 6.88 (2H, d, J = 8.8 Hz, H-3′ and H-5′), 5.38 (1H, d, J = 7.2 Hz, H-1″), 3.19 (1H, m, H-2″), 3.22 (1H, m, H-3″), 3.06 (1H, t, J = 8.8 Hz, H-4″), 3.25 (1H, m, H-5″), 3.69 (1H, br d, J = 11.0 Hz, H-6″a), 3.33 (1H, dd, J = 11.0, 6.0 Hz, H-6″b), 4.69 (1H, d, J = 2.5 Hz, H-1‴), 3.78 (1H, d, J = 2.5 Hz, H-2‴), 3.62 (1H, d, J = 9.5 Hz, H-4‴a), 3.45 (1H, d, J = 9.5 Hz, H-4‴b), 3.18 (2H, s, H-5‴).

¹³C-NMR (DMSO-d_6, 100 MHz): δ 156.4 (C-2), 133.1 (C-3), 177.2 (C-4), 161.1 (C-5), 98.8 (C-6), 164.3 (C-7), 93.7 (C-8), 156.5 (C-9), 103.9 (C, C-10), 120.9 (C-1′), 130.8 (C-2′, C-6′), 115.1 (C-3′, C-5′), 159.8 (C-4′), 100.9 (C-1″), 74.2 (C-2″), 75.8 (C-3″), 69.9 (C-4″), 75.9 (C-5″), 67.3 (C-6″), 109.2 (C-1‴), 76.4 (C-2‴), 78.7 (C-3‴), 73.2 (C-4‴), 65.5 (C-5‴).

HMQC: δ 6.19 (H-6) *vs* δ 98.8 (C-6), δ 6.41 (H-8) *vs* δ 93.7 (C-8), δ 8.00 (H-2′/H-6′) *vs* δ 130.8 (C-2′/C-6′), δ 6.88 (H-3′/5′) *vs* δ 115.1 (C-3′/C-5′), δ 5.38 (H-1″) *vs* δ 100.9 (C-1″), δ 3.19 (H-2″) *vs* δ 74.2 (C-2″), δ 3.22 (H-3″) *vs* δ 75.8 (C-3″), δ 3.06 (H-4″) *vs* δ 69.9 (C-4″), δ 3.25 (H-5″) *vs* δ 75.9 (C-5″), δ 3.69 (H-6″a) and 3.33 (H-6″b) *vs* δ 67.3 (C-6″), δ 4.69 (H-1‴) *vs* δ 109.2 (C-1‴), δ 3.78 (H-2‴) *vs* δ 76.4 (C-2‴), δ 3.62 (H-4‴a) and 3.45 (H-4‴b) *vs* δ 73.2 (C-4‴), δ 3.18 (H-5‴) *vs* δ 65.5 (C-5‴).

ESI-MS: *m/z* 581 [M + H]⁺, 449 [M + H − Api]⁺, 287 [M + H − Api − Glc]⁺.

HR-ESI-MS: *m/z* 581.1479 ([M + H]⁺, Calcd. for $C_{26}H_{29}O_{15}$, 581.1501).

Reference

Wu B, Takahashi T, Kashiwagi T, Tebayashi S, Kim C-S. (2007). New flavonoid glycosides from the leaves of *Solidago altissima*. *Chem Pharm Bull* **55**: 815–816.

Khonklonginol A [3,5-Dihydroxy-4'-methoxy-6'',6''-dimethylpyrano(2'',3'':7,6)-8-(3''',3'''-dimethylallyl)flavanone]

IUPAC name: (2*R*,3*R*)-3,5-Dihydroxy-2-(4-methoxyphenyl)-8,8-dimethyl-10-(3-methylbut-2-en-1-yl)-2,3-dihydropyrano[3,2-*g*]chromen-4(8*H*)-one

Sub-class: Prenylated pyrano-fused flavanone

Chemical structure

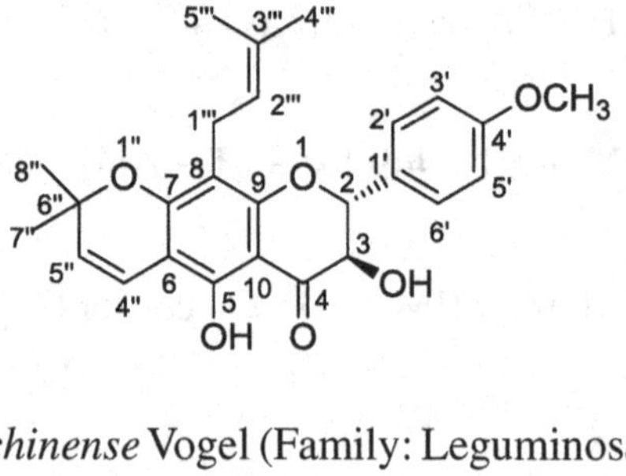

Source: *Eriosema chinense* Vogel (Family: Leguminosae-Papilionoideae); roots

Molecular formula: $C_{26}H_{28}O_6$

Molecular weight: 436

State: Yellow liquid

Bioactivity studied: Antimicrobial

160

Specific rotation: $[\alpha]_D$ +18.9°(CHCl$_3$, c 0.5)

IR (KBr): v_{max} 3467 (OH), 2975, 2926, 1626, 1516, 1463, 1379, 1289, 1251, 1191, 1126, 1026, 943, 907, 830, 737, 641, 575 cm^{-1}

^{1}H-NMR (CDCl$_3$, 400 MHz): δ 4.98 (1H, d, J = 12.0 Hz, H-2β), 4.42 (1H, d, J = 12.0 Hz, H-3$_{ax}$), 7.46 (2H, d, J = 8.8 Hz, H-2′ and H-6′), 6.96 (2H, d, J = 8.8 Hz, H-3′ and H-5′), 6.62 (1H, d, J = 10.0 Hz, H-4″), 5.51 (1H, d, J = 10.0 Hz, H-5″), 1.44 (3H, s, H-7″), 1.43 (3H, s, H-8″), 3.16 (2H, d, J = 6.8 Hz, H-1‴), 5.25 (1H, dt, J = 6.4, 1.6 Hz, H-2‴), 1.63 (3H, s, H-4‴), 1.59 (3H, s, H-5‴), 3.83 (3H, s, 4′-OCH_3), 11.41 (1H, s, 5-OH).

^{13}C-NMR (CDCl$_3$, 100 MHz): δ 82.9 (C-2), 72.6 (C-3), 196.4 (C-4), 156.1 (C-5), 103.2 (C-6), 160.7 (C-7), 109.3 (C-8), 159.5 (C-9), 100.4 (C-10), 128.8 (C-1′, C-2′ and C-6′), 114.0 (C-3′ and C-5′), 160.3 (C-4′), 115.4 (C-4″), 126.2 (C-5″), 78.5 (C-6″), 28.4 (C-7″ and C-8″), 21.4 (C-1‴), 122.3 (C-2‴), 131.3 (C-3‴), 17.8 (C-4‴), 25.7 (C-5‴), 55.3 (4′-OCH$_3$).

HR-EIMS: m/z 436.1873 ([M$^+$], Calcd. for C$_{26}$H$_{28}$O$_6$, 436.1886).

Reference

Sutthivaiyakit S, Thongnak O, Lhinhatrakool T, Yodchun O, Srimark R, Dowtaisong P, Chuankamnerdkarn M. (2009). Cytotoxic and antimycobacterial prenylated flavonoids from the roots of *Eriosema chinense*. *J Nat Prod* **72**: 1092–1096.

Khonklonginol E [3,5-Dihydroxy-4'-methoxy-6'',6''-dimethylpyrano(2'',3'':7,6)-8-(3''',3'''-dimethyl-2''',3'''-dihydroxypropyl)flavanone]

IUPAC name: (2R,3R)-10-(2,3-Dihydroxy-3-methylbutyl)-3,5-dihydroxy-2-(4-methoxyphenyl)-8,8-dimethyl-2,3-dihydropyrano[3,2-g]chromen-4(8H)-one

Sub-class: Prenylated pyrano-fused flavanone

Chemical structure

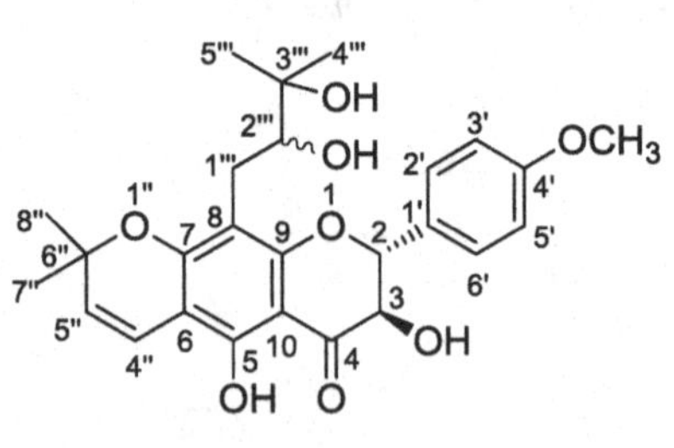

Source: *Eriosema chinense* Vogel (Family: Leguminosae-Papilionoideae); roots

Molecular formula: $C_{26}H_{30}O_6$

Molecular weight: 438

State: Yellow liquid

Specific rotation: $[\alpha]_D$ −1.9° (CHCl$_3$, c 1.8)

IR (KBr): ν_{max} 3401 (OH), 2975, 2932, 1645, 1633, 1588, 1516, 1463, 1381, 1288, 1251, 1192, 1127, 1030, 879, 831, 776, 736, 575, 520 cm^{-1}

^{1}H-NMR (CDCl$_3$, 400 MHz): δ 5.01 (1H, d, J = 11.9 Hz, H-2β), 4.50 (1H, d, J = 11.9 Hz, H-3$_{ax}$), 7.43 (2H, d, J = 8.7 Hz, H-2′ and H-6′), 6.96 (2H, d, J = 8.8 Hz, H-3′ and H-5′), 6.64 (1H, d, J = 10.0 Hz, H-4″), 5.52 (1H, d, J = 10.2 Hz, H-5″), 1.47 (3H, s, H-7″), 1.45 (3H, s, H-8″), 3.16 (2H, d, J = 6.9 Hz, H-1‴), 5.13 (1H, brt, J = 6.1 Hz, H-2‴), 1.17 (3H, s, H-4‴), 1.16 (3H, s, H-5‴), 3.82 (3H, s, 4′-OCH_3), 11.40 (1H, s, 5-OH).

^{13}C-NMR (CDCl$_3$, 100 MHz): δ 83.1 (C-2), 72.4 (C-3), 196.4 (C-4), 156.5 (C-5), 103.4 (C-6), 160.4 (C-7), 106.5 (C-8), 159.8 (C-9), 100.5 (C-10), 128.8 (C-1′), 128.6 (C-2′), 114.1 (C-3′), 160.3 (C-4′), 114.2 (C-5′), 128.6 (C-6′), 115.4 (C-4″), 126.2 (C-5″), 79.2 (C-6″), 28.6 (C-7″ and C-8″), 25.2 (C-1‴), 79.1 (C-2‴), 72.8 (C-3‴), 25.9 (C-4‴ and C-5‴), 55.3 (4′-OCH$_3$).

HR-EIMS: m/z 493.1830 ([M + Na]$^+$, Calcd. for C$_{26}$H$_{30}$O$_6$Na, 493.1833).

Reference

Sutthivaiyakit S, Thongnak O, Lhinhatrakool T, Yodchun O, Srimark R, Dowtaisong P, Chuankamnerdkarn M. (2009). Cytotoxic and antimycobacterial prenylated flavonoids from the roots of *Eriosema chinense*. *J Nat Prod* **72**: 1092–1096.

Khonklonginol F [3,5-Dihydroxy-4'-methoxy-6'',6'-dimethylpyrano(2'',3'':7,6)-8-(3''',3'''-dimethylallyl)flavone]

IUPAC name: 3,5-Dihydroxy-2-(4-methoxyphenyl)-8,8-dimethyl-10-(3-methylbut-2-en-1-yl)pyrano[3,2-*g*]chromen-4(8*H*)-one

Sub-class: Prenylated pyrano-fused flavone

Chemical structure

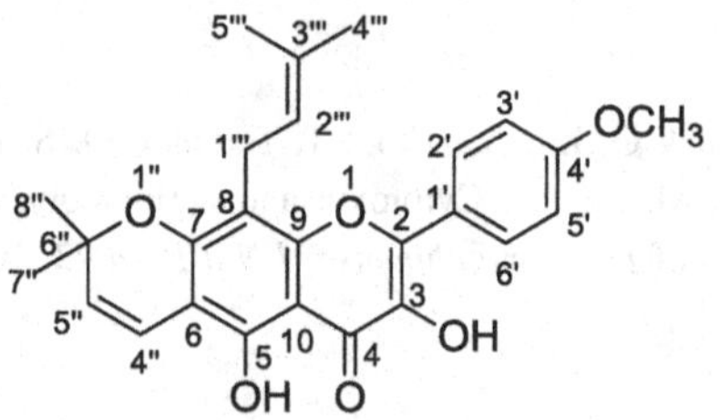

Source: *Eriosema chinense* Vogel (Family: Leguminosae/Papilionoideae); roots

Molecular formula: $C_{26}H_{26}O_6$

Molecular weight: 434

State: Yellow solid

Bioactivity studied: Antimicrobial

Specific rotation: $[\alpha]_D$ −19° (CHCl$_3$, c 1.6)

IR (KBr): ν_{max} 3316 (OH), 2924, 2852, 1736, 1651, 1620, 1596, 1552, 1483, 1428, 1359, 1303, 1256, 1187, 1159, 1125, 1091, 1052, 1035, 999, 899, 878, 834, 803, 772, 648, 585, 508 cm^{-1}

^{1}H-NMR (CDCl$_3$, 400 MHz): δ 8.15 (2H, d, J = 8.8 Hz, H-2′ and H-6′), 7.01 (2H, d, J = 8.8 Hz, H-3′ and H-5′), 6.62 (1H, d, J = 9.9 Hz, H-4″), 5.62 (1H, d, J = 10.1 Hz, H-5″), 1.45 (6H, s, H-7″ and H-8″), 3.49 (2H, d, J = 7.0 Hz, H-1‴), 5.21 (1H, dt, J = 7.0, 6.1 Hz, H-2‴), 1.67 (3H, s, H-4‴), 1.82 (3H, s, H-5‴), 3.87 (3H, s, 4′-OCH_3), 6.63 (1H, br s, 3-OH), 11.93 (1H, s, 5-OH).

^{13}C-NMR (CDCl$_3$, 100 MHz): δ 145.4 (C-2), 135.5 (C-3), 175.5 (C-4), 153.0 (C-5), 104.9 (C-6), 156.9 (C-7), 107.7 (C-8), 153.6 (C-9), 103.5 (C-10), 123.7 (C-1′), 129.3 (C-2′), 114.1 (C-3′), 161.0 (C-4′), 114.1 (C-5′), 129.3 (C-6′), 115.7 (C-4″), 128.1 (C-5″), 77.8 (C-6″), 28.3 (C-7″ and C-8″), 21.5 (C-1‴), 122.2 (C-2‴), 131.8 (C-3‴), 25.7 (C-4‴), 18.0 (C-5‴), 55.3 (4′-OCH_3).

HR-EIMS: m/z 435.1800 ([M + 1]$^+$, Calcd. for C$_{26}$H$_{27}$O$_6$, 435.1800).

Reference

Sutthivaiyakit S, Thongnak O, Lhinhatrakool T, Yodchun O, Srimark R, Dowtaisong P, Chuankamnerdkarn M. (2009). Cytotoxic and antimycobacterial prenylated flavonoids from the roots of *Eriosema chinense*. *J Nat Prod* 72: 1092–1096.

Kurzphenol B

IUPAC name: (2*R*,3*R*)-3,5,7-trihydroxy-6,8-bis(3-methylbut-2-en-1-yl)-2-phenylchroman-4-one

Sub-class: Prenylated flavanone

Chemical structure

Source: *Macaranga kurzii* (Family: Euphorbiaceae); twigs

Molecular formula: $C_{25}H_{28}O_5$

Molecular weight: 408

State: Yellow oil

Specific rotation: $[\alpha]^{26}_{D}$ −1.77° (MeOH, *c* 0.28)

UV-vis (MeOH): λ_{max} (log ε) 204 (4.68), 3.00 (4.27), 346 (3.74) nm

IR (KBr): ν_{max} 3421 (OH), 2968, 2916, 1631, 1452, 1374, 1276, 1230, 1182, 1112, 1003, 766, 698 cm^{-1}.

¹H-NMR (CDCl₃, 500 MHz): δ 5.02 (1H, d, J = 11.9 Hz, H-2β), 4.49 (1H, d, J = 11.9 Hz, H-3$_{ax}$), 7.54 (2H, d, J = 7.1 Hz, H-2′ and H-6′), 7.44 (2H, t, J = 7.1 Hz, H-3′ and H-5′), 7.41 (1H, t, J = 7.1 Hz, H-4′), 3.34 (2H, d, J = 7.0 Hz, H-1″), 5.22 (1H, t, J = 7.0 Hz, H-2″), 1.74 (3H, s, H-4″), 1.81 (3H, s, H-5″), 3.27 (2H, d, J = 7.1 Hz, H-1‴), 5.17 (1H, t, J = 7.1 Hz, H-2‴), 1.10 (3H, s, H-4‴), 1.66 (3H, s, H-5‴), 11.49 (1H, s, 5-OH), 6.49 (1H, s, 7-OH).

¹³C-NMR (CDCl₃, 125 MHz): δ 83.1 (C-2), 72.5 (C-3), 196.2 (C-4), 158.8 (C-5), 107.9 (C-6), 163.2 (C-7), 107.0 (C-8), 157.6 (C-9), 100.3 (C-10), 136.6 (C-1′), 127.3 (C-2′), 128.5 (C-3′), 129.1 (C-4′), 128.5 (C-5′), 127.3 (C-6′), 21.2 (C-1″), 121.4 (C-2″), 134.9 (C-3″), 25.8 (C-4″), 17.9 (C-5″), 21.7 (C-1‴), 121.5 (C-2‴), 134.4 (C-3‴), 25.8 (C-4‴), 17.8 (C-5‴).

HR-EIMS: m/z 408.1913 ([M]⁺, Calcd. for C₂₅H₂₈O₅, 408.1937).

Reference

Yang D-S, Wei J-G, Peng W-B, Wang S-M, Sun C, Yang Y-P, Liu K-C, Li X-L. (2014). Cytotoxic prenylated bibenzyls and flavonoids from *Macaranga kurzii*. *Fitoterapia* 99: 161–266.

Lawsochrysin (5-hydroxy-6-n-pentyl-7-n-pentyloxyflavone)

IUPAC name: 5-Hydroxy-6-pentyl-7-(pentyloxy)-2-phenyl-4H-chromen-4-one

Sub-class: Flavone

Chemical structure

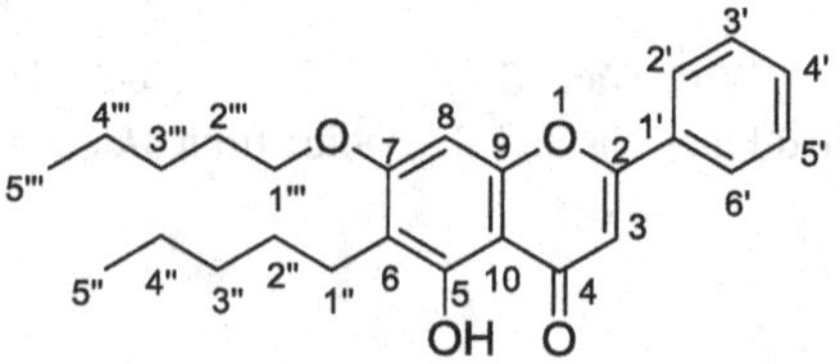

Source: *Lawsonia alba* Lam. (synonym: *Lawsonia inermis* Linn.) (Family: Lythraceae); leaves

Molecular formula: $C_{25}H_{30}O_4$

Molecular weight: 394

State: Yellow amorphous powder

UV: λ_{max} (CHCl$_3$) (log ε): 276 (4.75), 317 (4.20) nm

IR (film, CHCl$_3$): ν_{max} 3431 (OH), 2860, 1660 (α,β-unsaturated carbonyl), 1500-1444 (aromatic unsaturaion) cm^{-1}

^{1}H-NMR (CDCl$_3$, 500 MHz): δ 6.64 (1H, s, H-3), 6.46 (1H, s, H-8), 7.85 (2H, dd, J = 8.0, 1.5 Hz, H-2′ and H-6′), 7.48 (2H, m, H-3′ and H-5′), 7.50

(1H, m, H-4′), 2.64 (2H, t, J = 7.5 Hz, H-1″), 1.45-1.53 (2H, m, H-2″), 1.31-1.34 (2H, m, H-3″), 1.36-1.42 (2H, m, H-4″), 0.93 (3H, t, J = 7.0 Hz, H-5″), 4.02 (2H, t, J = 6.5 Hz, H-1‴), 1.84 (2H, quint, J = 6.5 Hz, H-2‴), 1.45-1.53 (2H, m, H-3‴), 1.31-1.34 (2H, m, H-4‴), 0.87 (3H, t, J = 7.0 Hz, H-5‴), 12.74 (1H, s, C_5-O*H*).

^{13}C-NMR (CDCl$_3$, 125 MHz): δ 163.0 (C-2), 105.9 (C-3), 182.4 (C-4), 158.5 (C-5), 114.4 (C-6), 163.4 (C-7), 90.0 (C-8), 156.1 (C-9), 105.4 (C-10), 131.5 (C-1′), 126.2 (C-2′ and C-6′), 129.0 (C-3′ and C-5′), 131.6 (C-4′), 22.3 (C-1″), 28.4 (C-2″), 28.2 (C-3″), 22.1 (C-4″), 14.0 (C-5″), 68.5 (C-1‴), 28.7 (C-2‴), 31.9 (C-3‴), 22.5 (C-4‴), 14.0 (C-5‴).

HSQC: δ 6.64 (H-3) *vs* δ 105.9 (C-3), δ 6.46 (H-8) *vs* δ 90.0 (C-8), δ 7.85 (H-2′/H-6′) *vs* δ 126.2 (C-2′/C-6′), δ 7.48 (H-3′/ H-5′) *vs* δ 129.0 (C-3′/C-5′), δ 7.50 (H-4′) *vs* δ 131.6 (C-4′), δ 2.64 (H-1″) *vs* δ 22.3 (C-1″), δ 1.45-1.53 (H-2″) *vs* δ 28.4 (C-2″), δ 1.31-1.34 (H-3″) *vs* δ 28.2 (C-3″), δ 1.36-1.42 (H-4″) *vs* δ 22.1 (C-4″), δ 0.93 (H-5″) *vs* δ 14.0 (C-5″), δ 4.02 (H-1‴) *vs* δ 68.5 (C-1‴), δ 1.84 (H-2‴) *vs* δ 28.7 (C-2‴), δ 1.45-1.53 (H-3‴) *vs* δ 31.9 (C-3‴), δ 1.31-1.34 (H-4‴) *vs* δ 22.5 (C-4‴), δ 0.87 (H-5‴) *vs* δ 14.0 (C-5‴).

HMBC: δ 6.64 (H-3) *vs* δ 163.0 (C-2) and 131.5 (C-1′), δ 6.46 (H-8) *vs* δ 114.4 (C-6), 163.4 (C-7), 90.0 156.1 (C-9) and 105.4 (C-10), δ 7.85 (H-2′/H-6′) *vs* δ 163.0 (C-2), 131.5 (C-1′), 129.0 (C-3′/C-5′) and 131.6 (C-4′), δ 7.48 (H-3′/ H-5′) *vs* δ 131.5 (C-1′), 126.2 (C-2′/C-6′) and 131.6 (C-4′), δ 7.50 (H-4′) *vs* δ 131.5 (C-1′), 126.2 (C-2′/C-6′) and 129.0 (C-3′/C-5′), δ 2.64 (H-1″) *vs* δ 158.5 (C-5), 114.4 (C-6), 163.4 (C-7) and 28.4 (C-2″), δ 1.45-1.53 (H-2″) *vs* δ 114.4 (C-6), δ 4.02 (H-1‴) *vs* δ 163.4 (C-7).

^{1}H-^{1}H COSY: δ 7.85 (H-2′/H-6′) *vs* δ 7.48 (H-3′/H-5′), δ 7.48 (H-3′/H-5′) *vs* δ 7.85 (H-2′/H-6′) and 7.50 (H-4′), δ 7.50 (H-4′) *vs* δ 7.48 (H-3′/H-5′), δ 2.64 (H-1″) *vs* δ 1.45-1.53 (H-2″), δ 1.45-1.53 (H-2″) *vs* δ 2.64 (H-1″) and 1.31-1.34 (H-3″), δ 1.31-1.34 (H-3″) *vs* δ 1.45-1.53 (H-2″) and 1.36-1.42 (H-4″), δ 1.36-1.42 (H-4″) *vs* δ 1.31-1.34 (H-3″) and 0.93 (H-5″), δ 0.93 (H-5″) *vs* δ 1.36-1.42 (H-4″), δ 4.02 (H-1‴) *vs* δ 1.84 (H-2‴), δ 1.84 (H-2‴) *vs* δ 4.02 (H-1‴) and 1.45-1.53 (H-3‴), δ 1.45-1.53 (H-3‴) *vs* δ 1.84 (H-2‴) and 1.31-1.34 (H-4‴), δ 1.31-1.34 (H-4‴) *vs* δ 1.45-1.53 (H-3‴) and 0.87 (H-5‴), δ 0.87 (H-5‴) *vs* δ 1.31-1.34 (H-4‴).

NOESY: δ 6.64 (H-3) *vs* δ 7.85 (H-6′), δ 4.02 (H-1‴) *vs* δ 6.46 (1H, s, H-8) (selected NOSEY correlations were shown).

EIMS: *m/z* (%rel.) 394 ([M]$^+$, 40), 351 (18), 337 (100), 281 (62), 267 (80), 252 (7.4), 165 (14.6).

HR-EIMS: *m/z* 394.2150 ([M]$^+$, Calcd. for $C_{25}H_{30}O_4$, 394.2144).

Reference

Uddin N, Siddiqui BS, Begum S, Bhatti HA, Khan A, Parveen S, Choudhary MI. (2011). Bioactive flavonoids from the leaves of *Lawsonia alba* (Henna). *Phytochemistry Lett* 4: 454–48.

Lawsonaringenin [4′,5-dihydroxy-7-(4″-pentenyloxy)flavanone]

IUPAC name: (R)-5-Hydroxy-2-(4-hydroxyphenyl)-7-(pent-4-en-1-yloxy)chroman-4-one

Sub-class: Flavanone

Chemical structure

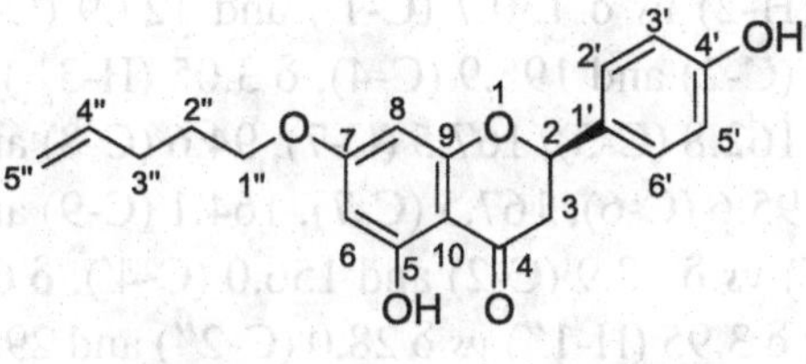

Source: *Lawsonia alba* Lam. (synonym: *Lawsonia inermis* Linn.) (Family: Lythraceae); leaves

Molecular formula: $C_{20}H_{20}O_5$

Molecular weight: 340

State: White amorphous powder

UV: λ_{max} (CHCl$_3$) (log ε): 287 (3.64), 326 (2.94) nm

IR (film, CHCl$_3$): ν_{max} 3226 (OH), 2943, 1639 (carbonyl), 1510-1446 (aromatic unsaturaion) cm^{-1}

^{1}H-NMR (CDCl$_3$, 500 MHz): δ 5.31 (1H, dd, J = 12.9, 3.0 Hz, H-2α), 2.73 (1H, dd, J = 17.1, 3.0 Hz, H-3$_{eq}$), 3.05 (1H, dd, J = 17.1, 12.9 Hz,

H-3$_{ax}$), 6.03 (1H, d, J = 2.1 Hz, H-6), 6.00 (1H, d, J = 2.1 Hz, H-8), 7.32 (2H, d, J = 8.5 Hz, H-2′ and H-6′), 6.88 (2H, d, J = 8.5 Hz, H-3′ and H-5′), 3.95 (2H, t, J = 6.6 Hz, H-1″), 1.84 (2H, quint, J = 6.6 Hz, H-2″), 2.17 (2H, q, J = 6.6 Hz, H-3″), 5.82 (1H, ddt, J = 17.1, 10.2, 6.6 Hz, H-4″), 5.06 (1H, ddd, J = 17.1, 3.3, 1.8 Hz, H-5″a), 4.97 (1H, ddd, J = 17.1, 3.3, 1.8 Hz, H-5″b), 11.98 (1H, s, C$_5$-OH).

^{13}C-NMR (CDCl$_3$, 125 MHz): δ 78.9 (C-2), 43.2 (C-3), 195.9 (C-4), 162.8 (C-5), 95.6 (C-6), 167.5 (C-7), 94.6 (C-8), 164.1 (C-9), 103.0 (C-10), 130.7 (C-1′), 127.9 (C-2′ and C-6′), 156.0 (C-3′ and C-5′), 156.0 (C-4′), 67.6 (C-1″), 28.0 (C-2″), 29.9 (C-3″), 137.3 (C-4″), 115.5 (C-5″).

HSQC: δ 5.31 (H-2) vs δ 78.9 (C-2), δ 2.73 (H-3$_{eq}$) and 3.05 (H-3$_{ax}$) vs δ 43.2 (C-3), δ 6.03 (H-6) vs δ 95.6 (C-6), δ 6.00 (H-8) vs δ 94.6 (C-8), δ 7.32 (H-2′/H-6′) vs δ 127.9 (C-2′/C-6′), δ 6.88 (H-3′/ H-5′) vs δ 156.0 (C-3′/C-5′), δ 3.95 (H-1″) vs δ 67.6 (C-1″), δ 1.84 (H-2″) vs δ 28.0 (C-2″), δ 2.17 (H-3″) vs δ 29.9 (C-3″), δ 5.82 (H-4″) vs δ 137.3 (C-4″), δ 5.06 (H-5″a) and 4.97 (H-5″b) vs δ 115.5 (C-5″).

HMBC: δ 5.31 (H-2) vs δ 130.7 (C-1′) and 127.9 (C-2′/C-6′), δ 2.73 (H-3$_{eq}$) vs δ 78.9 (C-2) and 195.9 (C-4), δ 3.05 (H-3$_{ax}$) vs δ 78.9 (C-2), δ 6.03 (H-6) vs δ 162.8 (C-5), 167.5 (C-7), 94.6 (C-8) and 103.0 (C-10), δ 6.00 (H-8) vs δ 95.6 (C-6), 167.5 (C-7), 164.1 (C-9) and 103.0 (C-10), δ 7.32 (H-2′/H-6′) vs δ 78.9 (C-2) and 156.0 (C-4′), δ 6.88 (H-3′/ H-5′) vs δ 130.7 (C-1′), δ 3.95 (H-1″) vs δ 28.0 (C-2″) and 29.9 (C-3″), δ 1.84 (H-2″) vs δ 67.6 (C-1″), 29.9 (C-3″) and 137.3 (C-4″), δ 2.17 (H-3″) vs δ 67.6 (C-1″), 28.0 (C-2″), 137.3 (C-4″) and 115.5 (C-5″), δ 5.82 (H-4″) vs δ 29.9 (C-3″), δ 5.06 (H-5″a) and 4.97 (H-5″b) vs δ 29.9 (C-3″)

^{1}H-^{1}H COSY: δ 5.31 (H-2α) vs δ 2.73 (H-3$_{eq}$) and 3.05 (H-3$_{ax}$), δ 7.32 (H-2′/H-6′) vs δ 6.88 (H-3′/H-5′), δ 3.95 (H-1″) vs δ 1.84 (H-2″), δ 1.84 (H-2″) vs δ 3.95 (H-1″) and 2.17 (H-3″), δ 2.17 (H-3″) vs δ 1.84 (H-2″) and 5.82 (H-4″), δ 5.82 (H-4″) vs δ 2.17 (H-3″), 5.06 (H-5″a) and 4.97 (H-5″b), δ 5.06 (H-5″a) and 4.97 (H-5″b) vs δ 5.82 (H-4″).

NOESY: δ 3.95 (H-1″) vs δ 6.00 (H-8), δ 3.95 (H-1″) vs δ 6.03 (H-6) (selected NOSEY correlations were shown).

EIMS: m/z (%rel.) 340 ([M]$^+$, 59), 272 (100), 221 (48), 153 (76), 120 (93)

HR-EIMS: m/z 340.1298 ([M]$^+$, Calcd. for $C_{20}H_{20}O_5$, 340.1311).

Reference

Uddin N, Siddiqui B S, Begum S, Bhatti H A, Khan A, Parveen S, Choudhary M I. (2011). Bioactive flavonoids from the leaves of *Lawsonia alba* (Henna). *Phytochemistry Lett.* 4: 454–48.

Licoflavanone-4′-*O*-methyl ether

IUPAC name: (*S*)-5,7-Dihydroxy-2-(4-methoxy-3-(3-methylbut-2-en-1-yl)phenyl)chroman-4-one

Sub-class: Prenylated flavanone

Chemical structure

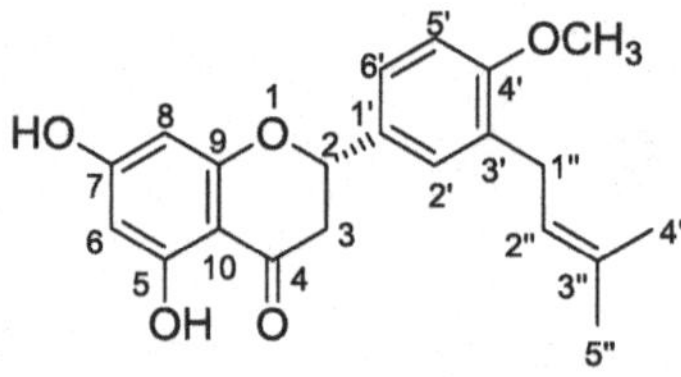

Source: *Erythrina mildbraedii* (Family: Leguminosae); root bark

Molecular formula: $C_{21}H_{22}O_5$

Molecular weight: 354

State: Amorphous gummy substance

Bioactivity studied: *In vitro* protein tyrosine phosphatase 1B (PTP1B) inhibitory activity ($IC_{50} = 29.6 \pm 2.5$ µM)

Specific rotation: $[\alpha]^{25}_D$ −21.5° (MeOH, *c* 0.3)

UV: λ_{max} (MeOH) (log ε): 289 (4.15), 321 (3.81) nm

^{1}H-NMR (CDCl₃, 400 MHz): δ 5.35 (1H, dd, *J* = 13.2, 2.8 Hz, H-2β), 2.78 (1H, dd, *J* = 17.2, 2.8 Hz, H-3$_{eq}$), 3.12 (1H, dd, *J* = 17.1, 13.2 Hz, H-3$_{ax}$), 5.98 (1H, d, *J* = 2.4 Hz, H-6), 5.99 (1H, d, *J* = 2.4 Hz, H-8), 7.20

(1H, d, *J* = 2.0 Hz, H-2'), 6.88 (1H, d, *J* = 8.4 Hz, H-5'), 7.26 (1H, dd, *J* = 8.4, 2.0 Hz, H-6'), 3.34 (2H, br d, *J* = 7.2 Hz, H-1''), 5.30 (1H, m, H-2''), 1.75 (3H, br s, H-4''), 1.71 (3H, br s, H-5''), 3.86 (3H, s, 4'-OCH_3).

^{13}C-NMR (CDCl$_3$, 100 MHz): δ 79.5 (C-2), 43.3 (C-3), 196.5 (C-4), 163.6 (C-5), 96.8 (C-6), 164.5 (C-7), 95.6 (C-8), 164.6 (C-9), 103.5 (C-10), 130.1 (C-1'), 127.8 (C-2'), 131.0 (C-3'), 158.1 (C-4'), 110.5 (C-5'), 125.3 (C-6'), 28.7 (C-1''), 122.1(C-2''), 133.3 (C-3''), 26.0 (C-4''), 18.0 (C-5''), 55.7 (4'-OCH$_3$).

HMBC: δ 5.35 (H-2) *vs* δ 196.5 (C-4) and 125.3 (C-6'), δ 3.12 (H-3$_{ax}$) *vs* δ 130.1 (C-1'), δ 7.20 (H-2') *vs* δ 158.1 (C-4'), δ 3.34 (H-1'') *vs* δ 158.1 (C-4') and 122.1(C-2''), δ 1.71 (H-5'') *vs* δ 122.1(C-2''), δ 3.86 (4'-OCH$_3$) *vs* δ 55.7 (4'-OCH$_3$) (selected 2D-correlations were shown).

EIMS: *m/z* (%rel.) 406 ([M]$^+$, 100), 323 (8), 285 (7), 202 (23), 189 (83), 179 (17), 153 (24), 147 (11).

HR-EIMS: *m/z* 354.1465 ([M]$^+$, Calcd. for C$_{21}$H$_{22}$O$_5$, 354.1467).

Reference

Jang J, Na M, Thuong P T, Njamen D, Mbafor J T, Fomum Z T, Woo E-R, Oh W K. (2008). Prenylated flavonoids with PTP1B inhibitory activity from the root bark of *Erythrina mildbraedii*. *Chem Pharm Bull* **56**: 85–88.

Liquiritigeninyl-(I-3,II-3)-naringenin

IUPAC name: (2S,2'S,3R,3'R)-5,7,7'-trihydroxy-2,2'-bis(4-hydroxyphenyl)-[3,3'-bichroman]-4,4'-dione.

Sub-class: Biflavonoid (bis-flavanone)

Chemical structure

Source: *Ormocarpum kirkii* S. Moore (Family: Papilionaceae); roots

Molecular formula: $C_{30}H_{22}O_9$

Molecular weight: 526

State: Yellow amorphous powder

Bioactivity studied: Antiplasmodial

Specific rotation: $[\alpha]_D^{20}$ +97° (MeOH, *c* 0.7)

CD (MeOH): λ_{max} ($\Delta\varepsilon$) 329 (+3.14 × 10⁴), 295 (−5.1 × 10⁴), 234 (+3.7 × 10⁴), 219 (+7.9 × 10⁴) nm

UV (MeOH): λ_{max} 205, 215, 290 nm

^{1}H-NMR (CD$_3$OD, 400 MHz): δ 5.84 (1H, d, J = 12.2 Hz, H-2β), 2.69 (1H, br d, J = 12.2 Hz, H-3α), 7.68 (1H, d, J = 8.7 Hz, H-5), 6.48 (1H, dd, J = 8.7, 2.2 Hz, H-6), 6.25 (1H, d, J = 2.2 Hz, H-8), 6.90 (2H, d, J = 8.5 Hz, H-2′ and H-6′), 6.76 (2H, d, J = 8.5 Hz, H-3′ and H-5′), 5.77 (1H, d, J = 12.2 Hz, H-2″β), 2.72 (1H, br d, J = 12.2 Hz, H-3″β), 5.87 (1H, d, J = 2.2 Hz, H-6″), 5.79 (1H, d, J = 2.2 Hz, H-8″), 6.87 (2H, d, J = 8.5 Hz, H-2‴ and H-6‴), 6.75 (2H, d, J = 8.5 Hz, H-3‴ and H-5‴).

^{13}C-NMR (CD$_3$OD, 100 MHz): δ 85.5 (C-2), 52.3 (C-3), 193.4 (C-4), 130.1 (C-5), 111.8 (C-6), 166.7 (C-7), 103.4 (C-8), 165.0 (C-9), 115.1 (C-10), 129.4 (C-1′), 130.2 (C-2′), 116.5 (C-3′), 159.5 (C-4′), 116.5 (C-5′), 130.2 (C-6′), 85.0 (C-2″), 51.5 (C-3″), 198.3 (C-4″), 165.5 (C-5″), 97.2 (C-6″), 168.2 (C-7″), 96.0 (C-8″), 164.4 (C-9″), 103.6 (C-10″), 129.2 (C-1‴), 130.2 (C-2‴), 116.5 (C-3‴), 159.5 (C-4‴), 116.5 (C-5‴), 130.2 (C-6‴).

HR-QTOF-MS: m/z 527.1342 ([M + H]$^+$), Calcd. for C$_{30}$H$_{23}$O$_9$, 527.4982).

Reference

Dhooghe L, Maregesi S, Mincheva I, Ferreira D, Marais J PJ, Lemière F, Matheeussen A, Cos P, Maes L, Vlietinck A, Apers S, Pieters L. (2010). Antiplasmodial activity of (I-3,II-3)-biflavonoids and other constituents from *Ormocarpum kirkii*. *Phytochemistry* **71**: 785–791.

Luteolin 7-*O*-[6″-(3‴-hydroxy-4‴-methoxy cinnamoyl)]-β-D-glucopyranoside

IUPAC name: (*E*)-((2*R*,3*S*,5*R*,6*S*)-6-((2-(3,4-dihydroxyphenyl)-5-hydroxy-4-oxo-4*H*-chromen-7-yl)oxy)-3,4,5-trihydroxytetrahydro-2*H*-pyran-2-yl) methyl 3-(3-hydroxy-4-methoxyphenyl)acrylate

Sub-class: Flavone glycoside

Chemical structure

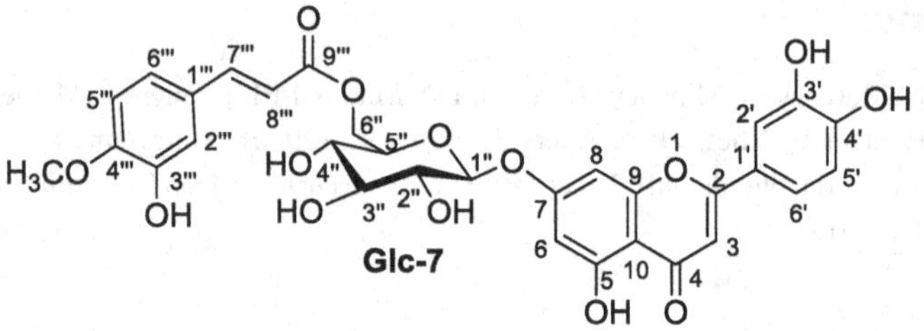

Source: *Elsholtzia bodinieri* Vaniot (Family: Labiatae); whole plants

Molecular formula: $C_{31}H_{28}O_{14}$

Molecular weight: 624

State: Yellow amorphous powder

Specific rotation: $[\alpha]^{22}_{D}$ −100.0° (pyridine, *c* 0.09)

IR (KBr): v_{max} 3425 (OH), 2924, 1689 (α,β-ester carbonyl), 1657 (α,β-unsaturated carbonyl), 1608, 1515, 1499, 1374 (aromatic ring), 1261, 1179, 1081 cm^{-1}

^{1}H-NMR (Pyridine-d_5, 400 MHz): δ 6.83 (1H, s, H-3), 6.95 (1H, d, J = 2.5 Hz, H-6), 6.90 (1H, d, J = 2.5 Hz, H-8), 7.86 (1H, d, J = 2.2 Hz, H-2′), 7.12 (1H, d, J = 8.8 Hz, H-5′), 7.48 (1H, dd, J = 8.8, 2.2 Hz, H-6′), 5.79 (1H, d, J = 7.8 Hz, H-1″), 4.26-4.89 (4H, m, H-2″, H-3″, H-4″ and H-5″), 5.20 (1H, d, J = 11.8 Hz, H-6″a), 4.86 (1H, dd, J = 11.8, 5.3 Hz, H-6″b), 6.97 (1H, s, H-2‴), 7.30 (1H, d, J = 4.0 Hz, H-5‴), 7.13 (1H, overlapped, H-6‴), 7.88 (1H, d, J = 15.9 Hz, H-7‴), 6.68 (1H, d, J = 15.9 Hz, H-8‴), 13.5 (1H, s, 5-OH), 3.80 (3H, s, 4‴-OCH_3).

^{13}C-NMR (Pyridine-d_5, 100 MHz): δ 164.0 (C-2), 104.0 (C-3), 183.0 (C-4), 162.8 (C-5), 100.7 (C-6), 165.5 (C-7), 95.7 (C-8), 157.8 (C-9), 106.7 (C-10), 122.6 (C-1′), 114.7 (C-2′), 149.1 (C-3′), 152.0 (C-4′), 116.9 (C-5′), 119.8 (C-6′), 101.9 (C-1″), 74.8 (C-2″), 78.5 (C-3″), 71.6 (C-4″), 75.8 (C-5″), 64.6 (C-6″), 125.2 (C-1‴), 106.7 (C-2‴), 149.1 (C-3‴), 140.7 (C-4‴), 115.2 (C-5‴), 124.0 (C-6‴), 146.2 (C-7‴), 115.2 (C-8‴), 167.6 (C-9‴), 56.0 (4‴-OCH_3).

HMQC: δ 6.83 (H-3) *vs* δ 104.0 (C-3), δ 6.95 (H-6) *vs* δ 100.7 (C-6), δ 6.90 (H-8) *vs* δ 95.7 (C-8), δ 7.86 (H-2′) *vs* δ 114.7 (C-2′), δ 7.12 (H-5′) *vs* δ 116.9 (C-5′), δ 7.48 (H-6′) *vs* δ 119.8 (C-6′), δ 5.79 (H-1″) *vs* δ 101.9 (C-1″), δ 4.26-4.89 (H-3″/H-4″/H-5″) *vs* δ 78.5 (C-3″)/71.6 (C-4″)/75.8 (C-5″), δ 5.20 (H-6″a)/4.86 (H-6″b) *vs* δ 64.6 (C-6″), δ 6.97 (H-2‴) *vs* δ 106.7 (C-2‴), δ 7.30 (H-5‴) *vs* δ 115.2 (C-5‴), δ 7.13 (H-6‴) *vs* δ 124.0 (C-6‴), δ 7.88 (H-7‴) *vs* δ 146.2 (C-7‴), δ 6.68 (H-8‴) *vs* δ 115.2 (C-8‴), δ 3.80 (4‴-OCH_3) *vs* δ 56.0 (4‴-OCH_3).

HR-FAB-MS: m/z 623.125 ([M − H]$^-$, Calcd. for $C_{31}H_{27}O_{14}$, 623.100).

Reference

Li R-T, Li J-T, Wang J-K, Han Q-B, Zhu Z-Y, and Sun H-D. (2008). Three new flavonoid glycosides isolated from *Elsholtzia bodinieri*. *Chem Pharm Bull* **56**: 592–594.

Lyratin B

IUPAC name: (3*R*,4*S*)-3-(5-Hydroxy-2,2-dimethyl-2*H*-chromen-6-yl) chroman-4,7-diol

Sub-class: Pyrano-fused isoflavan

Chemical structure

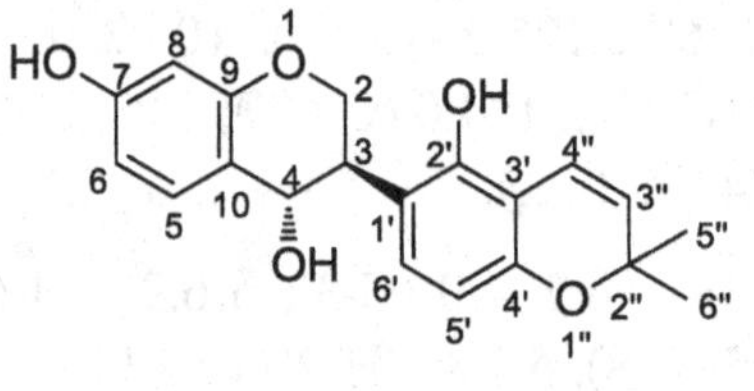

Source: *Solanum lyratum* Thunb (Family: Solanaceae); whole plants

Molecular formula: $C_{20}H_{20}O_5$

Molecular weight: 340

State: Brown paste

Bioactivity studied: *In vitro* anti-inflammatory

Specific rotation: $[\alpha]^{25}_{D}$ −106.2° (CHCl$_3$, *c* 0.32)

UV: λ_{max} (CHCl$_3$): 211, 298 nm

IR (KBr): 3431 (OH), 1634, 1610, 1568, 1457, 1010 cm^{-1}

^{1}H-NMR (CDCl$_3$, 400 MHz): δ 3.61 (1H, dd, *J* = 11.0, 10.9 Hz, H-2a), 4.23 (1H, dd, *J* = 10.9, 4.9 Hz, H-2b), 3.56 (1H, m, H-3α), 5.58 (1H, d, *J* = 6.2 Hz, H-4β), 7.29 (1H, d, *J* = 8.4 Hz, H-5), 6.48 (1H, dd, *J* = 8.4,

2.2 Hz, H-6), 6.27 (1H, d, J = 2.2 Hz, H-8), 6.28 (1H, d, J = 8.0 Hz, H-5′), 7.06 (1H, d, J = 8.0 Hz, H-6′), 5.68 (1H, d, J = 9.9 Hz, H-3″), 6.37 (1H, d, J = 9.9 Hz, H-4″), 1.35 (3H, s, H-5″), 1.33 (3H, s, H-6″), 9.64 (1H, s, 7-OH), 8.30 (1H, s, 2′-OH).

^{13}**C-NMR (CDCl$_3$, 100 MHz):** δ 65.9 (C-2), 39.0 (C-3), 78.5 (C-4), 132.1 (C-5), 109.7 (C-6), 158.8 (C-7), 102.8 (C-8), 156.4 (C-9), 111.1 (C-10), 119.5 (C-1′), 154.9 (C-2′), 105.1 (C-3′), 152.9 (C-4′), 107.9 (C-5′), 124.3 (C-6′), 75.7 (C-2″), 129.9 (C-3″), 115.8 (C-4″), 27.4 (C-5″ and C-6‴).

HMBC: δ 3.61 (H-2a) and 4.23 (H-2b) vs δ 39.0 (C-3) and 78.5 (C-4), δ 3.56 (H-3α) vs δ 119.5 (C-1′) and 154.9 (C-2′), δ 5.58 (H-4β) vs δ 132.1 (C-5), 156.4 (C-9) and 119.5 (C-1′), δ 7.29 (H-5) vs δ 158.8 (C-7), δ 6.48 (H-6) vs δ 102.8 (C-8) and 111.1 (C-10), δ 6.28 (H-5′) vs δ 119.5 (C-1′), δ 7.06 (H-6′) vs δ 154.9 (C-2′) and 152.9 (C-4′), δ 5.68 (H-3″) vs δ 105.1 (C-3′), δ 6.37 (H-4″) vs δ 154.9 (C-2′) and 152.9 (C-4′), δ 1.35 (H-5″) vs δ 129.9 (C-3″), δ 1.33 (H-6″) vs δ 75.7 (C-2″), δ 9.64 (7-OH) vs δ 158.8 (C-7), δ 8.30 (2′-OH) vs δ 154.9 (C-2′) (selected 2D-correlations were shown).

ESIMS: m/z (%rel.) 341.2 [M + H]$^+$

HR-ESIMS: m/z 341.1385 ([M + H]$^+$, Calcd. for $C_{20}H_{21}O_5$, 341.1389).

Reference

Zhang D-W, Li G-H, Yu Q-Y, Dai S-J. (2010). New anti-inflammatory 4-hdroxyisoflavans from *Solanum lyratum*. *Chem Pharm Bull* **56**: 840–842.

Malvidin 3-*O*-[6-*O*-(3-hydroxy-3-methylglutaryl)-β-glucopyranoside]

Sub-class: Anthocyanin glycoside

Chemical structure

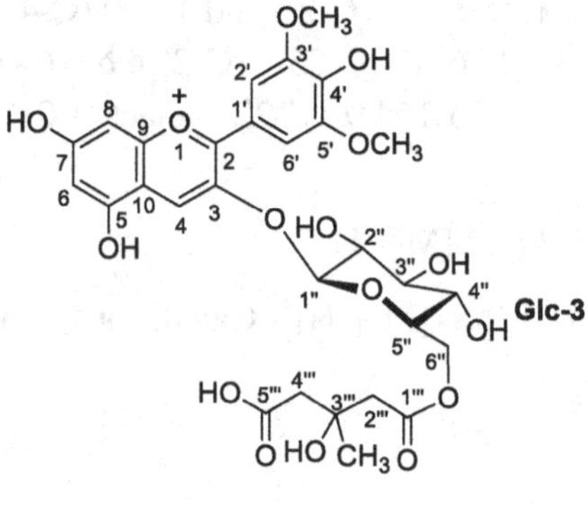

Source: *Impatiens textori* Miq. (Family: Balsaminaceae); flowers

Molecular formula: $C_{29}H_{33}O_{16}{}^{+}$

Molecular weight: 637

State: Pigment

UV (0.1% HCl-MeOH): λ_{max} 540, 274 nm

¹H-NMR [DCl-CD₃OD (1:9; v/v), 500 MHz]: δ 8.96 (1H, s, H-4), 6.70 (1H, d, *J* = 11.9 Hz, H-6), 7.01 (1H, d, *J* = 1.9 Hz, H-8), 7.96 (2H, s, H-2′ and H-6′), 4.00 (6H, s, 3′-OC*H*₃ and 5′-OC*H*₃), 5.37 (1H, d, *J* = 7.6 Hz,

H-1″), 3.64 (1H, t, J = 8.6 Hz, H-2″), 3.50 (1H, t, J = 8.9 Hz, H-3″), 3.41 (1H, t, J = 8.9 Hz, H-4″), 3.79 (1H, ddd, J = 8.9, 7.2, 2.1 Hz, H-5″), 4.21 (1H, dd, J = 12.2, 7.2 Hz, H-6″a), 4.46 (1H, dd, J = 12.2, 7.2 Hz, H-6″b), 2.53 (1H, d, J = 4.3 Hz, H-2‴a or H-4‴a), 2.53 (1H, d, J = 4.3 Hz, H-2‴b or H-4‴b), 2.62 (1H, d, J = 14.7 Hz, H-4‴a or H-2‴a), 2.68 (1H, d, J = 14.7 Hz, H-4‴b or H-2‴b), 1.25 (3H, s, 3‴-CH_3).

^{13}**C-NMR [DCl-CD$_3$OD (1:9; v/v), 150 MHz]:** δ 163.4 (C-2), 148.0 (C-3), 137.5 (C-4), 156.3 (C-5), 103.8 (C-6), 170.9 (C-7), 98.3 (C-8), 158.1 (C-9), 113.6 (C-10), 119.7 (C-1′), 110.8 (C-2′ and C-6′), 149.8 (C-3′ and C-5′), 145.6 (C-4′), 57.4 (3′-OCH$_3$ and 5′-OCH$_3$), 103.8 (C-1″), 75.0 (C-2″), 77.9 (C-3″), 71.4 (C-4″), 76.0 (C-5″), 64.5 (C-6″), 172.3 (C-1‴, COO), 46.1 (C-2‴ or C-4‴), 70.7 (C-3‴), 71.6 (C-4‴ or C-2‴), 173.3 (C-5‴, COOH), 27.8 (3‴-CH$_3$).

HR-FAB-MS: m/z 637.1780 ([M]$^+$, Calcd. for $C_{29}H_{33}O_{16}{}^+$, 637.1769.

Reference

Tatsuzawa F, Saito N, Mikanagi Y, Shinoda K, Toki K, Shigihara A, Honda T. (2009). An unusual acylated malvidin 3-glucoside from flowers of *Impatiens textori* Miq. (Balsaminaceae). *Phytochemistry* **70**: 672–674.

3-Methoxy-(3″,4″-dihydro-3″, 4″-diacetoxy)-2″,2″-dimethylpyrano-(7,8:5″,6″)-flavone

IUPAC name: 3-Methoxy-8,8-dimethyl-4-oxo-2-phenyl-4,8,9,10-tetrahydropyrano[2,3-*f*]chromene-9,10-diyl diacetate

Sub-class: Pyranoflavone

Chemical structure

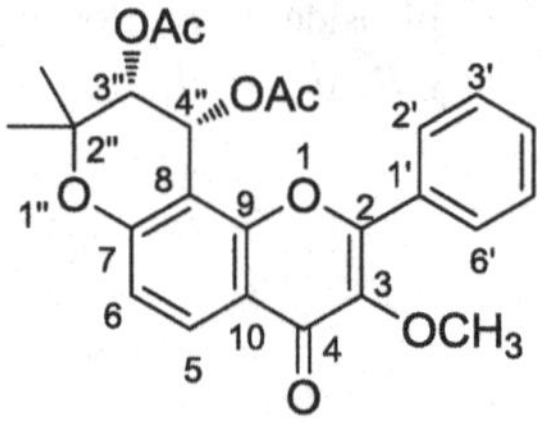

(Ac = COCH₃)

Source: *Derris indica* (Lam.) Bennet [Synonyms: *Pongamia pinnata* (Linn.) Pierre, *P. pinnata* (L.) Merr., *P. glabra* Vent., *Cytisus pinnaus* (L.)] (Family: Leguminosae); stems and roots

Molecular formula: $C_{25}H_{24}O_8$

Molecular weight: 452

State: Pale yellow solid

Melting point: 104–106 °C

Specific rotation: $[\alpha]^{24}_D$ −24° (CHCl$_3$, c 0.71)

Bioactivity studied: Antimycobacterial (MIC 25 μg/mL against *Mycobacterium tuberculosis*)

UV: λ_{max} (CHCl$_3$) (log ε): 238 (3.75), 256 (3.74), 313 (3.97) nm

IR (neat): ν_{max} 1746 (OAc), 1641 (α,β-unsaturated carbonyl), 1600 (aromatic unsaturaion) cm^{-1}

^{1}H-NMR (CDCl$_3$, 500 MHz): δ 8.18 (1H, d, J = 9.0 Hz, H-5), 6.93 (1H, d, J = 9.0 Hz, H-6), 7.98 (2H, m, H-2′ and H-6′), 7.49 (3H, m, H-3′, H-4′ and H-5′), 5.32 (1H, d, J = 5.0 Hz, H-3″), 6.66 (1H, d, J = 5.0 Hz, H-4″), 3.89 (3H, s, 3-OCH_3), 2.11 (3H, s, 3″-OCOCH_3), 1.88 (3H, s, 4″-OCOCH_3), 1.46 (3H, s, 2″-CH_3-1), 1.48 (3H, s, 2″-CH_3-2).

^{13}C-NMR (CDCl$_3$, 125 MHz): δ 155.0 (C-2), 141.6 (C-3), 174.3 (C-4), 127.9 (C-5), 116.0 (C-6), 157.9 (C-7), 106.2 (C-8), 154.8 (C-9), 118.3 (C-10), 130.5 (C-1′), 128.2 (C-2′, C-6′), 128.5 (C-3′, C-5′), 130.7 (C-4′), 77.3 (C-2″), 70.7 (C-3″), 61.1 (C-4″), 60.2 (C$_3$-OCH$_3$), 169.9 and 20.6 (3″-OCOCH$_3$), 170.4 and 20.5 (4″-OCOCH$_3$), 25.6 (2″-CH$_3$-1), 21.7 (2″-CH$_3$-2).

HMQC: δ 8.18 (H-5) *vs* δ 127.9 (C-5), δ 6.93 (H-6) *vs* δ 116.0 (C-6), δ 7.98 (H-2′/H-6′) *vs* δ 128.2 (C-2′,6′), δ 7.49 (H-3′/H-4′/H-5′) *vs* δ 128.5 (C-3′/C-5′) and δ 130.7 (C-4′), δ 5.32 (H-3″) *vs* δ 70.7 (C-3″), δ 6.66 (H-4″) *vs* δ 61.1 (C-4″), δ 3.89 (C$_3$-OCH$_3$) *vs* δ 60.2 (C$_3$-OCH$_3$), δ 2.11 (3″-OCOCH_3) *vs* δ 20.6 (3″-OCOCH_3), δ 1.88 (4″-OCOCH_3) *vs* δ 20.5 (4″-OCOCH_3), 1.46 (2″-CH_3-1) *vs* δ 25.6 (2″-CH_3-1), 1.48 (2″-CH_3-2) *vs* δ 21.7 (2″-CH_3-2).

HMBC: δ 8.18 (H-5) *vs* δ 174.3 (C-4) and δ 157.9 (C-7), δ 6.93 (H-6) *vs* δ 106.2 (C-8) and δ 118.3 (C-10), δ 5.32 (H-3″) *vs* δ 169.9 (3″-OCOCH$_3$) and δ 25.6 (2″-CH_3-1), δ 6.66 (H-4″) *vs* δ 170.4 (4″-OCOCH$_3$), 1.46 (2″-CH_3-1) *vs* δ 77.3 (C-2″), 1.48 (2″-CH_3-2) *vs* δ 77.3 (C-2″) (selected correlations are shown).

^{1}H-^{1}H COSY: δ 8.18 (H-5) *vs* δ 6.93 (H-6), δ 7.98 (H-2′/H-6′) *vs* δ 7.49 (H-3′/H-4′/H-5′),), δ 5.32 (H-3″) *vs* δ 6.66 (H-4″).

NOE: Irradiation of the methoxy protons (−OCH_3) at δ 3.89 enhanced the signals of H-2′ and 6′, thereby supporting the placement of the methoxy group at C-3.

EIMS: *m/z* (%rel.) 452 ([M]$^+$, 20), 450 (100), 332 (44), 294 (31).

HR-EIMS: *m/z* 452.1466 ([M]$^+$, Calcd. for $C_{25}H_{24}O_8$, 452.1471).

Reference

Koysomboon S, van Altena I, Kato S, Chantrapromma K. (2006). Antimycobacterial flavonoids from *Derris indica*. *Phytochemistry* **67**: 1034–1040.

7-Methoxy-2″-*O*-(2‴-methylbutyryl) orientin

IUPAC name: (2*R*,3*S*,4*R*,5*R*,6*S*)-2-(2-(3,4-dihydroxyphenyl)-5-hydroxy-7-methoxy-4-oxo-4*H*-chromen-8-yl)-4,5-dihydroxy-6-(hydroxymethyl) tetrahydro-2*H*-pyran-3-yl 2-methylbutanoate

Sub-class: Flavone acylated *C*-glycoside

Chemical structure

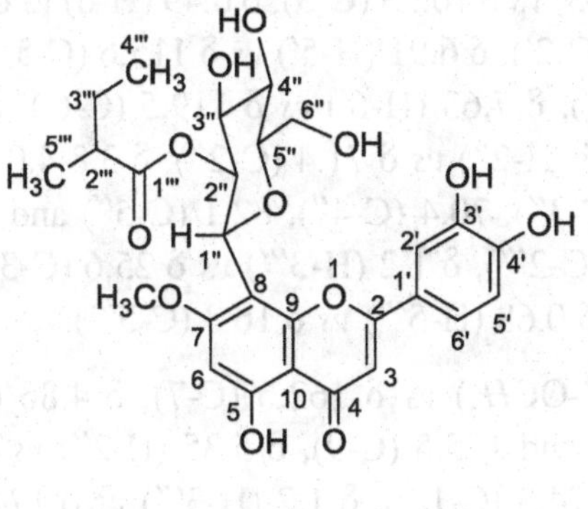

Source: *Trollius ledebouri* Reichb (Family: Ranunculaceae); flowers

Molecular formula: $C_{27}H_{30}O_{12}$

Molecular weight: 546

State: Yellow powder

Melting point: 155–157 °C

Specific rotation: $[\alpha]_D^{20}$ −60.0° (MeOH, *c* 0.06)

Bioactivity studied: Showed *in vivo* anti-inflammatory effect on TPA-induced ear edema with 58.6% inhibitory rate at a dose of 10 mg/kg body weight

UV (MeOH): λ_{max} 233, 290, 331 nm

FT-IR (KBr): ν_{max} 3409 (OH), 1623 (α,β-unsaturated carbonyl), 1504 cm^{-1}

^{1}H-NMR (DMSO-d_6, 600 MHz): δ 6.71 (1H, s, H-3), 6.49 (1H, s, H-6), 7.54 (1H, d, J = 2.2 Hz, H-2'), 6.91 (1H, d, J = 8.4 Hz, H-5'), 7.63 (1H, dd, J = 8.4, 2.2 Hz, H-6'), 3.86 (3H, s, 7-OCH$_3$), 4.86 (1H, d, J = 9.8 Hz, H-1''), 5.35 (1H, t, J = 9.8 Hz, H-2''), 3.8-4.0 (5H, m, H-3'', H-4'', H-5'' and H-6''), 2.0 (1H, m, H-2'''), 1.2 (2H, m, H-3'''), 0.58 (3H, t, J = 7.5 Hz, H-4'''), 0.68 (3H, d, J = 7.0 Hz, H-5''').

^{13}C-NMR (DMSO-d_6, 150 MHz): δ 164.4 (C-2), 102.3 (C-3), 182.0 (C-4), 161.6 (C-5), 94.4 (C-6), 162.6 (C-7), 103.1 (C-8), 155.5 (C-9), 104.1 (C-10), 121.7 (C-1'), 114.0 (C-2'), 145.7 (C-3'), 149.7 (C-4'), 115.6 (C-5'), 119.5 (C-6'), 56.4 (7-OCH$_3$), 70.6 (C-1''), 71.4 (C-2''), 75.6 (C-3''), 70.4 (C-4''), 82.1 (C-5''), 61.8 (C-6''), 174.3 (C-1'''), 40.1 (C-2'''), 25.6 (C-3'''), 11.0 (C-4'''), 16.3 (C-5''').

HMQC: δ 6.71(H-3) *vs* δ 102.3 (C-3), δ 6.49 (H-6) *vs* δ 94.4 (C-6), δ 7.54 (H-2') *vs* δ 114.0 (C-2'), δ 6.91 (H-5') *vs* δ 115.6 (C-5'), δ 3.86 (7-OCH$_3$) *vs* δ 56.4 (7-OCH$_3$), δ 7.63 (H-6') *vs* δ 119.5 (C-6'), δ 4.86 (H-1'') *vs* δ 70.6 (C-1''), δ 5.35 (H-2'') *vs* δ 71.4 (C-2''), δ 3.8-4.0 (H-3''/H-4''/H-5''/H-6'') *vs* δ 75.6 (C-3''), 70.4 (C-4''), 82.1 (C-5'') and 61.8 (C-6''), δ 2.0 (H-2''') *vs* δ 40.1 (C-2'''), δ 1.2 (H-3''') *vs* δ 25.6 (C-3'''), δ 0.58 (H-4''') *vs* δ 11.0 (C-4'''), δ 0.68 (H-5''') *vs* δ 16.3 (C-5''').

HMBC: δ 3.86 (7-OCH$_3$) *vs* δ 162.6 (C-7), δ 4.86 (H-1'') *vs* δ 162.6 (C-7), 103.1 (C-8) and 155.5 (C-9), δ 5.35 (H-2'') *vs* δ 174.3 (C-1'''), δ 5.35 (H-2'') *vs* δ 174.3 (C-1'''), δ 1.2 (H-3''') *vs* δ 174.3 (C-1'''), δ 0.68 (H-5''') *vs* δ 174.3 (C-1''') (selected 2D-correlations were shown).

EIMS: *m/z* 546 ([M]$^+$), 461, 426, 366, 329, 247, 195, 85.

HR-ESI-MS: *m/z* 546.1733 ([M]$^+$, Calcd. for C$_{27}$H$_{30}$O$_{12}$, 546.1737).

Reference

Wu X-A, Zhao Y-M, Yu N-J. (2006). Flavone *C*-glycosides from *Trollius ledebouri* reichb. *J Asian Nat Prod* **8**: 541–544.

6-Methoxykaempferol-3-*O*-β-D-gentiobioside [6-methoxy-3,5,7, 4'-tetrahydroxyflavon 3-*O*-β-D-glucopyranosyl(1→6)-*O*-β-D-glucopyranoside]

IUPAC name: 5,7-Dihydroxy-2-(4-hydroxyphenyl)-6-methoxy-3-(((2*S*,3*R*, 4*S*,5*S*,6*R*)-3,4,5-trihydroxy-6-((((2*R*,3*R*,4*S*,5*S*,6*R*)-3,4,5-trihydroxy-6-(hydroxymethyl)tetrahydro-2*H*-pyran-2-yl)oxy)methyl) tetrahydro-2*H*-pyran-2-yl)oxy)-4*H*-chromen-4-one

Sub-class: Flavonol glycoside

Chemical structure

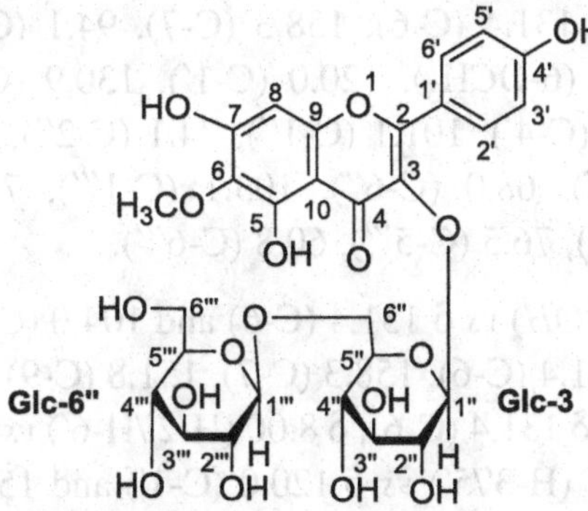

189

Source: *Chenopodium foliosum* Asch (Family: Amaranthaceae); aerial parts

Molecular formula: $C_{28}H_{32}O_{17}$

Molecular weight: 640

State: Pale yellow crystalline powder (from MeOH-H_2O)

Melting point: 230–231 °C

Specific rotation: $[\alpha]^{21.9}_D$ −138° (DMSO, *c* 0.1)

UV (MeOH): λ_{max} (log ε) 270 (4.25) and 342 (4.30) nm

FT-IR (ATR): v_{max} 3458–3210 (OH), 1615 (α,β-unsaturated carbonyl), 1563, 1471 cm^{-1}

^{1}H-NMR (DMSO-d_6, 600 MHz): δ 6.46 (1H, s, H-8), 12.64 (1H, br s, 5-O*H*), 3.73 (3H, s, 6-OC*H*$_3$), 8.00 (2H, d, *J* = 8.9 Hz, H-2′ and H-6′), 6.87 (2H, d, *J* = 8.9 Hz, H-3′ and H-5′), 5.35 (1H, d, *J* = 7.6 Hz, H-1″), 3.17 (1H, dd, *J* = 9.0, 7.6 Hz, H-2″), 3.22 (1H, t, *J* = 9.0 Hz, H-3″), 3.13 (1H, dd, *J* = 9.6, 9.0 Hz, H-4″), 3.29 (1H, ddd, *J* = 9.6, 5.8, 1.5 Hz, H-5″), 3.85 (1H, d, *J* = 11.8 Hz, H-6″a), 3.44 (1H, dd, *J* = 11.8, 5.8 Hz, H-6″b), 4.02 (1H, d, *J* = 7.7 Hz, H-1‴), 2.82 (1H, m, H-2‴), 2.93 (1H, t, *J* = 8.8 Hz, H-3‴), 2.96 (1H, dd, *J* = 9.2, 8.8 Hz, H-4‴), 2.80 (1H, m, H-5‴), 3.51 (1H, dd, *J* = 11.7, 1.8 Hz, H-6‴a), 3.34 (1H, dd, *J* = 11.7, 5.7 Hz, H-6‴b).

^{13}C-NMR (DMSO-d_6, 125 MHz): δ 156.4 (C-2), 132.8 (C-3), 177.4 (C-4), 152.2 (C-5), 131.4 (C-6), 158.3 (C-7), 94.1 (C-8), 151.8 (C-9), 104.0 (C-10), 59.9 (6-OCH$_3$), 120.0 (C-1′), 130.9 (C-2′, C-6′), 115.1 (C-3′, C-5′), 159.9 (C-4′), 101.1 (C-1″), 74.1 (C-2″), 76.2 (C-3″), 69.7 (C-4″), 76.3 (C-5″), 68.0 (C-6″), 103.1 (C-1‴), 73.4 (C-2‴), 76.6 (C-3‴), 69.7 (C-4‴), 76.5 (C-5‴), 60.8 (C-6‴).

HMBC: δ 12.64 (5-O*H*) *vs* δ 131.4 (C-6) and 104.0 (C-10), δ 6.46 (H-8) *vs* δ 177.4 (C-4), 131.4 (C-6), 158.3 (C-7), 151.8 (C-9) and 104.0 (C-10), δ 3.73 (6-OC*H*$_3$) *vs* δ 131.4 (C-6), δ 8.00 (H-2′/H-6′) *vs* δ 156.4 (C-2) and 159.9 (C-4′), δ 6.87 (H-3′/5′) *vs* δ 120.0 (C-1′) and 159.9 (C-4′), δ 5.35 (H-1″) *vs* δ 132.8 (C-3) and 76.2 (C-3″), δ 3.17 (H-2″) *vs* δ 101.1 (C-1″)

and 76.2 (C-3″), δ 3.22 (H-3″) *vs* δ 74.1 (C-2″) and 69.7 (C-4″), δ 3.13 (H-4″) *vs* δ 76.2 (C-3″), 76.3 (C-5″) and 68.0 (C-6″), δ 3.29 (H-5″) *vs* δ 69.7 (C-4″), δ 3.85 (H-6″a) and 3.44 (H-6″b) *vs* δ 69.7 (C-4″), 76.3 (C-5″) and 103.1 (C-1‴), δ 4.02 (H-1‴) *vs* δ 68.0 (C-6″), 73.4 (C-2‴) and 76.6 (C-3‴), δ 2.82 (H-2‴) *vs* δ 103.1 (C-1‴) and 76.6 (C-3‴), δ 2.93 (H-3‴) *vs* δ 73.4 (C-2‴) and 69.7 (C-4‴), δ 2.96 (H-4‴) *vs* δ 76.6 (C-3‴), 76.5 (C-5‴) and 60.8 (C-6‴), δ 3.51 (H-6‴a) and 3.34 (H-6‴b) *vs* δ 76.5 (C-5‴).

ESI-MS: *m/z* 641 [M + H]$^+$

HRLSI-MS: *m/z* 641.1738 ([M + H]$^+$, Calcd. for $C_{28}H_{33}O_{15}$, 641.1718).

Reference

Kokanova-Nedialkova Z, Bücherl D, Nikolov S, Heilmann J, Nedialkov PT. (2011). Flavonol glycosides from *Chenopodium foliosum* Asch. *Phytochemistry Lett* **4**: 367–371.

5-Methoxyl-4,2'-epoxy-3-(4',5'-dihydroxyphenyl)-linear pyranocoumarin

IUPAC name: 9,10-Dihydroxy-13-methoxy-3,3-dimethylbenzofuro[3,2-*c*] pyrano[3,2-*g*]chromen-7(3*H*)-one

Sub-class: Prenylcoumarin

Chemical structure

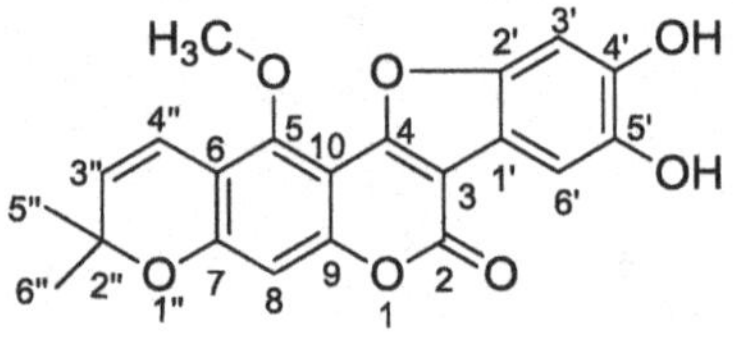

Source: *Ficus hirta* (Family: Moraceae); roots

Molecular formula: $C_{21}H_{16}O_7$

Molecular weight: 380

State: Yellow amorphous powder

Melting point: 280–282 °C

UV (MeOH): λ_{max} 206, 220, 250, 267, 361 nm

FT-IR (KBr): v_{max} 3408 (OH), 1728 (lactone carbonyl), 1627, 1606, 1580, 1513, 1480 (aromatic unsaturation) cm^{-1}.

^{1}H-NMR (DMSO-d_6, 400 MHz): δ 6.72 (1H, s, H-8), 7.18 (1H, s, H-3'),
7.25 (1H, s, H-6'), 5.91 (1H, d, J = 10.1 Hz, H-3''), 6.64 (1H, d, J = 10.1
Hz, H-4''), 1.41 (6H, s, H-5'' and H-6''), 3.91 (3H, s, 5-OCH_3), 9.62 (1H,
s, 4'-OH), 9.45 (1H, s, 5'-OH).

^{13}C-NMR (DMSO-d_6, 100 MHz): δ 157.5 (C-2), 103.5 (C-3), 157.7
(C-4), 150.9 (C-5), 112.4 (C-6), 156.2 (C-7), 101.0 (C-8), 154.0 (C-9),
101.8 (C-10), 113.8 (C-1'), 149.5 (C-2'), 99.0 (C-3'), 146.1 (C-4'), 144.8
(C-5'), 104.9 (C-6'), 78.0 (C-2''), 131.9 (C-3''), 115.1 (C-4''), 27.9 (C-5''
and C-6''), 63.4 (5-OCH$_3$).

HR-ESI-MS: m/z 379.0838 ([M − H]$^-$, Calcd. for $C_{21}H_{15}O_7$, 379.0823).

Reference

Ya J, Zhang X-Q, Wang Y, Zhang Q-W, Chen J-X, Ye W-C. (2010). Two new phenolic
compounds from the roots of *Ficus hirta*. *Nat. Prod. Res.* 7: 621–625.

(2*R*,3*R*)-6-Methylaromadendrin 3-*O*-β-D-glucopyranoside

IUPAC name: (2*R*,3*R*)-5,7-Dihydroxy-2-(4-hydroxyphenyl)-6-methyl-3-(((2*R*,3*S*,4*R*,5*R*,6*S*)-3,4,5-trihydroxy-6-(hydroxymethyl)tetrahydro-2*H*-pyran-2-yl)oxy)chroman-4-one

Sub-class: Flavanone glycoside

Chemical structure

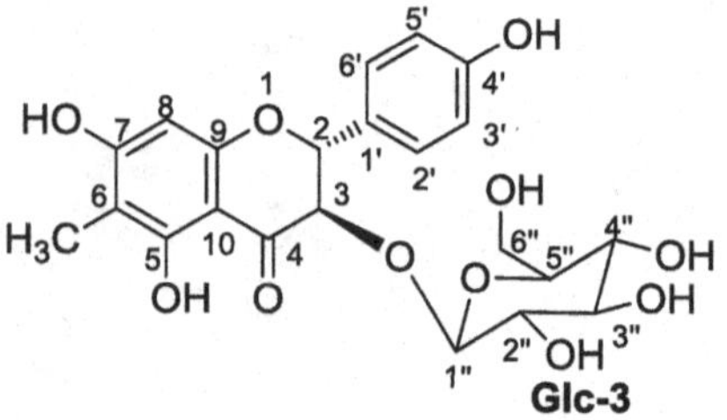

Glc-3

Source: *Cephalotaxus koreana* Nakai (Family: Cephalotaxaceae); aerial parts

Molecular formula: $C_{22}H_{24}O_{11}$

Molecular weight: 464

State: Yellowish powder

Melting point: 166–168 °C

Specific rotation: $[\alpha]^{24}_D$ −19.3° (MeOH, *c* 0.1)

UV (MeOH): λ_{max} (log ε): 213 (4.94), 256 (4.79), 355 (sh, 4.66) nm

IR (KBr): v_{max} 3364 (OH), 1653 (C=O), 1604, 1504, 1359, 1299, 1199 cm^{-1}

^{1}H-NMR (CD$_3$OD, 400 MHz): δ 5.23 (1H, d, *J* = 10.1 Hz, H-2β), 4.96 (1H, d, *J* = 10.1 Hz, H-3α), 5.91 (1H, s, H-8), 7.35 (2H, d, *J* = 8.6 Hz, H-2' and H-6'), 6.80 (2H, d, *J* = 8.6 Hz, H-3' and H-5'), 3.78 (1H, d, J = 7.7 Hz, H-1''), 2.90-3.80 (5H, m, glucosyl protons), 1.94 (3H, s, 6-C*H*$_3$).

^{13}C-NMR (CDCl$_3$, 125 MHz): δ 84.4 (C-2), 78.1 (C-3), 196.9 (C-4), 162.6 (C-5), 103.1 (C-6), 168.2 (C-7), 96.3 (C-8), 163.4 (C-9), 106.6 (C-10), 129.5 (C-1'), 131.3 (C-2'), 117.0 (C-3'), 160.1 (C-4'), 117.0 (C-5'), 131.3 (C-6'), 7.8 (6-C*H*$_3$), 103.4 (C-1''), 75.4 (C-2''), 79.1 (C-3''), 72.0 (C-4''), 78.4 (C-5''), 63.4 (C-6'').

HRFABMS: *m/z* 487.1216 ([M + Na])$^+$, Calcd. for C$_{22}$H$_{24}$O$_{11}$Na, 487.1217).

Reference

Yoon K D, Jeong D G, Hwang Y H, Ryu J M, Kim J. (2007). Inhibitors of osteoclast differentiation from *Cephalotaxus koreana*. *J Nat Prod* **70**: 2029–2032.

3′,4′-Methylenedioxy- [2″,3″:7,8]-furanoflavonol

IUPAC name: 2-(Benzo[*d*][1,3]dioxol-5-yl)-3-hydroxy-4*H*-furo[2,3-*h*] chromen-4-one

Sub-class: Furanoflavonol

Chemical structure

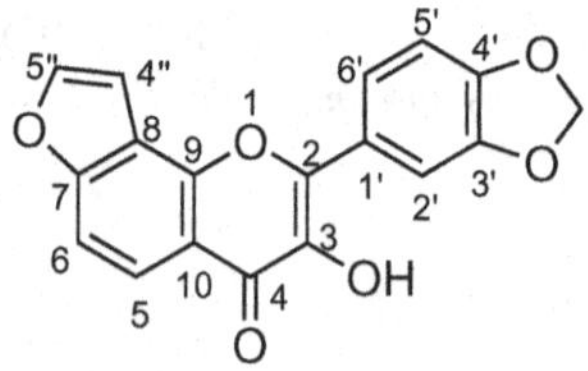

Source: *Millettia erythrocalyx* Gagnep. (Family: Ligumonisae); pods

Molecular formula: $C_{18}H_{10}O_6$

Molecular weight: 322

State: Yellow powder

UV (MeOH): λ_{max} (log ε): 246 (3.58), 324 (3.29) nm

IR (film): ν_{max} 3401 (OH), 1655 (C=O) cm^{-1}

^{1}H-NMR (CDCl$_3$, 500 MHz): δ 8.20 (1H, d, *J* = 8.8 Hz, H-5), 7.58 (1H, dd, *J* = 8.8, 0.9 Hz, H-6), 7.43 (1H, d, *J* = 1.8 Hz, H-2′), 6.98 (1H, d, *J* = 8.2 Hz, H-5′), 7.53 (1H, dd, *J* = 8.2, 1.8 Hz, H-6′), 6.09 (2H, s,

3′,4′-O-CH$_2$-O-), 7.11 (1H, dd, J = 2.4, 0.9 Hz, H-4″), 7.75 (1H, d, J = 2.4 Hz, H-5″).

^{13}C-NMR (CDCl$_3$, 125 MHz): δ 159.3 (C-2), 141.9 (C-3), 172.9 (C-4), 122.3 (C-5), 110.7 (C-6), 158.4 (C-7), 116.9 (C-8), 150.1 (C-9), 117.7 (C-10), 125.0 (C-1′), 109.4 (C-2′), 147.7 (C-3′), 150.0 (C-4′), 108.4 (C-5′), 124.6 (C-6′), 101.9 (3′,4′-O-CH$_2$-O-), 104.2 (C-4″), 146.0 (C-5″).

HR-ESI-TOFMS: m/z 323.0567 ([M + H])$^+$, Calcd. for C$_{18}$H$_{11}$O$_6$, 323.0555).

Reference

Sritularak B, Likhitwitayawuid K. (2006). Flavonoids from the pods of *Millettia erythrocalyx*. *Phytochemistry* **67**: 812–817.

N-(8-Methylquercetin-
3-*O*-[α-L-rhamnopyranosyl-(1→2)-
[α-L-rhamnopyranosyl-(1→6)]-
β-D-galactopyranosyl])-3-
hydroxypiperidin-2-one

IUPAC name: 1-(((3-((((2*S*,3*R*,4*S*,5*R*,6*R*)-4,5-Dihydroxy-3-((((2*S*,3*R*,4*R*,5*R*, 6*S*)-3,4,5-trihydroxy-6-methyltetrahydro-2*H*-pyran-2-yl)oxy)- 6-(((((2*R*,3*R*,4*R*,5*R*,6*S*)-3,4,5-trihydroxy-6-methyltetrahydro-2*H*-pyran-2-yl) oxy)methyl)tetrahydro-2*H*-pyran-2-yl)oxy)-2-(3,4-dihydroxyphenyl)-5,7-di- hydroxy-4-oxo-4*H*-chromen-8-yl)methyl)-3-hydroxypiperidin-2-one

Sub-class: Glycosylated flavonol alkaloid

Chemical structure

Source: *Astragalus monspessulanus* Linn. (Montpellier Milk Vetch; Family: Fabaceae); aerial parts

Molecular formula: $C_{39}H_{49}NO_{22}$

Molecular weight: 883

State: Orange amorphous powder

UV (MeOH): λ_{max} (log ε) 251 (4.13), 263 (sh, 3.97), 293 (sh, 3.60), 355 (3.96) nm

¹H-NMR (Pyridine-d_5, 400 MHz): δ 6.68 (1H, brs, H-6), 8.48 (1H, d, J = 2.0 Hz, H-2′), 7.31 (1H, d, J = 8.4 Hz, H-5′), 8.56 (1H, dd, J = 8.4, 2.0 Hz, H-6′), 4.53 (1H, d, J = 14.1 Hz, H-11a), 4.39 (1H, d, J = 14.1 Hz, H-11b), 3.77 (1H, t, J = 8.5 Hz, H-13), 2.31 (1H, m, H-14a), 2.21 (1H, m, H-14b), 1.82 (1H, m, H-15a), 1.74 (1H, m, H-15b), 3.50 (1H, m, H-16a), 2.81 (1H, q, J = 8.1 Hz, H-16b), 6.40 (1H, d, J = 7.7 Hz, H-1″), 5.00 (1H, t, J = 8.5 Hz, H-2″), 4.33 (1H, m, H-3″), 4.29 (1H, s, H-4″), 4.08 (1H, m, H-5″), 4.28 (1H, m, H-6″a), 5.13 (1H, dd, J = 10.1, 6.0 Hz, H-6″b), 5.13 (1H, brs, H-1‴), 4.29 (1H, m, H-2‴), 4.27 (1H, m, H-3‴), 4.09 (1H, m, H-4‴), 4.13 (1H, m, H-5‴), 1.40 (3H, d, J = 6.3 Hz, H-6‴), 6.34 (1H, brs, H-1‴‴), 4.90 (1H, m, H-2‴‴), 4.89 (1H, m, H-3‴‴), 4.32 (1H, m, H-4‴‴), 5.06 (1H, m, H-5‴‴), 1.65 (3H, d, J = 6.4 Hz, H-6‴‴).

^{13}**C-NMR (Pyridine-d_5, 100 MHz):** δ 157.2 (C-2), 134.4 (C-3), 178.7 (C-4), 162.6 (C-5), 99.4 (C-6), 165.1 (C-7), 99.6 (C-8), 154.5 (C-9), 105.1 (C-10), 49.7 (C-11), 174.9 (C-12), 67.6 (C-13), 29.7 (C-14), 23.6 (C-15), 53.8 (C-16), 122.5 (C-1′), 117.0 (C-2′), 147.1 (C-3′), 150.4 (C-4′), 116.5 (C-5′), 123.5 (C-6′), 101.0 (C-1″), 76.6 (C-2″), 75.9 (C-3″), 70.1 (C-4″), 75.1 (C-5″), 66.1 (C-6″), 102.0 (C-1‴), 72.1 (C-2‴), 72.5 (C-3‴), 73.8 (C-4‴), 69.6 (C-5‴), 18.4 (C-6‴), 102.5 (C-1⁗), 72.8 (C-2⁗), 72.7 (C-3⁗), 74.3 (C-4⁗), 69.9 (C-5⁗), 18.5 (C-6⁗).

HREIMS: m/z 882.2695 ([M − H]⁻, Calcd. for $C_{39}H_{48}NO_{22}$, 882.2673).

Reference

Krasteva I, Bratkov V, Bucar F, Kunert O, Kollroser M, Kondeva-Burdina M, Ionkova I. (2015). Flavoalkaloids and Flavonoids from *Astragalus monspessulanus*. *J Nat Prod* **78**: 2565–2571.

Murrmeranzin

IUPAC name: 8-(3-Hydroxy-2-((1-hydroxy-1-(7-methoxy-2-oxo-2*H*-chromen-8-yl)-3-methylbut-3-en-2-yl)oxy)-3-methylbutyl)-7-methoxy-2*H*-chromen-2-one

Sub-class: Biscoumarin ether

Chemical structure

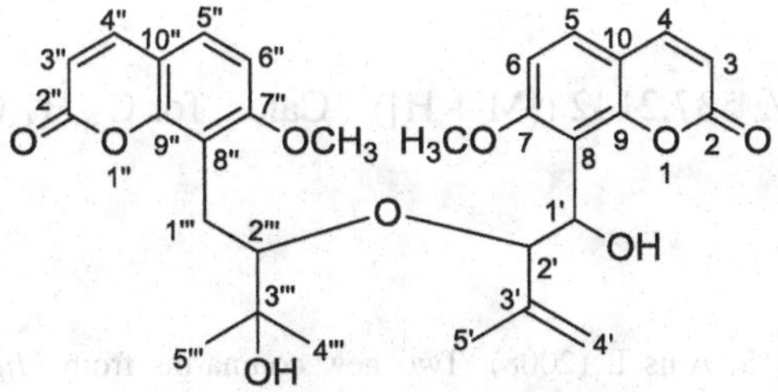

Source: *Murraya paniculata* Dunn. (Family: Rutaceae); aerial parts

Molecular formula: $C_{30}H_{32}O_9$

Molecular weight: 536

State: Colorless oil

UV (MeOH): λ_{max} (log ε) 221 (4.17), 320 (4.29) nm

FT-IR (KBr): ν_{max} 3343 (OH), 1720 (lactone carbonyl), 1607, 1566 (aromatic unsaturation) cm^{-1}

^{1}H-NMR (CD$_3$OD, 500 MHz): δ 6.23 (1H, d, *J* = 9.4 Hz, H-3), 7.86 (1H, d, *J* = 9.4 Hz, H-4), 7.52 (1H, d, *J* = 8.7 Hz, H-5), 7.03 (1H, d, *J* = 8.7 Hz,

H-6), 3.94 (3H, s, 7-OCH_3), 5.34 (1H, d, J = 8.7 Hz, H-1′), 4.83 (1H, d, J = 8.7 Hz, H-2′), 4.63 (1H, s, H-4′a), 4.53 (1H, s, H-4′b), 1.63 (3H, s, H-5′), 6.22 (1H, d, J = 9.4 Hz, H-3″), 7.85 (1H, d, J = 9.4 Hz, H-4″), 7.45 (1H, d, J = 8.7 Hz, H-5″), 7.02 (1H, d, J = 8.6 Hz, H-6″), 3.93 (3H, s, 7″-OCH_3), 3.02 (1H, dd, J = 13.5, 9.7 Hz, H-1‴a), 2.99 (1H, dd, J = 13.5, 3.1 Hz, H-1‴b), 3.66 (1H, dd, J = 9.7, 3.1 Hz, H-2‴), 1.27 (3H, s, H-4‴), 1.29 (3H, s, H-4‴).

^{13}C-NMR (CD$_3$OD, 100 MHz): δ 162.5 (C-2), 113.0 (C-3), 146.4 (C-4), 128.5 (C-5), 109.0 (C-6), 163.7 (C-7), 117.2 (C-8), 154.7 (C-9), 114.3 (C-10), 56.7 (7-OCH_3), 69.9 (C-1′), 78.8 (C-2′), 146.3 (C-3′), 113.7 (C-4′), 17.6 (C-5′), 162.2 (C-2″), 113.3 (C-3″), 146.2 (C-4″), 130.3 (C-5″), 109.6 (C-6″), 162.7 (C-7″), 117.9 (C-8″), 154.2 (C-9″), 114.4 (C-10″), 56.8 (7″-OCH_3), 26.3 (C-1‴), 79.2 (C-2‴), 74.1 (C-3‴), 25.5 (C-4‴), 25.6 (C-5‴).

EI-MS: m/z (rel%): 478 [M – C$_3$H$_7$O]$^+$ (2), 277 (26), 261 (71), 259 (5), 244 (10), 219 (41), 205 (29), 189 (93), 177(100), 175 (14), 144 (10), 131 (38), 59 (31).

HRFABMS: m/z 537.2142 ([M + H])$^+$, Calcd. for C$_{30}$H$_{33}$O$_9$, 537.2125).

Reference

Saied S, Nizami S S, Anis I. (2008). Two new coumarins from *Murraya paniculata*. *J Asian Nat Prod Res* **10**: 515–519.

Naringenin 5-methyl ether

IUPAC name: (S)-7-Hydroxy-2-(4-hydroxyphenyl)-5-methoxychroman-4-one

Sub-class: Flavanone

Chemical structure

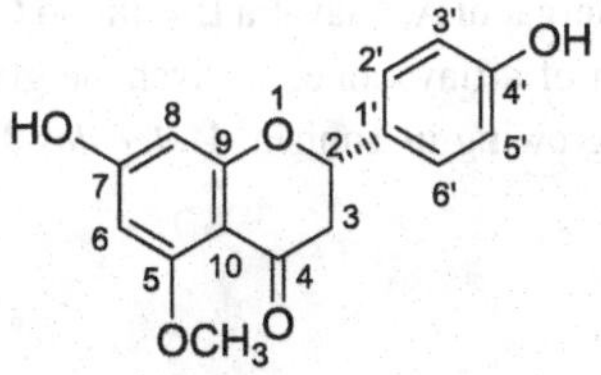

Source: *Echiochilon fruticosum* (Family: Boraginaceae); aerial parts

Molecular formula: $C_{16}H_{14}O_5$

Molecular weight: 286

State: Oil

Specific rotation: $[\alpha]^{22}_D$ −14.0° (CHCl₃, c 0.17)

IR (KBr): v_{max} 3366 (OH), 1642 (α,β-unsaturated carbonyl) cm⁻¹

¹H-NMR (CDCl₃, 300 MHz): δ 5.38 (1H, dd, J = 13.1, 2.9 Hz, H-2), 3.12 (1H, dd, J = 17.1, 2.9 Hz, H-3a), 2.85 (1H, dd, J = 17.1, 13.1 Hz, H-3b), 6.09 (1H, d, J = 2.17 Hz, H-6), 6.02 (1H, d, J = 2.3 Hz, H-8), 7.35 (2H, d, J = 8.4 Hz, H-2′ and H-6′), 6.90 (2H, d, J = 8.4 Hz, H-3′ and H-5′), 3.82 (3H, s, C₅-OCH₃).

^{13}**C-NMR (CDCl$_3$, 75 MHz):** δ 78.6 (C-2), 52.9 (C-3), 196.4 (C-4), 168.0 (C-5), 93.8 (C-6), 162.9 (C-7), 96.3 (C-8), 164.0 (C-9), 102.0 (C-10), 129.9 (C-1′), 127.7 (C-2′ and C-6′), 115.4 (C-3′ and C-5′), 156.4 (C-4′), *no data* (OCH$_3$).

HMQC: δ 5.38 (H-2) *vs* δ 78.6 (C-2), δ 3.12 (H-3a) and 2.85 (H-3b) *vs* δ 52.9 (C-3), δ 6.09 (H-6) *vs* δ 93.8 (C-6), δ 6.02 (H-8) *vs* δ 96.3 (C-8), δ 7.35 (H-2′/H-6′) *vs* δ 127.7 (C-2′/C-6′), δ 6.90 (H-3′/H-5′) *vs* δ 115.4 (C-3′/C-5′).

^{1}**H-^{1}H COSY:** δ 5.38 (H-2) *vs* δ 3.12 (H-3a), δ 7.35 (H-2′/H-6′) *vs* δ 6.90 (H-3′/H-5′).

ESMS: *m/z* 287 ([M + H]$^+$)

Reference

Hammami S, Ben Jannet H, Bergaoui A, Ciavatta L, Cimino G, Mighri Z. (2004). Isolation and structure elucidation of a flavanone, a flavanone glycoside and vomifoliol from *Echiochilon fruticosum* growing in Tunisia. *Molecules* **9**: 602–608.

Nervilifordin A [Rhamnocitrin-3-*O*-β-D-xylopyranosyl-(1→4)-β-D-glucopyranoside]

IUPAC name: 3-((((2S,4R,5S,6R)-3,4-Dihydroxy-6-(hydroxymethyl)-5-((((2R,3S,4R,5S)-3,4,5-trihydroxytetrahydro-2*H*-pyran-2-yl)oxy)tetrahydro-2*H*-pyran-2-yl)oxy)-5-hydroxy-2-(4-hydroxyphenyl)-7-methoxy-4*H*-chromen-4-one

Sub-class: Flavonol glycoside

Chemical structure

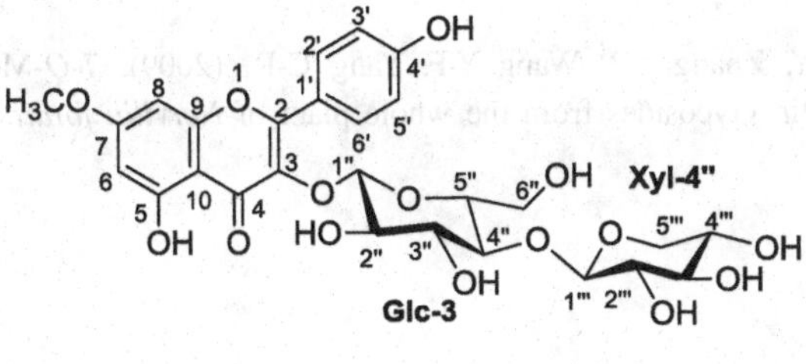

Source: *Nervilia fordii* (Hance) Schltr. (Family: Orchidaceae); whole plants

Molecular formula: $C_{27}H_{30}O_{15}$

Molecular weight: 594

State: Yellow amorphous powder

Specific rotation: $[\alpha]^{24}_{D}$ −9.8° (MeOH, *c* 0.09)

UV (MeOH): λ_{max} (log ε) 266 (4.16), 350 (4.10) nm

FT-IR (KBr): v_{max} 3374 (OH), 2895, 2768, 1665 (α,β-unsaturated carbonyl), 1593, 1494, 1445, 1355 (aromatic unsaturaion), 830, 804 cm^{-1}.

^{1}H-NMR (DMSO-d_6, 500 MHz): δ 6.31 (1H, d, J = 2.1 Hz, H-6), 6.67 (1H, d, J = 2.1 Hz, H-8), 8.05 (2H, d, J = 8.6 Hz, H-2' and H-6'), 6.88 (2H, d, J = 8.6 Hz, H-3' and H-5'), 3.84 (3H, s, 7-OCH_3), 5.50 (1H, d, J = 7.5 Hz, H-1''), 3.11 (1H, m, H-2''), 3.31 (1H, m, H-3''), 3.32 (1H, m, H-4''), 3.19 (1H, m, H-5''), 3.51 (1H, m, H-6''a), 3.43 (1H, m, H-6''b), 4.23 (1H, d, J = 7.7 Hz, H-1'''), 2.90 (1H, m, H-2'''), 3.00 (1H, m, H-3'''), 3.21 (1H, m, H-4'''), 3.67 (1H, dd, J = 11.0, 5.5 Hz, H-5'''a), 3.07 (1H, m, H-5'''b).

^{13}C-NMR (DMSO-d_6, 125 MHz): δ 156.7 (C-2), 133.4 (C-3), 177.5 (C-4), 160.9 (C-5), 97.9 (C-6), 165.2 (C-7), 92.3 (C-8), 156.3 (C-9), 105.0 (C-10), 120.7 (C-1'), 131.0 (C-2', C-6'), 115.2 (C-3', C-5'), 160.1 (C-4'), 56.1 (7-OCH_3), 100.6 (C-1''), 74.1 (C-2''), 74.3 (C-3''), 78.7 (C-4''), 75.5 (C-5''), 59.8 (C-6''), 103.5 (C-1'''), 73.3 (C-2'''), 76.5 (C-3'''), 69.4 (C-4'''), 65.8 (C-5''').

HR-ESI-MS: m/z 593.1505 ([M $-$ H]$^-$, Calcd. for $C_{27}H_{29}O_{15}$, 593.1506).

Reference

Tian L-W, Pei Y, Zhang Y-J, Wang Y-F, Yang C-R. (2009). 7-O-Methylkaempferol and -quercetin glycosides from the whole plant of *Nervilia fordii*. *J Nat Prod* **72**: 1057–1060.

Nervilifordin E [4′-*O*-β-D-glucopyranosylrhamnetin-3-*O*-β-D-glucopyranosyl-(4→1)-β-D-glucopyranoside]

IUPAC name: 3-((((2*S*,4*R*,5*S*,6*R*)-3,4-Dihydroxy-6-(hydroxymethyl)-5-((((2*R*,3*S*,4*R*,5*R*,6*S*)-3,4,5-trihydroxy-6-(hydroxymethyl)tetrahydro-2*H*-pyran-2-yl)oxy)tetrahydro-2*H*-pyran-2-yl)oxy)-5-hydroxy-2-(3-hydroxy-4-((((2*R*,3*S*,4*R*,5*R*,6*S*)-3,4,5-trihydroxy-6-(hydroxymethyl)tetrahydro-2*H*-pyran-2-yl)oxy)phenyl)-7-methoxy-4*H*-chromen-4-one

Sub-class: Flavonol glycoside

Chemical structure

Source: *Nervilia fordii* (Hance) Schltr. (Family: Orchidaceae); whole plants

Molecular formula: $C_{34}H_{42}O_{22}$

Molecular weight: 802

State: Yellow amorphous powder

Specific rotation: $[\alpha]^{24}_D$ −93.3° (MeOH, c 0.15)

UV (MeOH): λ_{max} (log ε) 268 (4.16), 344.5 (4.15) nm

FT-IR (KBr): v_{max} 3419 (OH), 2919, 1597, 1072, 803, 642, 579 cm^{-1}

^{1}H-NMR (DMSO-d_6, 500 MHz): δ 6.37 (1H, s, H-6), 6.76 (1H, s, H-8), 7.64 (1H, s, H-2′), 7.19 (1H, d, J = 8.4 Hz, H-5′), 7.61 (1H, d, J = 8.4 Hz, H-6′), 3.84 (3H, s, 7-OCH_3), 5.50 (1H, d, J = 7.3 Hz, H-1″), 3.27 (1H, m, H-2″), 3.38 (1H, m, H-3″), 3.36 (1H, m, H-4″), 3.26 (1H, m, H-5″), 3.63 (1H, m, H-6″a), 3.48 (1H, m, H-6″b), 4.25 (1H, d, J = 6.4 Hz, H-1‴), 2.97 (1H, m, H-2‴), 3.14 (1H, m, H-3‴), 3.03 (1H, t, J = 9.0 Hz, H-4‴), 3.18 (1H, m, H-5‴), 3.70 (1H, m, H-6‴a), 3.40 (1H, m, H-6‴b), 4.86 (1H, d, J = 6.4 Hz, H-1⁗), 3.12 (1H, m, H-2⁗), 3.31 (1H, m, H-3⁗), 3.18 (1H, m, H-4⁗), 3.38 (1H, m, H-5⁗), 3.69 (1H, m, H-6⁗a), 3.42 (1H, m, H-6⁗b).

^{13}C-NMR (DMSO-d_6, 125 MHz): δ 156.0 (C-2), 134.1 (C-3), 177.6 (C-4), 161.9 (C-5), 98.0 (C-6), 165.3 (C-7), 92.3 (C-8), 156.4 (C-9), 105.1 (C-10), 124.4 (C-1′), 116.7 (C-2′), 146.6 (C-3′), 147.8 (C-4′), 115.5 (C-5′), 120.8 (C-6′), 56.2 (7-OCH$_3$), 100.5 (C-1″), 73.9 (C-2″), 74.9 (C-3″), 80.3 (C-4″), 75.5 (C-5″), 60.3 (C-6″), 103.2 (C-1‴), 73.3 (C-2‴), 76.5 (C-3‴), 70.1 (C-4‴), 76.8 (C-5‴), 61.1 (C-6‴), 101.5 (C-1⁗), 73.3 (C-2⁗), 75.9 (C-3⁗), 69.8 (C-4⁗), 73.8 (C-5⁗), 60.8 (C-6⁗).

HR-ESI-MS: m/z 801.2089 ([M − H]$^-$, Calcd. for $C_{34}H_{41}O_{22}$, 801.2089).

Reference

Tian L-W, Pei Y, Zhang Y-J, Wang Y-F, Yang C-R. (2009). 7-O-Methylkaempferol and -quercetin glycosides from the whole plant of *Nervilia fordii*. *J Nat Prod* **72**: 1057–1060.

Noidesol A

IUPAC name: (2R,3R)-2-(3,4-Dihydroxyphenyl)-3,5-dihydroxy-7-methoxy-8-((2R,3S,4S,5R,6S)-3,4,5-trihydroxy-6-(hydroxymethyl)tetrahydro-2H-pyran-2-yl)chroman-4-one

Sub-class: Flavanone C-glycoside

Chemical structure

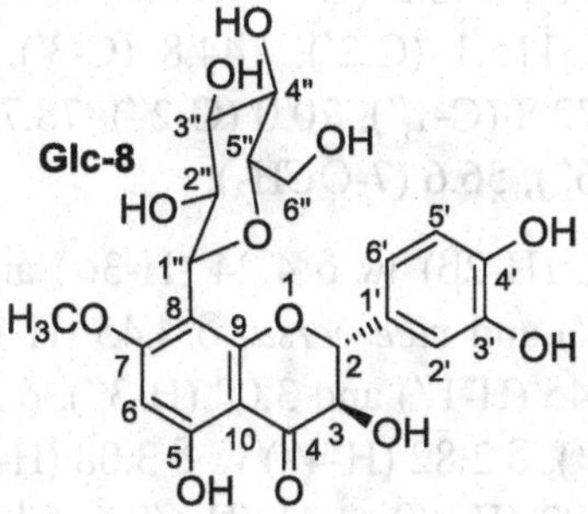

Source: *Gnetum gnemonoides* (Family: Gnetaceae); barks

Molecular formula: $C_{22}H_{24}O_{12}$

Molecular weight: 480

State: Colorless platelets

Melting point: 178 °C

Specific rotation: $[\alpha]^{26}_{D}$ −31° (MeOH, c 0.5)

UV (MeOH): λ_{max} (log ε) 204 (4.67), 291 (4.22), 328 (3.57) nm

CD (MeOH): λ_{ext} 215 (θ −22 000), 230 (+9700), 296 (−35 000), 333 (+8700) nm

FT-IR (KBr): 3410 (OH), 2871, 1636 (aromatic unsaturation), 1205, 1115 cm^{-1}

^{1}H-NMR (DMSO-d_6, 600 MHz): δ 5.02 (1H, d, J = 11.3 Hz, H-2β), 4.24 (1H, dd, J = 11.3, 6.6 Hz, H-3α), 6.21 (1H, s, H-6), 6.94 (1H, d, J = 2.0 Hz, H-2′), 6.70 (1H, d, J = 8.2 Hz, H-5′), 6.84 (1H, dd, J = 8.2, 1.9 Hz, H-6′), 4.45 (1H, d, J = 9.7 Hz, H-1″), 3.77 (1H, m, H-2″), 3.08 (1H, m, H-3″), 2.82 (1H, ddd, J = 9.2, 9.2, 5.2 Hz, H-4″), 3.10 (1H, m, H-5″), 3.41 (1H, m, H-6″a), 3.73 (1H, ddd, J = 11.6, 5.8, 1.7 Hz, H-6″b), 3.82 (3H, s, 7-OCH_3), 5.88 (1H, d, J = 6.6 Hz, 3-OH), 12.09 (1H, s, 5-OH), 8.80 (1H, s, 3′-OH), 8.94 (1H, s, 4′-OH), 4.57 (1H, d, J = 5.4 Hz, 2″-OH), 4.78 (1H, d, J = 3.7 Hz, 3″-OH), 4.79 (1H, d, J = 5.3 Hz, 4″-OH), 4.49 (1H, dd, J = 5.8, 5.7 Hz, 6″-OH).

^{13}C-NMR (DMSO-d_6, 150 MHz): δ 82.1 (C-2), 72.7 (C-3), 198.9 (C-4), 163.1 (C-5), 92.4 (C-6), 166.2 (C-7), 106.8 (C-8), 161.0 (C-9), 101.6 (C-10), 128.7 (C-1′), 115.1 (C-2′), 144.8 (C-3′), 145.2 (C-4′), 115.2 (C-5′), 118.2 (C-6′), 72.8 (C-1″), 70.0 (C-2″), 78.7 (C-3″), 70.7 (C-4″), 81.6 (C-5″), 62.2 (C-6″), 56.6 (7-OCH_3),

^{1}H-^{1}H COSY: δ 5.02 (H-2β) vs δ 4.24 (H-3α) and vice versa, δ 6.70 (H-5′) vs δ 6.84 (H-6′) and vice versa, δ 4.45 (H-1″) vs δ 3.77 (H-2″), δ 3.77 (H-2″) vs δ 4.45 (H-1″) and 3.08 (H-3″), δ 3.08 (H-3″) vs δ 3.77 (H-2″) and 2.82 (H-4″), δ 2.82 (H-4″) vs δ 3.08 (H-3″) vs δ 3.10 (H-5″), δ 3.10 (H-5″) vs δ 2.82 (H-4″), 3.41 (H-6″a) and 3.73 (H-6″b), δ 3.41 (H-6″a) and 3.73 (H-6″b) vs δ 3.10 (H-5″).

HMBC: δ 5.02 (H-2β) vs δ 198.9 (C-4), 161.0 (C-9), 115.1 (C-2′) and 118.2 (C-6′), δ 4.24 (H-3α) vs δ 128.7 (C-1′), δ 6.84 (H-6′) vs δ 163.1 (C-5), 166.2 (C-7), 106.8 (C-8) and 101.6 (C-10), δ 6.94 (H-2′) vs δ 82.1 (C-2), δ 6.70 (H-5′) vs δ 128.7 (C-1′) and 144.8 (C-3′), δ 6.84 (H-6′) vs δ 82.1 (C-2) and 145.2 (C-4′), δ 4.45 (H-1″) vs δ 166.2 (C-7), 106.8 (C-8), 78.7 (C-3″) and 81.6 (C-5″), δ 2.82 (H-4″) vs δ 78.7 (C-3″), 81.6 (C-5″) and 62.2 (C-6″), δ 3.82 (7-OCH_3) vs δ 166.2 (C-7) (selected 2D-correlations were shown).

ESIMS: m/z 503 [M + Na]$^+$

HRESI-TOFMS: m/z 503.1167 [M + Na]$^+$ (Calcd. for $C_{22}H_{24}O_{12}Na$, 503.1165).

Reference

Shimokawa Y, Akao Y, Hirasawa Y, Awang K, Hadi A H A, Sato S, Aoyama C, Takeo J, Shiro M, Morita H. Gneyulins A and B, stilbene trimers, and Noidesols A and B, dihydroflavonol-C-glucosides, from the bark of *Gnetum gnemonoides*. *J Nat Prod* **73**: 763–767.

Ormosinol [5,7,2′,4′-tetrahydroxyl-6,8,5′-tri-(γ,γ-dimethylallyl) isoflavanone]

IUPAC name: 3-(2,4-Dihydroxy-5-(3-methylbut-2-en-1-yl)phenyl)-5,7-dihydroxy-6,8-bis(3-methylbut-2-en-1-yl)chroman-4-one

Sub-class: Isoflavanone

Chemical structure

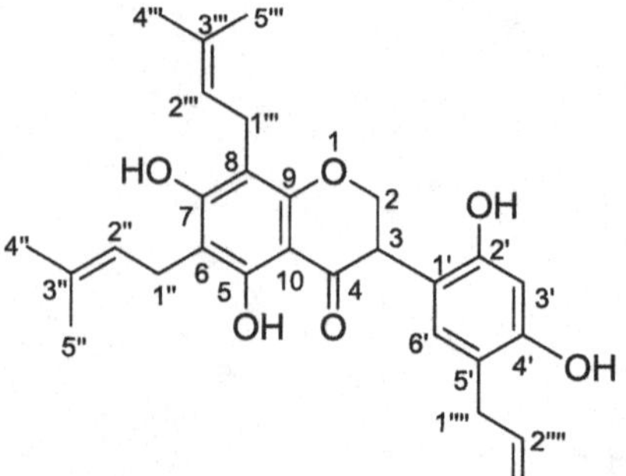

Source: *Ormosia henryi* (Family: Leguminosae); root barks

Molecular formula: $C_{30}H_{36}O_6$

Molecular weight: 492

State: White powder

Melting point: 99–104 °C

Bioactivity studied: Anti-oxidative activity against DPPH radicals (IC_{50} 28.5 µM

Specific rotation: $[\alpha]_D^{20}$ +1.0° (MeOH, c 2.0)

UV (MeOH): λ_{max} (log ε) 203 (2.90), 295 (2.40) nm

FT-IR (KBr): 3380 (OH), 2969, 2913, 1700 (C=O), 1631 (aromatic unsaturation), 1509, 1442 and 1378 cm^{-1}

^{1}H-NMR (DMSO-d_6, 400 MHz): δ 4.40 (1H, dd, J = 10.4, 4.8 Hz, H-2a), 4.40 (1H, dd, J = 10.4, 5.6 Hz, H-2b), 4.08 (1H, dd, J = 5.6, 4.4 Hz, H-3), 6.34 (1H, s, H-3'), 6.62 (1H, s, H-6'), 12.61 (1H, s, 5-OH), 9.56 (1H, br s, 7-OH), 9.14 (1H, s, 2'-OH), 9.24 (1H, s, 4'-OH), 3.19 (2H, d, J = 7.6 Hz, H-1''), 5.09 (1H, m, H-2''), 1.67 (3H, s, H-4''), 1.69 (3H, s, H-5''), 3.21 (2H, d, J = 7.6 Hz, H-1'''), 5.09 (1H, m, H-2'''), 1.61 (3H, s, H-4'''), 1.59 (3H, s, H-5'''), 3.02 (2H, d, J = 6.8 Hz, H-1''''), 5.15 (1H, t, J = 3.6 Hz, H-2''''), 1.59 (3H, s, H-4''''), 1.54 (3H, s, H-5'''').

^{13}C-NMR (DMSO-d_6, 100 MHz): δ 69.8 (C-2), 46.2 (C-3), 198.3 (C-4), 158.9 (C-5), 107.9 (C-6), 157.7 (C-7), 107.0 (C-8), 161.2 (C-9), 102.3 (C-10), 117.8 (C-1'), 154.8 (C-2'), 102.5 (C-3'), 153.6 (C-4'), 112.1 (C-5'), 130.3 (C-6'), 21.4 (C-1''), 122.8 (C-2''), 130.3 (C-3''), 17.7 (C-4''), 21.0 (C-5''), 17.5 (C-1'''), 123.3 (C-2'''), 130.7 (C-3'''), 25.5 (C-4'''), 27.2 (C-5'''), 17.7 (C-1''''), 123.0 (C-2''''), 130.3 (C-3''''), 25.4 (C-4''''), 27.1 (C-5'''').

HMBC: δ 6.62 (H-6') vs δ 46.2 (C-3), δ 9.56 (7-OH) vs δ 157.7 (C-7) and 107.0 (C-8), δ 9.14 (2'-OH) vs δ 117.8 (C-1') and 102.5 (C-3'), δ 9.24 (4'-OH) vs δ 102.5 (C-3') and 112.1 (C-5'), δ 3.19 (H-1'') vs δ 158.9 (C-5) and 157.7 (C-7), δ 3.21 (H-1''') vs δ 157.7 (C-7) and 161.2 (C-9), δ 5.09 (H-2''') vs δ 107.0 (C-8), δ 3.02 (H-1'''') vs δ 112.1 (C-5') and 130.3 (C-6') (selected 2D-correlations were shown).

HR-ESIMS: m/z 493.2609 ([M + H]$^+$), Calcd. for $C_{30}H_{37}O_6$, 493.2584).

Reference

Feng S, Hao J, Xu Z, Chen T, Qiu SX (Shengxiang). (2012). Polyprenylated isoflavanone and isoflavonoids from *Ormosia henryi* and their cytotoxicity and anti-oxidation activity. *Fitoterapia* **83**: 161–165.

Ormosinoside [Isoprunetin-7-*O*-β-D-xylopyranosyl-(1→6)-β-D-glucopyranoside

IUPAC name: 3-(2,4-Dihydroxy-5-(3-methylbut-2-en-1-yl)phenyl)-5,7-dihydroxy-6,8-bis(3-methylbut-2-en-1-yl)chroman-4-one

Sub-class: Isoflavone glycoside

Chemical structure

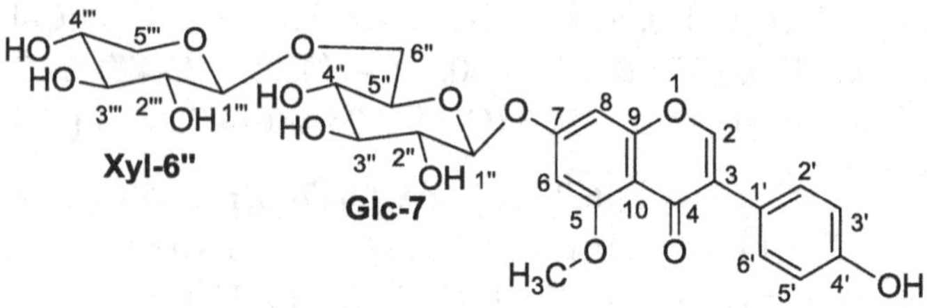

Source: *Ormosia henryi* (Family: Leguminosae); root barks

Molecular formula: $C_{27}H_{30}O_{14}$

Molecular weight: 578

State: White powder

Melting point: 225–228 °C

Specific rotation: $[\alpha]_D^{20}$ −57° (MDSO, *c* 1.0)

UV (MeOH): λ_{max} (log ε) 202 (3.69), 250 (3.75) nm

FT-IR (KBr): 3390(OH), 2919, 1637(aromatic unsaturation), 1515, 1461, 1428 and 1375 cm^{-1}

^{1}H-NMR (DMSO-d_6, 400 MHz): δ 8.11 (1H, s, H-2), 6.55 (1H, d, J = 2.0 Hz, H-6), 6.73 (1H, d, J = 2.0 Hz,, H-8), 7.30 (2H, d, J = 8.0 Hz, H-2′ and H-6′), 6.78 (2H, d, J = 8.0 Hz, H-3′ and H-5′), 3.67 (3H, s, 5-OCH_3), 5.04 (1H, d, J = 6.8 Hz, H-1″), 4.16 (1H, d, J = 7.2 Hz, H-1‴), 2.0-5.0 (10H, m, sugar protons).

^{13}C-NMR (DMSO-d_6, 100 MHz): δ 150.8 (C-2), 124.9 (C-3), 173.9 (C-4), 160.8 (C-5), 97.0 (C-6), 161.3 (C-7), 95.7 (C-8), 157.1 (C-9), 109.6 (C-10), 56.2 (5-OCH$_3$), 122.6 (C-1′), 130.2 (C-2′), 114.9 (C-3′), 158.9 (C-4′), 114.9 (C-5′), 130.2 (C-6′), 99.7 (C-1″), 76.4 (C-2″), 76.6 (C-3″), 69.8 (C-4″), 75.7 (C-5″), 65.7 (C-6″), 104.2 (C-1‴), 73.1 (C-2‴), 73.4 (C-3‴), 69.5 (C-4‴), 68.8 (C-5‴).

HR-ESIMS: m/z 577.1556 ([M − H]$^-$), Calcd. for $C_{27}H_{29}O_{14}$, 577.1562).

Reference

Feng S, Hao J, Xu Z, Chen T, Qiu SX (Shengxiang). (2012). Polyprenylated isoflavanone and isoflavonoids from *Ormosia henryi* and their cytotoxicity and anti-oxidation activity. *Fitoterapia* **83**: 161–165.

Oxytropisoflavan A

IUPAC name: (4-(2-Hydroxy-1-((R)-7-hydroxy-3-(2-hydroxy-3,4-dimethoxyphenyl)chroman-6-yl)-3-(4-hydroxyphenyl)propyl)benzene-1,3-diol

Sub-class: Isoflavan

Chemical structure

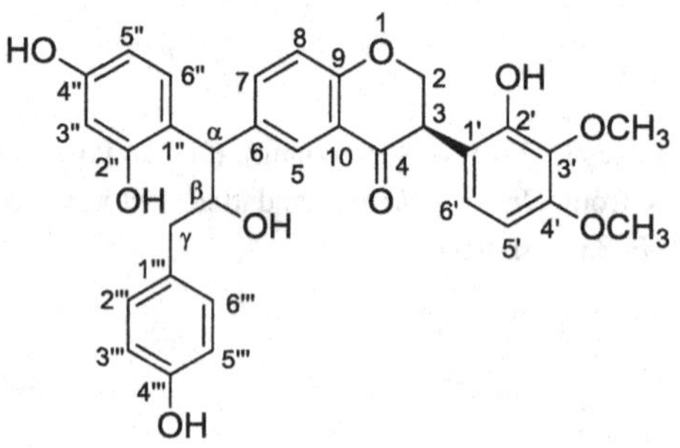

Source: *Oxytropis falcata* Bunge (Family: Leguminosae); aerial parts and roots

Molecular formula: $C_{32}H_{32}O_9$

Molecular weight: 560

State: Pale brown oil

Specific rotation: $[\alpha]^{20}_D$ −28° (CH_3COCH_3, c 0.7)

UV: λ_{max} (MeOH) (log ε): 212 (1.63), 282 (0.27) nm

IR(KBr): ν_{max} 3363 (OH), 2922, 1510, 1463, 1096 cm^{-1}

^{1}H-NMR (CD$_3$COCD$_3$, 400 MHz): δ 4.23 (1H, dd, J = 10.0, 3.2 Hz, H-2a), 3.96 (1H, t, J = 10.0 Hz, H-2b), 3.48 (1H, m, H-3α), 2.90 (1H, dd, J = 15.6, 11.2 Hz, H-4a), 2.78 (1H, dd, J = 15.6, 4.0 Hz, H-4b), 6.85 (1H, s, H-5), 6.33 (1H, s, H-8), 6.50 (1H, d, J = 8.8 Hz, H-5′), 6.83 (1H, d, J = 8.8 Hz, H-6′), 4.40 (1H, d, J = 5.2 Hz, H-α), 4.68 (1H, m, H-β), 2.73 (1H, dd, J = 14.0, 4.4 Hz, H-γ$_1$), 2.61 (1H, dd, J = 14.0, 7.2 Hz, H-γ$_2$), 6.36 (1H, d, J = 2.4 Hz, H-3″), 6.24 (1H, dd, J = 8.4, 2.4 Hz, H-5″), 7.02 (1H, d, J = 8.4, Hz, H-6″), 7.02 (2H, d, J = 8.4, Hz, H-2‴ and H-6‴), 6.72 (2H, d, J = 8.4 Hz, H-3‴ and H-5‴), 3.78 (3H, s, 3′-OCH_3), 3.81 (3H, s, 4′-OCH_3), 9.63 (1H, s, 7-OH), 7.94 (1H, s, 2′-OH), 5.13 (1H, s, β-OH), 8.44 (1H, s, 2″-OH), 8.01 (1H, s, 4″-OH), 8.07 (1H, s, 4‴-OH).

^{13}C-NMR (CD$_3$COCD$_3$, 100 MHz): δ 69.7 (C-2), 32.6 (C-3), 30.4 (C-4), 132.9 (C-5), 120.1 (C-6), 155.2 (C-7), 104.1 (C-8), 154.1 (C-9), 113.5 (C-10), 121.1 (C-1′), 148.4 (C-2′), 136.3 (C-3′), 152.0 (C-4′), 103.7 (C-5′), 121.7 (C-6′), 47.7 (C-α), 74.4 (C-β), 41.4 (C-γ), 120.4 (C-1″), 155.2 (C-2″), 102.9 (C-3″), 156.8 (C-4″), 106.9 (C-5″), 130.9 (C-6″), 130.5 (C-1‴), 130.5 (C-2‴), 115.0 (C-3‴), 155.8 (C-4‴), 115.0 (C-5‴), 130.5 (C-6‴), 60.1 (3′-OCH$_3$), 55.4 (4′-OCH$_3$).

HMBC: δ 4.23 (H-2a) and 3.96 (H-2b) vs δ 30.4 (C-4) and 154.1 (C-9), δ 2.90 (H-4a) and 2.78 (H-4b) vs δ 69.7 (C-2), 32.6 (C-3), 132.9 (C-5), 154.1 (C-9), 113.5 (C-10) and 121.1 (C-1′), δ 6.85 (H-5) vs δ 30.4 (C-4), 155.2 (C-7), 154.1 (C-9) and 47.7 (C-α), δ 6.33 (H-8) vs δ 120.1 (C-6), 155.2 (C-7), 154.1 (C-9) and 113.5 (C-10), δ 6.50 (H-5′) vs δ 121.1 (C-1′), 136.3 (C-3′) and 152.0 (C-4′), δ 6.83 (H-6′) vs δ 148.4 (C-2′) and 152.0 (C-4′), δ 4.40 (H-α) vs δ 132.9 (C-5), 120.1 (C-6), 155.2 (C-7), 120.4 (C-1″), 155.2 (C-2″), 130.9 (C-6″) and 41.4 (C-γ), δ 4.68 (H-β) vs δ 120.4 (C-1″) and 130.5 (C-1‴), δ 2.73 (H-γ$_1$) and 2.61 (H-γ$_2$) vs δ 47.7 (C-α), 74.4 (C-β),130.5 (C-2‴) and 130.5 (C-6‴), δ 6.36 (H-3″) vs δ 120.4 (C-1″), 155.2 (C-2″), 156.8 (C-4″) and 106.9 (C-5″), δ 6.24 (H-5″) vs δ 120.4 (C-1″) and 102.9 (C-3″), δ 7.02 (H-6″) vs δ 47.7 (C-α), 155.2 (C-2″) and 156.8 (C-4″), δ 7.02 (H-2‴) vs δ 41.4 (C-γ), 115.0 (C-3‴), 155.8 (C-4‴) and 130.5 (C-6‴), δ 6.72 (H-3‴) vs δ 130.5 (C-1‴), 130.5 (C-2‴), 155.8 (C-4‴) and 115.0 (C-5‴), δ 6.72 (H-5‴) vs δ 130.5 (C-1‴), 115.0 (C-3‴), 155.8 (C-4‴) and 130.5 (C-6‴), δ 7.02 (H-6‴) vs δ 41.4 (C-γ), 130.5 (C-2‴), 155.8 (C-4‴) and 115.0 (C-5‴), δ 3.78

(3'-OCH$_3$) vs δ 136.3 (C-3'), δ 3.81 (4'-OCH$_3$) vs δ 152.0 (C-4'), δ 9.63 (7-OH) vs δ 120.1 (C-6), 155.2 (C-7) and 104.1 (C-8), δ 7.94 (2'-OH) vs δ 121.1 (C-1'), 148.4 (C-2') and 136.3 (C-3'), δ 5.13 (1H, s, β-OH) vs δ 47.7 (C-α), 74.4 (C-β) and 41.4 (C-γ), δ 8.44 (1H, s, 2''-OH) vs δ 120.4 (C-1''), 155.2 (C-2''), 102.9 (C-3''), δ 8.01 (4''-OH) vs δ 102.9 (C-3''), 156.8 (C-4'') and 106.9 (C-5''), δ 8.07 (4'''-OH) vs δ 115.0 (C-3'''), 155.8 (C-4''') and 115.0 (C-5''').

EIMS: m/z (%rel.) 552 (0.3), 450 (4), 432 (11), 423 (2), 302 (59), 240 (33), 239 (32), 180 (retro-Diels-Alder fragmented ion, 100), 107 (46).

HR-EIMS: m/z 583.1944 ([M + Na]$^+$, Calcd. for C$_{32}$H$_{32}$O$_9$Na, 583.1939).

Reference

Chen W-H, Wang R, Shi Y-P. (2010). Flavonoids in the poisonous plant *Oxytropis falcata*. *J Nat Prod* **73**: 1398–1403.

Pavietin

IUPAC name: (S)-7-Hydroxy-6-(2-(hydroxymethyl)butoxy)-4-methyl-2H-chromen-2-one

Sub-class: Coumarin

Chemical structure

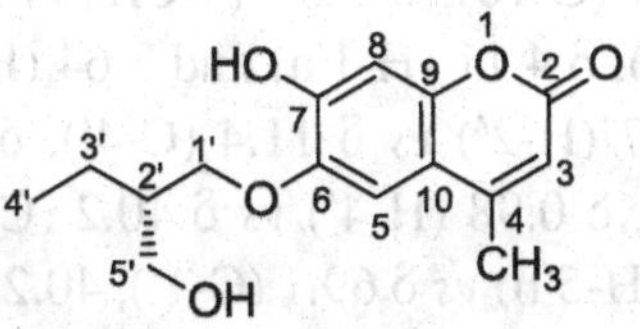

Source: *Aesculus pavia* (Family: Sapindaceae); leaves

Molecular formula: $C_{15}H_{18}O_5$

Molecular weight: 278

State: Yellow amorphous solid

Bioactivity studied: Antifungal

Specific rotation: $[\alpha]^{25}_D$ −27.0° (CHCl₃, c 0.1)

UV (CHCl₃): λ_{max} (log ε) 226 (3.95), 293 (sh), 346 (4.26) nm

FT-IR (KBr): ν_{max} 3400 (OH), 1680 (C=O), 1450, 1295, 1162 cm⁻¹

¹H-NMR (CD₃OD, 500 MHz): δ 6.08 (1H, s, H-3), 7.04 (1H, s, H-5), 6.78 (1H, s, H-8), 2.39 (3H, s, 4-CH₃), 4.11 (1H, dd, J = 10.0, 3.0 Hz, H-1′a), 3.64 (1H, br d, J = 10.0, 3.0 Hz, H-1′b), 2.07 (1H, m, H-2′), 1.35

(2H, dq, J = 7.0, 3.0 Hz, H-3'), 0.98 (3H, t, J = 7.0 Hz, H-4'), 3.61 (1H, dd, J = 10.0, 3.0 Hz, H-5'a), 3.29 (1H, dd, J = 10.0, 3.0 Hz, H-5'b).

^{13}C-NMR (CD$_3$OD, 125 MHz): δ 164.5 (C-2), 111.3 (C-3), 156.0 (C-4), 110.0 (C-5), 149.7 (C-6), 151.9 (C-7), 103.7 (C-8), 144.6 (C-9), 113.5 (C-10), 69.1 (C-1'), 40.2 (C-2'), 24.9 (C-3'), 11.4 (C-4'), 63.9 (C-5'), 18.8 (4-CH_3).

HMQC: δ 6.08 (H-3) *vs* δ 111.3 (C-3), δ 7.04 (H-5) *vs* δ 110.0 (C-5), δ 6.78 (H-8) *vs* δ 103.7 (C-8), δ 4.11 (H-1'a) and 3.64 (H-1'b) *vs* δ 69.1 (C-1'), δ 2.07 (H-2') *vs* δ 40.2 (C-2'), δ 1.35 (H-3') *vs* δ 24.9 (C-3'), δ 0.98 (H-4') *vs* δ 11.4 (C-4'), δ 3.61 (H-5'a) and 3.29 (H-5'b) *vs* δ 63.9 (C-5'), δ 3.39 (4-CH_3) *vs* δ 18.8 (4-CH_3).

HMBC: δ 6.08 (H-3) *vs* δ 164.5 (C-2), 113.5 (C-10) and 18.8 (4-CH_3), δ 7.04 (H-5) *vs* δ 156.0 (C-4), 149.7 (C-6), 151.9 (C-7), 144.6 (C-9) and 113.5 (C-10), δ 6.78 (H-8) *vs* δ 164.5 (C-2), 149.7 (C-6), 151.9 (C-7), 144.6 (C-9) and 113.5 (C-10), δ 2.39 (4-CH_3) *vs* δ 111.3 (C-3), 156.0 (C-4) and 113.5 (C-10), δ 4.11 (H-1'a) and 3.64 (H-1'b) *vs* δ 149.7 (C-6) and 63.9 (C-5'), δ 2.07 (H-2') *vs* δ 11.4 (C-4'), δ 1.35 (H-3') *vs* δ 69.1 (C-1') and 63.9 (C-5'), δ 0.98 (H-4') *vs* δ 40.2 (C-2') and 24.9 (C-3'), δ 3.61 (H-5'a) and 3.29 (H-5'b) *vs* δ 69.1 (C-1'), 40.2 (C-2') and 24.9 (C-3').

HR-FABMS: *m/z* 279.1256 ([M + H]$^+$, Calcd. for $C_{15}H_{18}O_5$, 279.1233).

Reference

Curir P, Galeotti F, Dolci M, Barile E, Lanzotti V. (2007). Pavietin, a coumarin from *Aesculus pawia* with antifungal activity. *J Nat Prod* **70**: 1668–1671.

Pelargonidin 3-(2″-(6′″-*E*-sinapoyl-β-glucopyranosyl)-6″-(*E*-*p*-coumaroyl)-β-glucopyranoside)-5-β-glucopyranoside

Sub-class: Anthocyanin glycoside

Chemical structure

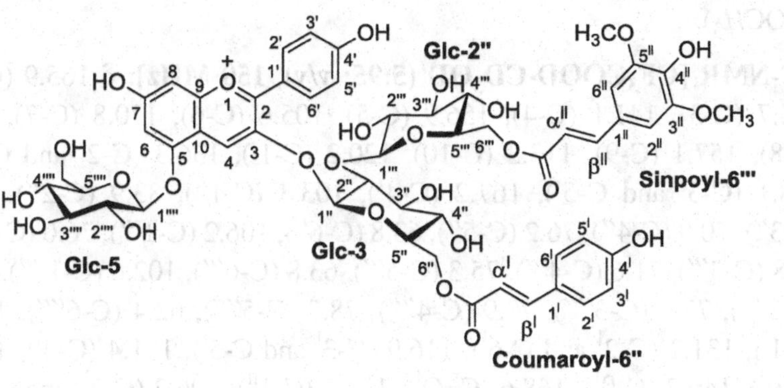

Source: *Cleme hassleriana* (spider flower; Family: Capparidaceae/Capparaceae); flowers

Molecular formula: $C_{35}H_{57}O_{26}^{+}$

Molecular weight: 1110

State: Pigment

UV: λ_{max} 510, 319 nm

^{1}H-NMR [CF$_3$COOD-CD$_3$OD (5:95; v/v), 600 MHz]: δ 8.99 (1H, d, J = 0.5 Hz, H-4), 7.94 (1H, d, J = 1.7 Hz, H-6), 6.73 (1H, dd, J = 1.8, 0.6 Hz, H-8), 8.56 (2H, d, J = 9.0 Hz, H-2′ and H-6′), 7.07 (2H, d, J = 9.0 Hz, H-3′ and H-5′), 5.29 (1H, d, J = 6.6 Hz, H-1″), 4.01 (1H, dd, J = 8.2, 6.7 Hz, H-2″), 3.82 (1H, t, J = 8.4 Hz, H-3″), 3.66 (1H, dd, J = 9.8, 8.9 Hz, H-4″), 3.71 (1H, ddd, J = 9.8, 7.2, 2.3 Hz, H-5″), 4.49 (1H, dd, J = 11.7, 2.3 Hz, H-6″a), 4.41 (1H, dd, J = 11.8, 6.9 Hz, H-6″b), 4.86 (1H, d, J = 7.3 Hz, H-1‴), 3.52 (1H, dd, J = 8.6, 7.2 Hz, H-2‴), 3.55 (1H, t, J = 8.8 Hz, H-3‴), 3.54 (1H, m, H-4‴), 3.64 (1H, m, H-5‴), 4.47 (1H, dd, J = 11.7, 2.3 Hz, H-6‴a), 4.43 (1H, dd, J = 11.9, 6.8 Hz, H-6‴b), 5.17 (1H, d, J = 7.8 Hz, H-1⁗), 3.79 (1H, dd, J = 8.4, 7.9 Hz, H-2⁗), 3.63 (1H, dd, J = 8.6, 8.4 Hz, H-3⁗), 3.57 (1H, dd, J = 9.1, 8.6 Hz, H-4⁗), 3.58 (1H, m, H-5⁗), 4.03 (1H, dd, J = 12.3, 7.4 Hz, H-6⁗a), 3.84 (1H, dd, J = 12.1, 2.6 Hz, H-6⁗b), 7.29 (2H, d, J = 8.6 Hz, H-2^l and H-6^l), 6.87 (2H, d, J = 8.6 Hz, H-3^l and H-5^l), 5.99 (1H, d, J = 15.8 Hz, H-α^l), 7.33 (1H, d, J = 15.8 Hz, H-β^l), 6.44 (2H, s, H-2^{ll} and H-6^{ll}), 5.95 (1H, d, J = 15.8 Hz, H-α^{ll}), 7.11 (1H, d, J = 15.8 Hz, H-β^{ll}), 3.79 (6H, s, 3^{ll}-OCH_3 and 5^{ll}-OCH_3).

^{13}C-NMR [CF$_3$COOD-CD$_3$OD (5:95; v/v), 150 MHz]: δ 165.9 (C-2), 144.7 (C-3), 141.1 (C-4), 156.9 (C-5), 105.6 (C-6), 170.8 (C-7), 97.5 (C-8), 157.1 (C-9), 113.3 (C-10), 120.2 (C-1′), 136.3 (C-2′ and C-6′), 118.1 (C-3′ and C-5′), 167.2 (C-4′), 103.3 (C-1″), 83.9 (C-2″), 77.4 (C-3″), 70.9 (C-4″), 76.2 (C-5″), 63.8 (C-6″), 106.2 (C-1‴), 76.0 (C-2‴), 77.8 (C-3‴), 71.6 (C-4‴), 75.3 (C-5‴), 63.8 (C-6‴), 102.6 (C-1⁗), 74.5 (C-2⁗), 77.9 (C-3⁗), 70.9 (C-4⁗), 78.7 (C-5⁗), 62.4 (C-6⁗), 126.5 (C-1^l), 131.2 (C-2^l and C-6^l), 116.9 (C-3^l and C-5^l), 161.4 (C-4^l), 114.4 (C-α^l), 146.7 (C-β^l), 168.6 (C=O^l), 125.7 (C-1^{ll}), 106.2 (C-2^{ll} and C-6^{ll}), 148.7 (C-3^{ll} and C-5^{ll}), 139.4 (C-4^{ll}), 149.1 (C-5^{ll}), 115.1 (C-α^{ll}), 146.6 (C-β^{ll}), 168.4 (C=O^{ll}), 56.6 (3^{ll}-OCH_3 and 5^{ll}-OCH_3).

HMBC: δ 8.99 (H-4) *vs* δ 165.9 (C-2), 144.7 (C-3), 156.9 (C-5), 157.1 (C-9) and 113.3 (C-10), δ 7.94 (H-6) *vs* δ 156.9 (C-5) and 170.8 (C-7), δ 6.73 (H-8) *vs* δ 170.8 (C-7), 157.1 (C-9) and 113.3 (C-10), δ 8.56 (H-2′/H-6′) *vs* δ 165.9 (C-2), 120.2 (C-1′), 118.1 (C-3′/C-5′) and 167.2

(C-4′), δ 7.07 (H-3′/H-5′) *vs* δ 167.2 (C-4′) and 136.3 (C-2′/C-6′), δ 5.29
(H-1″) *vs* δ 144.7 (C-3), δ 4.49 (H-6″a) *vs* δ 168.6 (C=O^l), δ 4.41 (H-6″b)
vs δ 168.6 (C=O^l), δ 5.99 (H-α^l) *vs* δ 168.6 (C=O^l) and 146.7 (C-β^l),
δ 7.33 (H-β^l) *vs* δ 114.4 (C-α^l) and 126.5 (C-1^l), δ 7.29 (H-2^l/H-6^l) *vs*
δ 126.5 (C-1^l) and 116.9 (C-3^l/C-5^l), δ 6.87 (H-3^l/H-5^l) *vs* δ 131.2 (C-2^l/
C-6^l) and 161.4 (C-4^l), 4.86 (H-1‴) *vs* δ 83.9 (C-2″), δ 4.47 (H-6‴a) *vs*
δ 168.4 (C=O^{ll}), δ 4.43 (H-6‴b) *vs* δ 168.4 (C=O^{ll}), δ 5.95 (H-α^{ll}) *vs*
δ 168.4 (C=O^{ll}) and 146.6 (C-β^{ll}), δ 7.11 (H-β^{ll}) *vs* δ 115.1 (C-α^{ll}) and
125.7 (C-1^{ll}), δ 6.44 (H-2^{ll}/H-6^{ll}) *vs* δ 125.7 (C-1^{ll}) and 139.4 (C-4^{ll}), 56.6
(3^{ll}-OCH$_3$/5^{ll}-OCH$_3$) *vs* δ 148.7 (C-3^{ll}/C-5^{ll}), δ 5.17 (H-1⁗) *vs* δ 156.9
(C-5) [selected HMBC correlations were shown].

HR-ESI-MS: *m/z* 1109.3118 ([M]$^+$, Calcd. for $C_{53}H_{57}O_{26}^+$, 1109.3138.

Reference

Jordheim M, Andersen ØM, Nozzolillo C, Amiguet VT. (2009). Acylated anthocyanins in
inflorescence of spider flower (*Cleome hassleriana*). *Phytochemistry* **70**: 740–745.

5,6,2′,5′,6′-Pentamethoxy-3′,4′-methylenedioxyflavone

IUPAC name: 5,6-Dimethoxy-2-(4,6,7-trimethoxybenzo[*d*][1,3]dioxol-5-yl)-4*H*-chromen-4-one

Sub-class: Flavone

Chemical structure

Source: *Struthiola argentea* (Family: Thymelaeaceae); whole plants

Molecular formula: $C_{21}H_{20}O_9$

Molecular weight: 416

State: Pale yellow solid

Bioactivity studied: Anthelmintic (EC_{50} 3.1 µg/ml)

UV (MeOH): λ_{max} (log ε) 203 (4.48), 230 (4.35), 329 (3.82) nm

IR (ZnSe film): ν_{max} 2928, 2842, 1641, 1479, 1418, 1355, 1283, 1047 cm^{-1}

^{1}**H-NMR (CDCl$_3$, 500 MHz):** δ 6.26 (1H, s, H-3), 7.28 (1H, d, J = 9.0 Hz, H-7), 7.19 (1H, d, J = 9.0 Hz, H-8), 6.00 (2H, s, 3',4'-OCH_2-O-), 3.99 (3H, s, 5-OCH_3), 3.92 (3H, s, 6-OCH_3), 3.90 (3H, s, 2'-OCH_3), 3.97 (3H, s, 5'-OCH_3), 3.79 (3H, s, 6'-OCH_3).

^{13}**C-NMR (CDCl$_3$, 125 MHz):** δ 158.2 (C-2), 114.8 (C-3), 177.9 (C-4), 148.0 (C-5), 149.8 (C-6), 119.1 (C-7), 113.5 (C-8), 152.4 (C-9), 119.0 (C-10), 113.6 (C-1'), 136.4 (C-2'), 141.6 (C-3' or C-4'), 134.4 (C-4' or C-3'), 133.2 (C-5'), 146.3 (C-6'), 101.9 (3',4'-OCH$_2$-O-), 61.9 (5-OCH$_3$), 57.3 (6-OCH$_3$), 60.4 (2'-OCH$_3$), 60.6 (5'-OCH$_3$), 62.1 (6'-OCH$_3$).

^{1}**H-^{1}H COSY:** δ 7.28 (H-7) *vs* δ 7.19 (H-8).

HMBC: δ 6.26 (H-3) *vs* δ 158.2 (C-2), 177.9 (C-4), 119.0 (C-10) and 113.6 (C-1'), δ 7.28 (H-7) *vs* δ 148.0 (C-5), 149.8 (C-6), 113.5 (C-8) and 152.4 (C-9), δ 7.19 (H-8) *vs* δ 177.9 (C-4), 149.8 (C-6) and 152.4 (C-9), δ 6.00 (3',4'-OCH$_2$-O-) *vs* δ 141.6 (C-3' or C-4') and 134.4 (C-4' or C-3'), δ 3.99 (5-OCH$_3$) *vs* δ 61.9 (5-OCH$_3$), δ 3.92 (6-OCH$_3$) *vs* δ 57.3 (6-OCH$_3$), δ 3.90 (2'-OCH$_3$) *vs* δ 60.4 (2'-OCH$_3$), δ 3.97 (5'-OCH$_3$) *vs* δ 60.6 (5'-OCH$_3$), δ 3.79 (6'-OCH$_3$) *vs* δ 62.1 (6'-OCH$_3$).

NOE: δ 6.26 (H-3) *vs* δ 3.90 (2'-OCH$_3$) and δ 3.79 (6'-OCH$_3$); δ 3.79 (6'-OCH$_3$) *vs* δ 6.26 (H-3) and δ 3.97 (5'-OCH$_3$).

HR-ESI-FTMS: m/z 417.1178 ([M + H]$^+$, Calcd. for C$_{21}$H$_{21}$O$_9$, 417.1185).

Reference

Ayers S, Zink D L, Mohn K, Powell J S, Brown C M, Murphy T, Brand R, Pretorius S, Stevenson D, Thompson D, Singh SB. (2008). Flavones from *Struthiola argentea* with anthelmintic activity *in vitro*. *Phytochemistry* **69**: 541–545.

Peregrinumin A [Acacetin-7-*O*-(2,3-*O*-diacetyl-α-L-rhamnosyl)-(1→6)-β-D-glucoside]

IUPAC name: (2*R*,3*R*,4*R*,5*S*,6*S*)-5-Hydroxy-6-methyl-2-(((2*R*,3*S*,4*S*,5*R*,6*S*)-3,4,5-trihydroxy-6-((5-hydroxy-2-(4-methoxyphenyl)-4-oxo-4*H*-chromen-7-yl)oxy)tetrahydro-2*H*-pyran-2-yl)methoxy)tetrahydro-2*H*-pyran-3,4-diyl diacetate

Sub-class: Flavone glycoside

Chemical structure

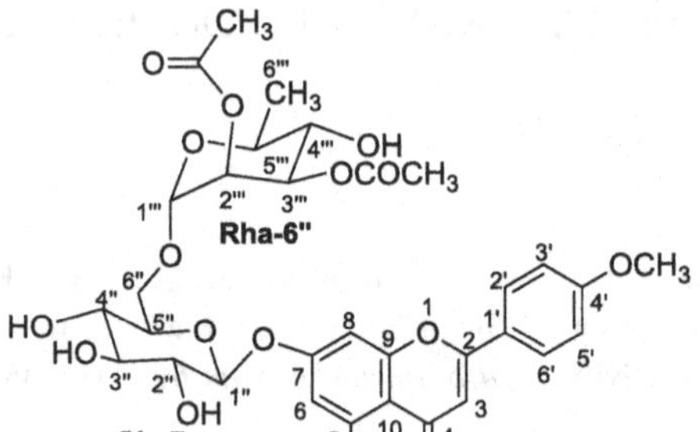

Source: *Dracocephalum peregrinum* (Family: Lamiaceae); whole plants

Molecular formula: $C_{32}H_{36}O_{16}$

Molecular weight: 676

State: Yellow amorphous powder

Specific rotation: $[\alpha]^{25}_{D}$ −78° (MeOH, c 0.1250)

Bioactivity studied: Anti-inflammatory

UV (MeOH): λ_{max} (log ε) 205 (3.49), 269 (1.93), 323 (2.37) nm

FT-IR (KBr): v_{max} 3421 (OH), 2923, 2852, 1741 (ester carbonyl), 1656 (α,β-unsaturated carbonyl), 1605 (aromatic unsaturation) cm^{-1}

^{1}H-NMR (DMSO-d_6, 600 MHz): δ 6.94 (1H, s, H-3), 6.43 (1H, d, J = 1.8 Hz, H-6), 6.83 (1H, d, J = 1.8 Hz, H-8), 8.05 (2H, d, J = 8.4 Hz, H-2′ and H-6′), 7.12 (2H, d, J = 8.0 Hz, H-3′ and H-5′), 12.9 (1H, br s, 5-OH), 3.88 (3H, br s, 4′-OCH_3), 5.16 (1H, d, J = 7.8 Hz, H-1″), 3.28 (1H, m, H-2″), 3.35 (1H, m, H-3″), 3.19 (1H, dd, J = 8.4, 8.4 Hz, H-4″), 3.70 (1H, dd, J = 8.4, 7.8 Hz, H-5″), 3.85 (1H, m, H-6″a), 3.60 (1H, dd, J = 3.6, 1.8 Hz, H-6″b), 4.67 (1H, s, H-1‴), 5.04 (1H, dd, J = 3.6, 1.8 Hz, H-2‴), 4.85 (1H, dd, J = 10.2, 3.6 Hz, H-3‴), 3.29 (1H, m, H-4‴), 3.64 (1H, dd, J = 15.6, 6.0 Hz, H-5‴), 1.14 (3H, m, H-6‴), 1.99 (3H, br s, 2‴-OCOCH_3), 1.82 (3H, br s, 3‴-OCOCH_3).

^{13}C-NMR (DMSO-d_6, 150 MHz): δ 163.8 (C-2), 103.7 (C-3), 182.0 (C-4), 161.1 (C-5), 99.6 (C-6), 162.8 (C-7), 94.4 (C-8), 156.7 (C-9), 105.2 (C-10), 122.7 (C-1′), 128.5 (C-2′), 114.6 (C-3′), 162.4 (C-4′), 114.6 (C-5′), 128.5 (C-6′), 55.6 (4′-OCH_3), 99.4 (C-1″), 73.0 (C-2″), 76.2 (C-3″), 69.5 (C-4″), 75.5 (C-5″), 65.8 (C-6″), 97.0 (C-1‴), 69.1 (C-2‴), 71.5 (C-3‴), 69.1 (C-4‴), 68.3 (C-5‴), 17.5 (C-6‴), 169.6 (2‴-OCOCH$_3$), 20.6 (2‴-OCOCH_3), 169.4 (3‴-OCOCH$_3$), 20.4 (3‴-OCOCH_3).

HMQC: δ 6.94 (H-3) vs δ 103.7 (C-3), δ 6.43 (H-6) vs δ 99.6 (C-6), δ 6.83 (H-8) vs δ 99.4 (C-8), δ 8.05 (H-2′/H-6′) vs δ 128.5 (C-2′/ C-6′), δ 7.12 (H-3′/H-5′) vs δ 114.6 (C-3′/ C-5′), δ 3.88 (4′-OCH_3) vs δ 55.6 (4′-OCH_3), δ 5.16 (H-1″) vs δ 99.4 (C-1″), δ 3.28 (H-2″) vs δ 73.0 (C-2″), δ 3.35 (H-3″) vs δ 76.2 (C-3″), δ 3.19 (H-4″) vs δ 69.5 (C-4″), δ 3.70 (H-5″) vs δ 75.5 (C-5″), δ 3.85 (H-6″a) and 3.60 (H-6″b) vs δ 65.8 (C-6″), δ 4.67 (H-1‴) vs δ 97.0 (C-1‴), δ 5.04 (H-2‴) vs δ 69.1 (C-2‴), δ 4.85 (H-3‴) vs δ 71.5 (C-3‴), δ 3.29 (H-4‴) vs δ 69.1 (C-4‴), δ 3.64 (H-5‴) vs δ 68.3 (C-5‴), δ 1.14 (H-6‴) vs δ 17.5 (C-6‴), δ 1.99 (2‴-OCOCH_3) vs δ 20.6 (2‴-OCOCH_3), δ 1.82 (3‴-OCOCH_3) vs δ 20.4 (3‴-OCOCH_3).

HMBC: δ 6.94 (H-3) *vs* δ 105.2 (C-10), δ 6.43 (H-6) *vs* δ 94.4 (C-8) and 105.2 (C-10), δ 6.83 (H-8) *vs* δ 94.6 (C-6) and 105.2 (C-10), δ 8.05 (H-2'/H-6') *vs* δ 162.4 (C-4'), δ 7.12 (H-3'/H-5') *vs* δ 122.7 (C-1'), δ 5.16 (H-1″) *vs* δ 162.8 (C-7) and 76.2 (C-3″), δ 3.28 (H-2″) *vs* δ 69.5 (C-4″), δ 3.35 (H-3″) *vs* δ 99.4 (C-1″) and 75.5 (C-5″), δ 3.19 (H-4″) *vs* δ 73.0 (C-2″), δ 3.70 (H-5″) *vs* δ 99.4 (C-1″) and 76.2 (C-3″), δ 3.85 (H-6″a) and 3.60 (H-6″b) *vs* δ 97.0 (C-1‴), δ 4.67 (H-1‴) *vs* δ 71.5 (C-3‴), δ 5.04 (H-2‴) *vs* 69.1 (C-4‴) and 169.6 (2‴-OCOCH₃), δ 4.85 (H-3‴) *vs* δ 97.0 (C-1‴) and 169.4 (3‴-OCOCH₃), δ 3.29 (H-4‴) *vs* δ 69.1 (C-2‴) and 17.5 (C-6‴), δ 1.99 (2‴-OCOC*H₃*) *vs* δ 169.6 (2‴-O*C*OCH₃), δ 1.82 (3‴-OCOC*H₃*) *vs* δ 169.4 (3‴-O*C*OCH₃) [selected 2D-correlations were shown].

HR-ESI-MS: *m/z* 699.18948 ([M + Na]⁺, Calcd. for $C_{32}H_{36}O_{16}Na$, 699.1901).

Reference

Fu P, Zhao C-C, Tang J, Shen Y-H, Xu X, Zhang W-D. (2009). New flavonoid glycosides and cyanogenic glycosides from *Dracocephalum peregrinum*. *Chem Pharm Bull* **57**: 207–210.

Phellamurin

IUPAC name: (2S,3S)-3,5-Dihydroxy-2-(4-hydroxyphenyl)-8-(3-methylbut-2-en-1-yl)-7-(((2S,3R,4S,5S,6R)-3,4,5-trihydroxy-6-(hydroxymethyl)tetrahydro-2H-pyran-2-yl)oxy)chroman-4-one

Sub-class: Flavanone glycoside

Chemical structure

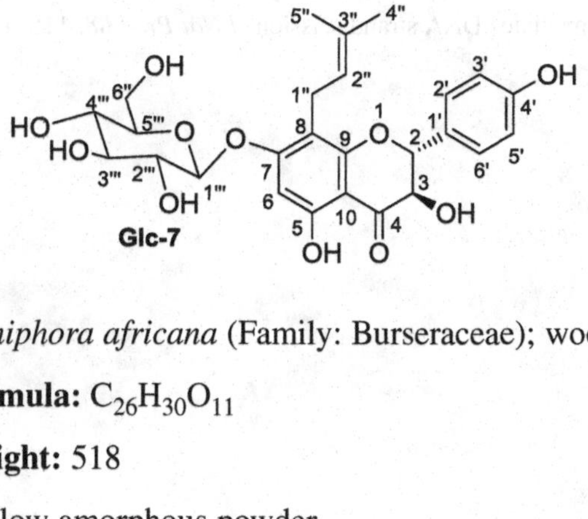

Glc-7

Source: *Commiphora africana* (Family: Burseraceae); wood stems

Molecular formula: $C_{26}H_{30}O_{11}$

Molecular weight: 518

State: Pale yellow amorphous powder

Melting point: 235–237 °C

Bioactivity studied: DNA cleavage activity

^{1}H-NMR (CD$_3$OD, 500 MHz): δ 4.99 (1H, d, J = 11.5 Hz, H-2β), 4.58 (1H, d, J = 11.5 Hz, H-3α), 6.28 (1H, s, H-6), 7.35 (2H, d, J = 8.5 Hz, H-2′ and H-6′), 6.82 (2H, d, J = 8.5 Hz, H-3′ and H-5′), 3.38 (2H, br d,

J = 7.0 Hz, H-1″), 5.22 (1H, br t, J = 7.0 Hz, H-2″), 1.64 (3H, s, H-4″), 1.78 (3H, s, H-5″), 5.01 (1H, d, J = 8.0 Hz, H-1‴), 3.26 (1H, m, H-2‴), 3.46 (1H, m, H-3‴), 3.64 (1H, m, H-4‴), 3.48 (1H, m, H-5‴), 3.88 (1H, dd, J = 11.5, 1.9 Hz, H-6‴a), 3.79 (1H, dd, J = 11.5, 6.0 Hz, H-6‴b).

^{13}C-NMR (CD$_3$OD, 125 MHz): δ 85.2 (C-2), 73.9 (C-3), 199.6 (C-4), 159.3 (C-5), 95.3 (C-6), 164.9 (C-7), 112.1 (C-8), 162.4 (C-9), 103.2 (C-10), 131.9 (C-1′), 130.4 (C-2′ and C-6′), 116.1 (C-3′ and C-5′), 161.4 (C-4′), 22.1 (C-1″), 123.7 (C-2″), 129.2 (C-3″), 18.0 (C-4″), 25.9 (C-5″), 101.4 (C-1‴), 74.8 (C-2‴), 78.0 (C-3‴), 71.2 (C-4‴), 78.3 (C-5‴), 62.4 (C-6‴).

Positive ion CI-MS (Methane): m/z (rel%) 519 [M + H]$^+$ (26), 518 [M]$^+$ (100), 356 (15), 355 (40), 337 (31).

Reference

Ma J, Jones S H, Hecht S M. (2005). A dihydroflavonol glucoside from *Commiphora africana* that mediates DNA strand scission. *J Nat Prod* **68**: 115–117.

(2*R*)-Phellodensin-F

IUPAC name: (*R*)-5-Hydroxy-2-(4-hydroxyphenyl)-8-(3-methylbut-2-en-1-yl)-7-(((2*S*,3*R*,4*S*,5*S*,6*R*)-3,4,5-trihydroxy-6-(hydroxymethyl)tetrahydro-2*H*-pyran-2-yl)oxy)chroman-4-one

Sub-class: Flavanone glycoside

Chemical structure

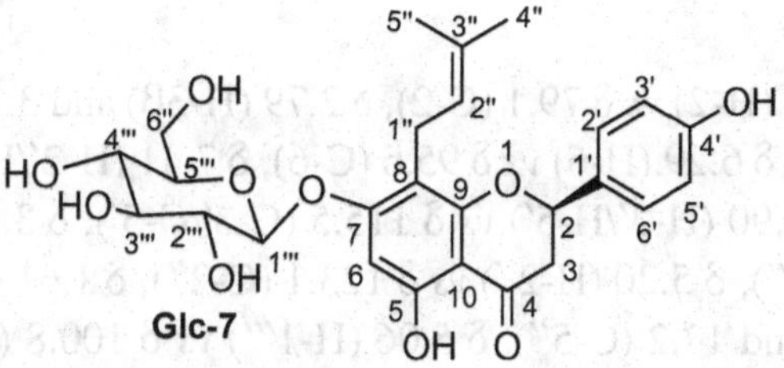

Source: *Phellodendron japonicum* Maxim. (Family: Rutaceae); leaves

Molecular formula: $C_{26}H_{30}O_{10}$

Molecular weight: 502

State: White powder

Melting point: 220–221 °C

Specific rotation: $[\alpha]_D^{25}$ –67.5° (MeOH, *c* 0.15)

UV (MeOH): λ_{max} 286, 342 nm

231

IR (KBr): v_{max} 3380 (OH), 2921, 1635 (α,β-unsaturated carbonyl), 1371, 1074 cm^{-1}

^{1}H-NMR (Acetone-d_6, 400 MHz): δ 5.48 (1H, dd, J = 12.8, 3.6 Hz, H-2), 2.79 (1H, m, H-3β), 6.29 (1H, s, H-6), 3.15-3.22 (2H, m, H-3α, 1″), 7.41 (2H, d, J = 8.4 Hz, H-2′ and H-6′), 6.90 (2H, d, J = 8.4 Hz, H-3′ and H-5′), 3.52-3.36 (1H, dd, J = 13.6, 7.6 Hz, H-1″), 5.20 (1H, br t, J = 7.6 Hz, H-2″), 1.61 (6H, s, H-4″ and H-5″), 5.06 (1H, d, J = 7.6 Hz, H-1‴), 3.52-3.54 (2H, m, H-2‴ and H-3‴), 3.46 (1H, m, H-4‴), 3.60 (1H, m, H-5‴), 3.71 (1H, m, H-6‴a), 3.98 (1H, dd, J = 8.4, 2.8 Hz, H-6‴b), 12.09 (1H, s, 5-OH), 8.52 (1H, br s, 4′-OH), 4.29 (1H, d, J = 4.4 Hz, 2‴-OH), 4.38 (1H, d, J = 1.6 Hz, 3‴-OH), 4.47 (1H, d, J = 2.8 Hz, 4‴-OH).

^{13}C-NMR (Acetone-d_6, 100 MHz): δ 79.1 (C-2), 42.9 (C-3), 197.6 (C-4), 162.3 (C-5), 95.6 (C-6), 163.7 (C-7), 109.6 (C-8), 159.6 (C-9), 103.7 (C-10), 130.2 (C-1′), 128.2 (C-2′ and C-6′), 115.5 (C-3′ and C-5′), 157.9 (C-4′), 21.8 (C-1″), 123.1 (C-2″), 130.5 (C-3″), 25.2 (C-4″), 17.2 (C-5″), 100.8 (C-1‴), 74.0 (C-2‴), 77.2 (C-3‴), 70.5 (C-4‴), 77.3 (C-5‴), 61.8 (C-6‴).

HMQC: δ 5.58 (H-2) *vs* δ 79.1 (C-2), δ 2.79 (H-3β) and 3.15-3.22 (H-3α) *vs* δ 42.9 (C-3), δ 6.29 (H-6) *vs* δ 95.6 (C-6), δ 7.41 (H-2′/H-6′) *vs* δ 128.2 (C-2′/C-6′), δ 6.90 (H-3′/H-5′) *vs* δ 115.5 (C-3′/C-5′), δ 3.15-3.22 (H-1″) *vs* δ 123.1 (C-1″), δ 5.20 (H-2″) *vs* δ 123.1 (C-2″), δ 1.61 (H-4″/H-5″) *vs* δ 25.2 (C-4″) and 17.2 (C-5″), δ 5.06 (H-1‴) *vs* δ 100.8 (C-1‴), δ 3.52-3.54 (H-2‴/H-3‴) *vs* δ 74.0 (C-2‴) and 77.2 (C-3‴), δ 3.60 (H-5‴) *vs* δ 77.3 (C-5‴), δ 3.71 (H-6‴a) and 3.98 (H-6‴b) *vs* δ 61.8 (C-6‴).

FAB-MS: *m/z* (%rel): 503 [M + H]$^+$ (32), 341 (100), 285 (38), 219 (22), 185 (71), 165 (58), 149 (29).

HR-FABMS: *m/z* 503.1919 ([M + H]$^+$, Calcd. for $C_{26}H_{31}O_{10}$, 503.1917).

Reference

Chiu C-Y, Li C-Y, Chiu C-C, Niwa M, Kitanaka S, Damu A G, Lee E-J, Wu T-S. (2005). Constituents of leaves of *Phellodendron japonicum* Maxim. and their antioxidant activity. *Chem Pharm Bull* **53**: 1118–1121.

Pisonianone [5,7,2'-Trihydroxy-6-methoxy-8-methylisoflavone]

IUPAC name: 5,7-Dihydroxy-3-(2-hydroxyphenyl)-6-methoxy-8-methyl-4*H*-chromen-4-one

Sub-class: Isoflavone

Chemical structure

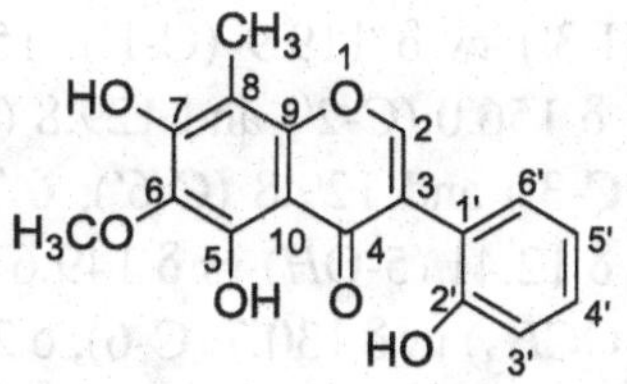

Source: *Pisonia aculeate* Linn. (Family: Nyctaginaceae); stems and roots

Molecular formula: $C_{17}H_{14}O_6$

Molecular weight: 314

State: Yellow needles (MeOH)

Melting point: 184–186 °C

Bioactivity studied: Anti-tubercular

UV (MeOH): λ_{max} (log ε) 267 (4.31), 340 (3.66), 395 (3.21) nm

FT-IR (KBr): ν_{max} 3352 (OH), 1652 (C=O), 1614, 1569, 1486 cm^{-1}

¹H-NMR (CDCl₃, 400 MHz): δ 8.11 (1H, s, H-2), 7.10 (1H, dd, J = 7.8, 1.2 Hz, H-3′), 7.36 (1H, td, J = 7.8, 1.8 Hz, H-4′), 7.00 (1H, td, J = 7.8, 1.2 Hz, H-5′), 7.17 (1H, dd, J = 7.8, 1.8 Hz, H-6′), 12.41 (1H, br s, 5-OH), 6.81(1H, br s, 7-OH), 8.20 (1H, br s, 2′-OH), 4.06 (3H, s, 6-OCH_3), 2.29 (3H, s, 8-CH_3).

¹³C-NMR (CDCl₃, 100 MHz): δ 155.9 (C-2), 122.7 (C-3), 182.6 (C-4), 149.6 (C-5), 130.7 (C-6), 154.2 (C-7), 102.3 (C-8), 151.3 (C-9), 105.6 (C-10), 119.9 (C-1′), 156.0 (C-2′), 119.6 (C-3′), 130.7 (C-4′), 121.2 (C-5′), 129.8 (C-6′), 60.9 (6-OCH_3), 7.6 (8-CH_3).

HMQC: δ 8.11 (H-2) *vs* δ 155.9 (C-2), δ 7.10 (H-3′) *vs* δ 119.6 (C-3′), δ 7.36 (H-4′) *vs* δ 130.7 (C-4′), δ 7.00 (H-5′) *vs* δ 121.2 (C-5′), δ 7.17 (H-6′) *vs* δ 129.8 (C-6′), δ 4.06 (6-OCH_3) *vs* δ 60.9 (6-OCH_3), δ 2.29 (8-CH_3) *vs* δ 7.6 (8-CH_3).

¹H-¹H COSY: δ 7.10 (H-3′) *vs* δ 7.36 (H-4′), δ 7.36 (H-4′) *vs* δ 7.10 (H-3′) and 7.00 (H-5′), δ 7.00 (H-5′) *vs* δ 7.36 (H-4′) and 7.17 (H-6′), δ 7.17 (H-6′) *vs* δ 7.00 (H-5′).

HMBC: δ 8.11 (H-2) *vs* δ 122.7 (C-3), 182.6 (C-4), 151.3 (C-9), and 119.9 (C-1′), δ 7.10 (H-3′) *vs* δ 119.9 (C-1′), 156.0 (C-2′), and 121.2 (C-5′), δ 7.36 (H-4′) *vs* δ 156.0 (C-2′) and 129.8 (C-6′), δ 7.30 (H-5′) *vs* δ 119.9 (C-1′), 119.6 (C-3′) and 129.8 (C-6′), δ 7.17 (H-6′) *vs* δ 122.7 (C-3) and 156.0 (C-2′), δ 12.41 (5-OH) *vs* δ 149.6 (C-5), 130.7 (C-6) and 105.6 (C-10), δ 4.06 (6-OCH_3) *vs* δ 130.7 (C-6), δ 2.29 (8-CH_3) *vs* δ 154.2 (C-7), 102.3 (C-8) and 151.3 (C-9).

NOSY: δ 7.10 (H-3′) *vs* δ 7.36 (H-4′), δ 7.36 (H-4′) *vs* δ 7.10 (H-3′) and 7.00 (H-5′), δ 7.00 (H-5′) *vs* δ 7.36 (H-4′) and 7.17 (H-6′), δ 7.17 (H-6′) *vs* δ 7.00 (H-5′) and 8.11 (H-2).

ESIMS: m/z 337 ([M + Na]⁺,

HR-ESI-MS: m/z 337.0690 ([M + Na]⁺, Calcd. for $C_{17}H_{14}O_6Na$, 337.0688).

Reference

Wu M-C, Peng C-F, Chen I-S, and Tsai I-L. (2011). Antitubercular chromones and flavonoids from *Pisonia aculeate*. *J Nat Prod* **74**: 976–982.

Pisonivanol [(2*R*,3*R*)-3,7-dihydroxy-5,6-dimethoxyflavanone]

IUPAC name: (2R,3R)-3,7-Dihydroxy-5,6-dimethoxy-2-phenylchroman-4-one

Sub-class: Hydroxyflavanone

Chemical structure

Source: *Pisonia* Linn. (Family: Nyctaginaceae); stems and roots

Molecular formula: $C_{17}H_{16}O_6$

Molecular weight: 316

State: Colorless needles (MeOH)

Melting point: 167–169 °C

Bioactivity studied: Anti-tubercular

Specific rotation: $[\alpha]^{24}_D$ +27.0° (MeOH, *c* 0.05)

UV (MeOH): λ_{max} (log ε) 236 sh (4.03), 279 (4.01), 318 (3.54) nm

FT-IR (KBr): ν_{max} 3391 (OH), 1676 (C=O), 1607, 1576, 1474 cm^{-1}

¹H-NMR (CDCl₃, 400 MHz): δ 5.03 (1H, d, J = 12.4 Hz, H-2β), 4.47 (1H, dd, J = 12.4, 1.9 Hz, H-3α), 6.39 (1H, s, H-8), 7.56-7.54 (2H, m, H-2′ and H-6′), 7.48-7.26 (3H, m, H-3′, H-4′ and H-5′), 3.96 (1H, d, J = 1.9 Hz, 3-O*H*), 6.64 (1H, br s, 7-O*H*), 3.98 (3H, s, 5-OC*H₃*), 3.94 (3H, s, 6-OC*H₃*).

¹³C-NMR (CDCl₃, 100 MHz): δ 83.3 (C-2), 73.1 (C-3), 191.3 (C-4), 152.9 (C-5), 135.5 (C-6), 156.7 (C-7), 99.5 (C-8), 159.8 (C-9), 106.4 (C-10), 136.4 (C-1′), 127.5 (C-2′), 128.7 (C-3′), 129.2 (C-4′), 128.7 (C-5′), 127.5 (C-6′), 61.6 (5-OCH₃ and 6-OCH₃).

HMBC: δ 5.03 (H-2β) *vs* δ 73.1 (C-3), 191.3 (C-4) and 127.5 (C-2′/C-6′), δ 5.03 (H-3α) *vs* δ 83.3 (C-2) and 136.4 (C-1′), δ 6.39 (H-8) *vs* δ 135.5 (C-6), 156.7 (C-7), 159.8 (C-9) and 106.4 (C-10), δ 3.98 (5-OC*H₃*) *vs* δ 152.9 (C-5), δ 3.94 (6-OC*H₃*) *vs* δ 135.5 (C-6), δ 6.64 (7-O*H*) *vs* δ 156.7 (C-7) and 99.5 (C-8) (selected 2D-correlations were shown).

ESIMS: *m/z* 339 ([M + Na]⁺,

HR-ESI-MS: *m/z* 339.0843 ([M + Na]⁺, Calcd. for $C_{17}H_{16}O_6Na$, 339.0845).

Reference

Wu M-C, Peng C-F, Chen I-S, and Tsai I-L. (2011). Antitubercular chromones and flavonoids from *Pisonia aculeate*. *J Nat Prod* **74**: 976–982.

Pisonivanone [(2*S*)-5,7,2'-trihydroxy-8-methylflavanone]

IUPAC name: (*S*)-5,7-Dihydroxy-2-(2-hydroxyphenyl)-8-methylchroman-4-one

Sub-class: Flavanone

Chemical structure

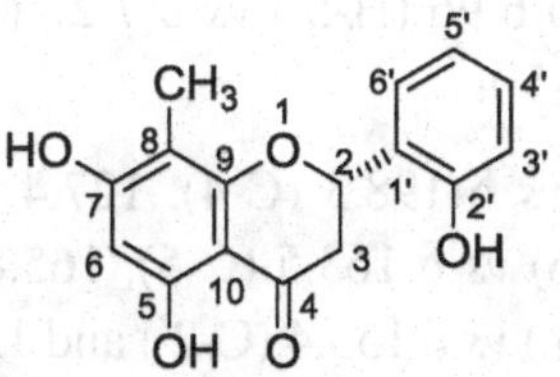

Source: *Pisonia aculeate* Linn. (Family: Nyctaginaceae); stems and roots

Molecular formula: $C_{16}H_{14}O_5$

Molecular weight: 286

State: Colorless needles (EtOAc)

Melting point: 192–194 °C

Bioactivity studied: Anti-tubercular

Specific rotation: $[\alpha]^{24}_{D}$ −102.0° (MeOH, *c* 0.1)

UV (MeOH): λ_{max} (log ε) 238 sh (3.69), 289 (4.12), 335 (3.45) nm

FT-IR (KBr): ν_{max} 3374 (OH), 1642 (C=O), 1600, 1505, 1457 cm^{-1}

¹H-NMR (Acetone-d_6, 400 MHz): δ 5.78 (1H, dd, J = 12.8, 3.1 Hz, H-2β), 2.88 (1H, dd, J = 17.1, 3.1 Hz, H-3a), 3.04 (1H, dd, J = 17.1, 12.8 Hz, H-3b), 6.05 (1H, s, H-6), 6.95 (1H, dd, J = 8.0, 1.0 Hz, H-3′), 7.22 (1H, td, J = 8.0, 1.7 Hz, H-4′), 6.96 (1H, td, J = 8.0, 1.0 Hz, H-5′), 7.59 (1H, dd, J = 8.0, 1.7 Hz, H-6′), 12.13 (1H, br s, 5-OH), 8.80 (1H, br s, 7-OH)*, 9.57 (1H, br s, 2′-OH)*, 2.00 (3H, s, 8-CH_3) (*interchangeable).

¹³C-NMR (Acetone-d_6, 100 MHz): δ 75.9 (C-2), 43.1 (C-3), 198.3 (C-4), 163.5 (C-5), 96.9 (C-6), 165.8 (C-7), 104.6 (C-8), 162.0 (C-9), 103.8 (C-10), 127.4 (C-1′), 155.4 (C-2′), 116.9 (C-3′), 130.7 (C-4′), 121.4 (C-5′), 128.0 (C-6′), 8.4 (8-CH_3).

HMQC: δ 5.78 (H-2β) vs δ 75.9 (C-2), δ 2.88 (H-3a) and 3.04 (H-3b) vs δ 43.1 (C-3), δ 6.05 (H-6) vs δ 96.9 (C-6), δ 6.95 (H-3′) vs δ 116.9 (C-3′), δ 7.22 (H-4′) vs δ 130.7 (C-4′), δ 6.96 (H-5′) vs δ 121.4 (C-5′), δ 7.59 (H-6′) vs δ 128.0 (C-6′), δ 2.00 (8-CH_3) vs δ 8.4 (8-CH_3).

¹H-¹H COSY: δ 6.95 (H-3′) vs δ 7.22 (H-4′), δ 7.22 (H-4′) vs δ 6.95 (H-3′) and 6.96 (H-5′), δ 6.96 (H-5′) vs δ 7.22 (H-4′) and 7.59 (H-6′), δ 7.59 (H-6′) vs δ 6.96 (H-5′).

HMBC: δ 5.78 (H-2β) vs δ 198.3 (C-4), 127.4 (C-1′), 155.4 (C-2′) and 128.0 (C-6′), δ 6.05 (H-6) vs δ 163.5 (C-5), 165.8 (C-7), 104.6 (C-8) and 103.8 (C-10), δ 6.95 (H-3′) vs δ 155.4 (C-2′) and 121.4 (C-5′), δ 7.22 (H-4′) vs δ 155.4 (C-2′), 116.9 (C-3′) and 128.0 (C-6′), δ 6.96 (H-5′) vs δ 127.4 (C-1′), 116.9 (C-3′) and 130.7 (C-4′), δ 7.59 (H-6′) vs δ 75.9 (C-2), 155.4 (C-2′), and 130.7 (C-4′), δ 12.13 (5-OH) vs δ 163.5 (C-5), 96.9 (C-6) and 103.8 (C-10), δ 2.00 (8-CH_3) vs δ 165.8 (C-7), 104.6 (C-8) and 162.0 (C-9).

NOSY: δ 5.78 (H-2β) vs δ δ 2.88 (H-3a) and 3.04 (H-3b), δ 6.95 (H-3′) vs δ 7.22 (H-4′), δ 6.96 (H-5′) vs δ 7.59 (H-6′).

ESIMS: m/z 309 ([M + Na]⁺,

HR-ESI-MS: m/z 309.0741 ([M + Na]⁺, Calcd. for $C_{16}H_{14}O_5Na$, 309.0739).

Reference

Wu M-C, Peng C-F, Chen I-S, and Tsai I-L. (2011). Antitubercular chromones and flavonoids from *Pisonia aculeate*. *J Nat Prod* **74**: 976–982.

Platyisoflavanone A [(*S*)-5,7-dihydroxy-2′,4′-dimethoxy-3′-(3″-methylbut-2″-enyl)-isoflavanone]

IUPAC name: (*S*)-3-(2,4-dimethoxy-3-(3-methylbut-2-en-1-yl)phenyl)-5,7-dihydroxychroman-4-one

Sub-class: Isoflavanone

Chemical structure

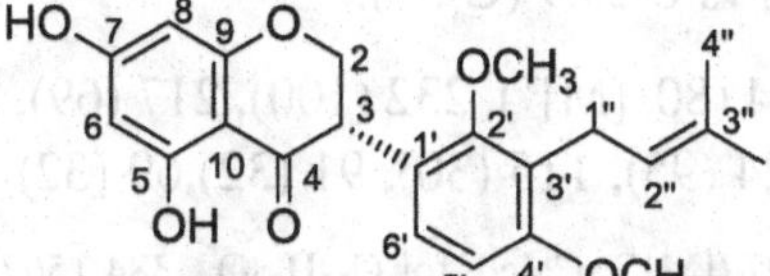

Source: *Platycelphium voënse* (Family: Leguminosae); barks

Molecular formula: $C_{22}H_{24}O_6$

Molecular weight: 384

State: White amorphous solid

Melting point: 160–164 °C

Bioactivity studied: Anti-TB activity against *Mycobacterium tuberculosis* (MIC = 23.7 μM); cytotoxic against vero cells (IC_{50} = 21.1 μM).

Specific rotation: $[\alpha]_D^{24}$ +14.1° (CH$_2$Cl$_2$, *c* 1% v/v)

UV (CH$_2$Cl$_2$): λ_{max} 288 nm

^{1}H-NMR (CD$_2$Cl$_2$, 600 MHz): δ 4.48 (1H, dd, J = 11.1, 11.2 Hz, H-2a), 4.46 (1H, dd, J = 11.1, 5.6 Hz, H-2b), 4.38 (1H, dd, J = 11.2, 5.6 Hz, H-3β), 5.97 (1H, d, J = 2.3 Hz, H-6), 5.95 (1H, d, J = 2.3 Hz, H-8), 6.92 (1H, d, J = 8.5 Hz, H-6'), 12.18 (1H, s, 5-O*H*), 3.71 (3H, s, 2'-OC*H$_3$*), 3.80 (3H, s, 4'-OC*H$_3$*), 3.32 (1H, dd, J = 14.3, 6.6 Hz, H-1″a), 3.38 (1H, dd, J = 14.4, 6.6 Hz, H-1″b), 5.25 (1H, t, J = 6.7 Hz, H-2″), 1.77 (3H, s, H-4″), 1.68 (3H, s, H-5″).

^{13}C-NMR (CD$_2$Cl$_2$, 150 MHz): δ 71.6 (C-2), 45.9 (C-3), 198.2 (C-4), 164.0 (C-5), 96.7 (C-6), 164.9 (C-7), 95.5 (C-8), 163.9 (C-9), 103.8 (C-10), 120.5 (C-1'), 157.9 (C-2'), 124.2 (C-3'), 158.9 (C-4'), 107.2 (C-5'), 127.4 (C-6'), 62.4 (2'-OCH$_3$), 56.0 (4'-OCH$_3$), 23.7 (C-1″), 123.1 (C-2″), 131.8 (C-3″), 18.1 (C-4″), 25.9 (C-5″).

HMQC: δ 4.48 (H-2a) and 4.46 (H-2b) vs δ 71.6 (C-2), δ 4.38 (H-3) vs δ 45.9 (C-3), δ 5.97 (H-6) vs δ 96.7 (C-6), δ 5.95 (H-8) vs δ 95.5 (C-8), δ 6.92 (H-6') vs δ 127.4 (C-6'), δ 3.71 (2'-OC*H$_3$*) vs δ 62.4 (2'-OCH$_3$), δ 3.80 (4'-OC*H$_3$*) vs δ 56.0 (4'-OCH$_3$), δ 3.32 (H-1″a) and 3.38 (H-1″b) vs δ 23.7 (C-1″), δ 5.25 (H-2″) vs δ 123.1 (C-2″), δ 1.77 (H-4″) vs δ 18.1 (C-4″), δ 1.68 (H-5″) vs δ 25.9 (C-5″).

EIMS: m/z (%rel) 384 (80, [M]$^+$), 232 (100), 217 (69), 205 (83), 189 (47), 177 (54), 152 (65),124 (95), 115 (30), 91 (32),69 (32), 51 (53), 45 (54).

HR-ESIMS: m/z 384.1597 ([M]$^+$), Calcd. for C$_{22}$H$_{24}$O$_6$, 384.1567).

Reference

Gumula I, Heydenreich M, Derese S, Ndiege IO, Yenesew A. (2012). Four isoflavanones from the stem bark of *Platycelphium voënse*. *Phytochemistry Lett* **5**: 150–154.

Polygonflavanol A

IUPAC name: (2S,3R,4S,5S,6R)-2-(2-((2S,3R,3aS,4R,5R)-5-(3,4-dihydroxyphenyl)-4,8-dihydroxy-2-(4-hydroxyphenyl)-3,3a,4,5-tetrahydro-2H-pyrano[4,3,2-de]chromen-3-yl)-4,6-dihydroxyphenoxy)-6-(hydroxymethyl)tetrahydro-2H-pyran-3,4,5-triol

Sub-class: Flavonostilbene glycoside

Chemical structure

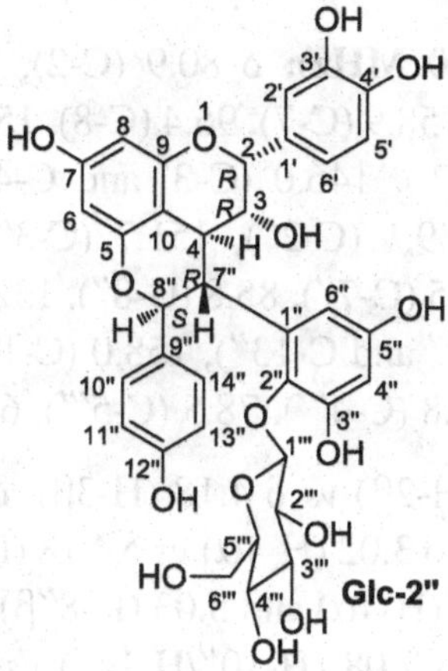

Source: *Polygonum multiflorum* Thunb. (Family: Polygonaceae); roots

Molecular formula: $C_{35}H_{34}O_{15}$

Molecular weight: 694

State: Brown oil

241

Bioactivity studied: Anti-inflammatory

Specific rotation: $[\alpha]_D^{19}$ +9.46° (MeOH, c 0.01)

CD (CH$_3$CN): λ_{max} ($\Delta\varepsilon$) 245.8 (−8.60), 286.7 (−5.07), 242 (−3.34) nm

UV (MeOH): λ_{max} (log ε) 205 (3.68), 280 (2.77) nm

IR (KBr): v_{max} 3444 (OH), 1618, 1508, 1462 cm^{-1}

^{1}H-NMR (CD$_3$OD, 400 MHz): δ 4.97 (1H, d, J = 3.8 Hz, H-2β), 4.18 (1H, dd, J = 7.0. 3.8 Hz, H-3β), 3.02 (1H, dd, J = 11.3, 7.0 Hz, H-4α), 5.93 (1H, d, J = 2.2 Hz, H-6), 6.01 (1H, d, J = 2.2 Hz, H-8), 6.89 (1H, s, H-2′), 6.71 (1H, d, J = 8.5 Hz, H-5′), 6.73 (1H, d, J = 8.5 Hz, H-6′), 6.09 (1H, d, J = 2.6 Hz, H-4″), 6.07 (1H, d, J = 2.6 Hz, H-6″), 4.30 (1H, t, J = 10.4 Hz, H-7″β), 5.03 (1H, d, J = 10.4 Hz, H-8″α), 7.08 (2H, d, J = 8.5 Hz, H-10″ and H-14″), 6.58 (2H, d, J = 8.5 Hz, H-11″ and H-13″), 4.46 (1H, d, J = 7.9 Hz, H-1‴), 3.46 (1H, overlapped signal, H-2‴), 3.34 (1H, overlapped signal, H-3‴), 3.43 (1H, overlapped signal, H-4‴), 3.37 (1H, overlapped signal, H-5‴), 3.79 (1H, d, J = 12.0 Hz, H-6‴a), 3.72 (1H, dd, J = 12.0, 4.3 Hz, H-6‴b).

^{13}C-NMR (CD$_3$OD, 100 MHz): δ 80.9 (C-2), 71.9 (C-3), 41.3 (C-4), 157.1 (C-5), 96.8 (C-6), 158.9 (C-7), 96.4 (C-8), 156.9 (C-9), 102.0 (C-10), 131.4 (C-1′), 115.7 (C-2′), 146.0 (C-3′ and C-4′), 116.0 (C-5′), 120.6 (C-6′), 136.8 (C-1″), 139.1 (C-2″), 151.7 (C-3″), 103.3 (C-4″), 155.9 (C-5″), 106.1 (C-6″), 40.5 (C-7″), 85.8 (C-8″), 132.0 (C-9″), 130.6 (C-10″ and C-14″), 115.5 (C-11″ and C-13″), 158.0 (C-12″), 107.7 (C-1‴), 75.4 (C-2‴), 78.3 (C-3‴), 70.8 (C-4‴), 78.5 (C-5‴), 62.3 (C-6‴).

^{1}H-^{1}H COSY: δ 4.97 (H-2β) *vs* δ 4.18 (H-3β), δ 4.18 (H-3β) *vs* δ 4.97 (H-2β) and 3.02 (H-4α), δ 3.02 (H-4α) *vs* δ 4.18 (H-3β) and 4.30 (H-7″β), δ 4.30 (H-7″β) *vs* δ 3.02 (H-4α) and 5.03 (H-8″β), δ 6.71 (H-5′) *vs* δ 6.73 (H-6′), and *vice versa*, δ 7.08 (H-10″/H-14″) *vs* δ 6.58 (H-11″/H-13″), and *vice versa*, δ 4.46 (H-1‴) *vs* δ 3.46 (H-2‴), δ 3.46 (H-2‴) *vs* δ δ 4.46 (H-1‴) and 3.34 (H-3‴), δ 3.34 (H-3‴) *vs* δ 3.46 (H-2‴) and 3.43 (H-4‴), δ 3.43 (H-4‴) *vs* δ 3.34 (H-3‴) and 3.37 (H-5‴), δ 3.37 (H-5‴) *vs* δ 3.43 (H-4‴), 3.79 (H-6‴a) and 3.72 (H-6‴b), δ 3.79 (H-6‴a) and 3.72 (H-6‴b) *vs* δ 3.37 (H-5‴).

HMBC: δ 4.97 (H-2β) *vs* δ 41.3 (C-4) and 120.6 (C-6′), δ 6.89 (H-2′) *vs* δ 80.9 (C-2), δ 5.03 (H-7″α) *vs* δ 106.1 (C-6″), δ 5.03 (H-8″α) *vs* δ 41.3 (C-4) and 130.6 (C-10″), δ 5.03 4.46 (H-1‴) *vs* δ 139.1 (C-2″) (selected HMBC correlations were shown).

ROSEY (NOE): δ 4.97 (H-2β) *vs* δ 4.18 (H-3β), δ 4.18 (H-3β) *vs* δ 4.97 (H-2β) and 4.30 (H-7″β), β), δ 3.02 (H-4α) *vs* δ 5.03 (H-8″α).

HR-ESIMS: *m/z* 717.1789 ([M + Na]$^+$, Calcd. for $C_{35}H_{34}O_{15}Na$, 717.1190).

Reference

Chen L-L, Huang X-J, Li M-M, Oua G-M, Zhao B-X, Chen M-F, Zhang Q-W, Wang Y, Ye W-C. (2012). Polygonflavanol A, a novel flavonostilbene glycoside from the roots of *Polygonum multiflorum*. *Phytochemistry Lett* **5**: 756–760.

Pongamone A [7-(γ,γ-Dimethylallyloxy)-8,4'-dimethoxy-isoflavone]

IUPAC name: 8-Methoxy-3-(4-methoxyphenyl)-7-((3-methylbut-2-en-1-yl)oxy)-4H-chromen-4-one

Sub-class: Isoflavone

Chemical structure

Source: *Pongamia pinnata* (Synonym: *P. pinnata*) (Family: Leguminosae); stems

Molecular formula: $C_{22}H_{22}O_5$

Molecular weight: 366

State: Amorphous powder

Bioactivity studied: Antiviral [IC_{50} >10 μg/mL against Duck hepatitis B virus: DNA polymerase and reverse transcriptase (DHBV RCs DNAP) and HIV-1 RT]

UV (MeOH): λ_{max} 257, 305 nm

IR (KBr): v_{max} 3029, 1642 (α,β-unsaturated carbonyl), 1603 (aromatic unsaturation), 1556, 1510, 1449, 1374, 1281, 1178, 1054 cm^{-1}

^{1}H-NMR (CDCl$_3$, 500 MHz): δ 8.02 (1H, s, H-2), 8.03 (1H, d, J = 9.0 Hz, H-5), 7.07 (1H, d, J = 9.0 Hz, H-6), 7.52 (2H, d, J = 8.5 Hz, H-2′ and H-6′), 7.00 (2H, d, J = 8.5 Hz, H-3′ and H-5′), 4.75 (2H, d, J = 7.5 Hz, H-1″), 5.55 (1H, t, J = 7.5 Hz, H-2″), 1.81 (3H, s, 4″-CH_3), 1.84 (3H, s, 5″-CH_3), 4.03 (3H, s, 8-OCH_3), 3.87 (3H, s, 4′-OCH_3).

^{13}C-NMR (CDCl$_3$, 125 MHz): δ 152.2 (C-2), 124.0 (C-3), 176.0 (C-4), 121.5 (C-5), 111.6 (C-6), 157.0 (C-7), 137.0 (C-8), 151.0 (C-9), 119.0 (C-10), 124.0 (C-1′), 130.2 (C-2′, C-6′), 114.0 (C-3′, C-5′), 160.0 (C-4′), 66.3 (C-1″), 119.1 (C-2″), 139.0 (C-3″), 25.8 (C-4″), 18.3 (C-5″), 61.5 (C$_8$-OCH_3), 56.4 (C$_4$′-OCH_3).

HMQC: δ 8.02 (H-2) vs δ 152.2 (C-2), δ 8.03 (H-5) vs δ 121.5 (C-5), δ 7.07 (H-6) vs δ 111.6 (C-6), δ 7.52 (H-2′/H-6′) vs δ 130.2 (C-2′/C-6′), δ 7.00 (H-3′/ H-5′) vs δ 114.0 (C-3′/C-5′), δ 4.75 (H-1″) vs δ 124.0 (C-1″), δ 5.55 (H-2″) vs δ 119.1 (C-2″), δ 1.81 (H-4″) vs δ 25.8 (C-4″), δ 1.84 (H-5″) vs δ 18.3 (C-5″), δ 4.03 (8-OCH_3) vs δ 61.5 (C$_8$-OCH_3), δ 3.87 (4′-OCH_3) vs δ 56.4 (C$_4$′-OCH_3).

EIMS: m/z 366 ([M]$^+$), 312, 298, 297, 283, 255, 252, 240, 213, 166, 149, 138, 132, 117.

HR-FABMS: m/z 367.1537 ([M + H]$^+$, Calcd. for C$_{22}$H$_{23}$O$_5$, 367.1545).

Reference

Li L, Li X, Shi C, Deng Z, Fu H, Proksch P, Lin W. (2006). Pongamone A–E, five flavonoids from the stems of a mangrove plant, *Pongamia pinnata*. *Phytochemistry* **67**: 1347–1352.

Pongamone C [(2*S*)-3', 4'-Methylenedioxy-6-γ, γ-dimethylallylpyrano [5'',6'':7,8]-flavanone]

IUPAC name: (*S*)-2-(Benzo[*d*][1,3]dioxol-5-yl)-8,8-dimethyl-6-(3-meth-ylbut-2-en-1-yl)-2,3-dihydropyrano[2,3-*f*]chromen-4(8*H*)-one

Sub-class: Pyrano-fused flavanone

Chemical structure

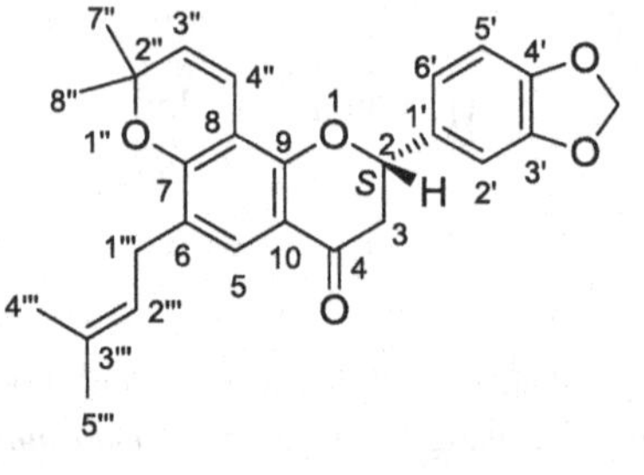

Source: *Pongamia pinnata* (Synonym: *P. pinnata*) (Family: Leguminosae); stems

Molecular formula: $C_{26}H_{26}O_5$

Molecular weight: 418

State: Pale yellow amorphous powder

Specific rotation: $[\alpha]^{20}_{D}$ −55° (CHCl$_3$, c 0.05)

Bioactivity studied: Antiviral [IC$_{50}$ >10 µg/mL against Duck hepatitis B virus: DNA polymerase and reverse transcriptase (DHBV RCs DNAP) and HIV-1 RT]

UV (MeOH): λ_{max} 234, 260, 340 nm

CD (MeCN): CE 307 (−6.68), 345 (+2.15) nm

IR (KBr): ν_{max} 3075, 1710 (carbonyl), 1673 (unsaturation), 1623 (aromatic unsaturation), 1465, 1413, 1378, 1196, 1094, 1066, 999, 903 cm^{-1}

^{1}H-NMR (CDCl$_3$, 500 MHz): δ 5.38 (1H, dd, J = 13.0, 3.0 Hz, H-2), 3.00 (1H, dd, J = 16.5, 13.0 Hz, H-3a), 2.79 (1H, dd, J = 16.5, 13.0 Hz, H-3b), 7.62 (1H, s, H-5), 7.01 (1H, d, J = 1.5 Hz, H-2′), 6.86 (1H, d, J = 8.0 Hz, H-5′), 6.92 (1H, dd, J = 1.5 Hz, H-6′), 5.59 (1H, d, J = 10.0 Hz, H-3″), 6.65 (1H, d, J = 10.0 Hz, H-4″), 1.46 (3H, s, 7″-CH_3), 1.49 (3H, s, 8″-CH_3), 3.38 (2H, d, J = 7.5 Hz, H-1‴), 5.26 (1H, t, J = 7.5 Hz, H-2‴), 1.75 (6H, s, 4‴-CH_3 and 5‴-CH_3), 6.03 (2H, s, −OCH_2O−).

^{13}C-NMR (CDCl$_3$, 125 MHz): δ 77.6 (C-2), 44.5 (C-3), 191.0 (C-4), 125.0 (C-5), 117.0 (C-6), 156.8 (C-7), 115.0 (C-8), 157.0 (C-9), 115.6 (C-10), 131.3 (C-1′), 107.0 (C-2′), 148.0 (C-3′, C-4′), 108.9 (C-5′), 120.0 (C-6′), 77.5 (C-2″), 129.0 (C-3″), 115.0 (C-4″), 28.4 (C-7″, C-8″), 28.0 (C-1‴), 121.0 (C-2‴), 132.0 (C-3‴), 27.0 (C-4‴), 17.8 (C-5‴), 101.5 (−OCH_2O−).

HMQC: δ 5.38 (H-2) *vs* δ 77.6 (C-2), δ 3.00 (H-3a) and 2.79 (H-3b) *vs* δ 44.5 (C-3), δ 7.62 (H-5) *vs* δ 125.0 (C-5), δ 7.01 (H-2′) *vs* δ 107.0 (C-2′), δ 6.86 (H-5′) *vs* δ 108.9 (C-5′), δ 6.92 (H-6′) *vs* δ 120.0 (C-6′), δ 5.59 (H-3″) *vs* δ 129.0 (C-3″), δ 6.65 (H-4″) *vs* δ 115.0 (C-4″), δ 1.46 (7″-CH_3) and 1.49 (8″-CH_3) *vs* δ 28.4 (C-7″/C-8″), δ 3.38 (H-1‴) *vs* δ 28.0 (C-1‴), 5.26 (H-2‴) *vs* δ 121.0 (C-2‴), δ 1.75 (4‴-CH_3/5‴-CH_3) *vs* δ 27.0 (C-4‴) and 17.8 (C-5‴), δ 6.03 (−OCH_2O−) *vs* δ 101.5 (−OCH_2O−).

HMBC: δ 7.62 (1H, s, H-5) *vs* δ 191.0 (C-4), 156.8 (C-7) and 28.0 (C-1‴), δ 7.01 (H-2′) *vs* δ 148.0 (C-3′, C-4′), δ 6.86 (H-5′) *vs* δ 148.0 (C-3′/C-4′), δ 3.38 (H-1‴) *vs* δ 156.8 (C-7), δ 6.65 (H-4″) *vs* δ 156.8

(C-7), δ 6.03 (–OCH_2O–) vs δ 148.0 (C-3′/C-4′) (selected correlations are shown).

EIMS: m/z 418 ([M]$^+$), 403, 377, 363, 255, 215, 148, 147, 115.

HR-FABMS: m/z 419.1855 ([M + H]$^+$, Calcd. for $C_{26}H_{27}O_5$, 419.1855).

Reference

Li L, Li X, Shi C, Deng Z, Fu H, Proksch P, Lin W. (2006). Pongamone A–E, five flavonoids from the stems of a mangrove plant, *Pongamia pinnata*. *Phytochemistry* **67**: 1347–1352.

Praeroside VI [*cis*-Khellactone-3'-*O*-β-D-apiofuranosyl (1→6)-β-D-glucopyranoside]

IUPAC name: (9*R*,10*R*)-9-((((2*S*,3*R*,4*S*,5*S*,6*R*)-6-(((((2*R*,3*R*,4*R*)-3,4-Dihydroxy-4-(hydroxymethyl)tetrahydrofuran-2-yl)oxy)methyl)-3,4,5-trihydroxytetrahydro-2H-pyran-2-yl)oxy)-10-hydroxy-8,8-dimethyl-9,10-dihydropyrano[2,3-*f*]chromen-2(8*H*)-one

Sub-class: Pyranocoumarin glycoside

Chemical structure

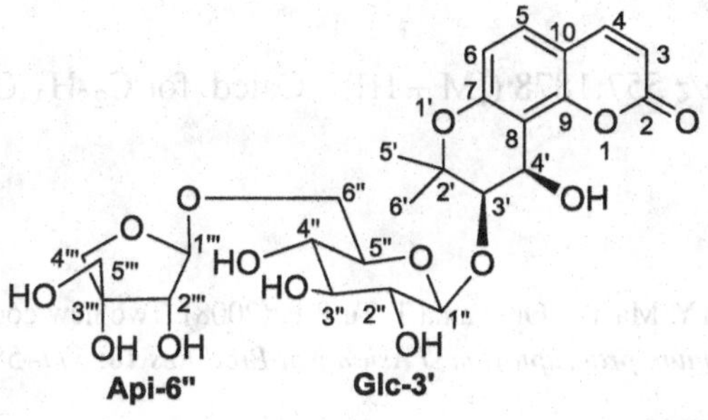

Source: *Peucedanum praeruptorum* Dunn. (Family: Apiaceae); roots

Molecular formula: $C_{25}H_{32}O_{14}$

Molecular weight: 556

State: White amorphous powder

Melting point: 109–111 °C

Specific rotation: $[\alpha]^{20}_D$ –46.5° (MeOH, c 0.85)

UV (MeOH): λ_{max} (log ε) 327 (3.04) nm

FT-IR (KBr): v_{max} 3380 (OH), 1715 (lactone carbonyl), 1600, 1585, 1500 (aromatic unsaturation) cm^{-1}

^{1}H-NMR (DMSO-d_6, 500 MHz): δ 6.25 (1H, d, J = 9.6 Hz, H-3), 7.87 (1H, d, J = 9.6 Hz, H-4), 7.48 (1H, d, J = 8.8 Hz, H-5), 6.78 (1H, d, J = 8.8 Hz, H-6), 4.02 (1H, d, J = 4.4 Hz, H-3'), 5.35 (1H, d, J = 4.4 Hz, H-4'), 1.48 (3H, s, H-5'), 1.49 (3H, s, H-6'), 4.64 (1H, d, J = 7.6 Hz, H-1''), 3.31 (1H, m, H-2''), 3.42 (1H, d, J = 9.2 Hz, H-3''), 3.33 (1H, m, H-4''), 3.43 (1H, m, H-5''), 3.62 (1H, dd, J = 11.3, 6.1 Hz, H-6''α), 3.98 (1H, dd, J = 13.2, 10.0 Hz, H-6''β), 5.00 (1H, d, J = 2.4 Hz, H-1'''), 3.86 (1H, d, J = 2.4 Hz, H-2'''), 3.73 (1H, d, J = 9.4 Hz, H-4'''a), 3.92 (1H, d, J = 9.4 Hz, H-4'''b), 3.53 (1H, brs, H-5''').

^{13}C-NMR (DMSO-d_6, 125 MHz): δ 162.8 (C-2), 112.6 (C-3), 145.7 (C-4), 130.1 (C-5), 115.4 (C-6), 157.4 (C-7), 111.4 (C-8), 155.4 (C-9), 113.5 (C-10), 80.5 (C-2'), 79.9 (C-3'), 60.1 (C-4'), 22.3 (C-5'), 26.7 (C-6'), 102.7 (C-1''), 75.4 (C-2''), 77.2 (C-3''), 71.4 (C-4''), 77.7 (C-5''), 68.4 (C-6''), 110.6 (C-1'''), 78.1 (C-2'''), 79.3 (C-3'''), 74.8 (C-4'''), 65.5 (C-5''').

HRFABMS: m/z 557.1878 ([M + H])$^+$, Calcd. for $C_{25}H_{33}O_{14}$, 557.1870).

Reference

Chang H-T, Okada Y, Ma T-J, Okuyama T, Tu P-F. (2008). Two new coumarin glycosides from *Peucedanum praeruptorum*. *J Asian Nat Prod Res* **10**: 577–581.

Protoapigenone

IUPAC name: 5,7-Dihydroxy-2-(1-hydroxy-4-oxocyclohexa-2,5-dien-1-yl)-4H-chromen-4-one

Sub-class: An unprecedented flavone

Chemical structure

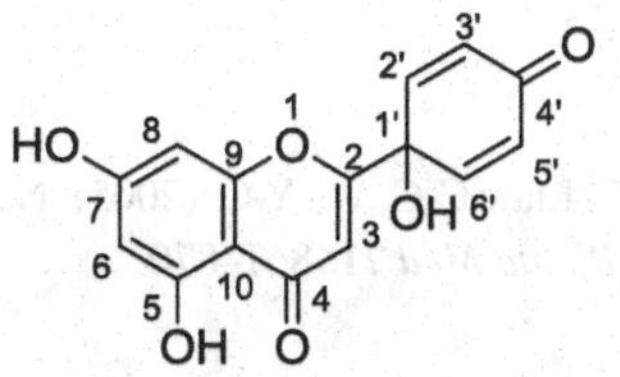

Source: *Thelyptris torresiana* (Gaud.) Alston (Synonym: *Macrothelypteris torresiana* (Gaud.) Ching.) (Family: Thelypteridaceae); whole plants

Molecular formula: $C_{15}H_{10}O_6$

Molecular weight: 286

State: Pale yellow needles (EtOAc–MeOH 2:1)

Melting point: 180–181 °C

Bioactivity studied: Cytotoxic [cancer cell line (IC_{50} expressed in µg/mL): Hep G2 (1.60 ± 0.33); Hep 3B (0.23 ± 0.01); MCF-7 (078 ± 0.02); A549 (3.88 ± 0.02); MDA-MB-231 (0.27 ± 0.02).

UV (MeOH): λ_{max} 325, 299, 259, 249, 230, 206 nm

IR (neat): ν_{max} 3224 (Ar-OH), 2928, 1656 (α,β-unsaturated carbonyl), 1621 (aromatic unsaturation), 1579, 1349, 1164 cm^{-1}

¹H-NMR (Pyridine-d_5, 200 MHz): δ 7.05 (1H, s, H-3), 6.59 (1H, d, J = 2.2 Hz, H-6), 6.72 (1 H, d, J = 2.2 Hz, H-8), 7.23 (2H, d, J = 10.2 Hz, H-2′ and H-6′), 6.52 (2H, d, J = 10.2 Hz, H-3′ and H-5′), 13.31 (1H, br s, C$_5$-OH).

¹³C-NMR (Pyridine-d_5, 50 MHz): δ 167.8 (C-2), 107.4 (C-3), 182.8 (C-4), 163.0 (C-5), 100.2 (C-6), 166.2 (C-7), 95.0 (C-8), 158.7 (C-9), 105.1 (C-10), 69.7 (C-1′), 148.3 (C-2′, C-6′), 129.5 (C-3′, C-5′), 185.2 (C-4′).

EIMS: m/z (%rel.) 286 ([M]$^+$, 100), 270 (39), 242 (18), 229 (31). 153 (33), 134 (12).

HR-EIMS: m/z 287.0553 ([M + H]$^+$, Calcd. for C$_{15}$H$_{11}$O$_4$, 287.0550).

Reference

Lin A-S, Chang F-R, Wu C-C, Liaw C-C, Wu Y-C. (2005). New cytotoxic flavonoids from *Thelypteris torresiana. Planta Med* **71**: 867–870.

Quercetin 3-*O*-[2‴,6‴-*O*-diacetyl-β-D-glucopyranosyl-(1→6)-β-D-glucopyranoside]

IUPAC name: ((2*R*,3*S*,4*S*,5*S*,6*R*)-5-Acetoxy-6-((((2*R*,3*S*,4*S*,6*S*)-6-((2-(3,4-dihydroxyphenyl)-5,7-dihydroxy-4-oxo-4*H*-chromen-3-yl)oxy)-3,4,5-trihydroxytetrahydro-2*H*-pyran-2-yl)methoxy)-3,4-dihydroxytetrahydro-2*H*-pyran-2-yl)methyl acetate

Sub-class: Acetylated flavonol glycoside

Chemical structure

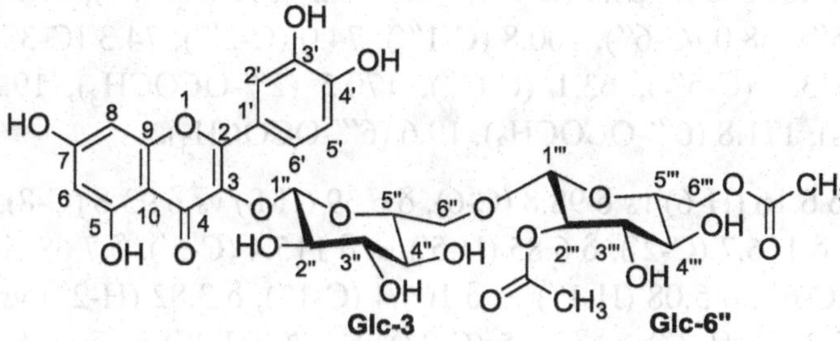

Source: *Meconopsis quintuplinervia* Regel (Family: Papaveraceae); aerial parts

Molecular formula: $C_{31}H_{34}O_{19}$

Molecular weight: 710

State: Yellow amorphous powder

Specific rotation: $[\alpha]^{20}_D$ +30.8° (MeOH, c 0.25)

UV (MeOH): λ_{max} (log ε) 206 (4.45), 257 (4.18), 270 (4.07), 296 (3.77), 363 (4.13) nm

FT-IR (KBr): ν_{max} 3419 (OH), 2908, 1732 (ester carbonyl), 1653 (α,β-unsaturated carbonyl), 1604, 1506, 1444, 1361 (aromatic unsaturaion), 1244, 1078 cm^{-1}

^{1}H-NMR (CD$_3$OD, 300 MHz): δ 6.16 (1H, d, J = 2.1 Hz, H-6), 6.39 (1H, d, J = 2.1 Hz, H-8), 8.07 (1H, d, J = 2.1 Hz, H-2'), 6.85 (1H, d, J = 8.5 Hz, H-5'), 7.68 (1H, dd, J = 8.5, 2.1 Hz, H-6'), 5.08 (1H, d, J = 7.8 Hz, H-1''), 3.82 (1H, dd, J = 7.8, 7.8 Hz, H-2''), 3.55 (1H, dd, J = 7.8, 7.5 Hz, H-3''), 3.70 (1H, dd, J = 7.8, 7.8 Hz, H-4''), 3.52 (1H, m, H-5''), 3.64(2H, m, H-6''), 4.16 (1H, d, J = 8.1 Hz, H-1'''), 4.37 (1H, dd, J = 8.4, 8.1 Hz, H-2'''), 2.98 (1H, dd, J = 9.3, 8.4 Hz, H-3'''), 3.15 (1H, dd, J = 9.3 Hz, H-4'''), 2.84 (1H, m, H-5'''), 4.20 (1H, dd, J = 12.3, 2.1 Hz, H-6'''a), 4.03 (1H, dd, J = 12.3, 5.5 Hz, H-6'''b), 1.59 (3H, s, 2'''-OCOCH$_3$), 1.98 (3H, s, 6'''-OCOCH$_3$).

^{13}C-NMR (CD$_3$OD, 75 MHz): δ 156.9 (C-2), 134.8 (C-3), 178.1 (C-4), 161.8 (C-5), 98.8 (C-6), 165.1 (C-7), 93.8 (C-8), 157.2 (C-9), 104.4 (C-10), 121.4 (C-1'), 116.7 (C-2'), 144.8 (C-3'), 149.2 (C-4'), 115.4 (C-5'), 121.6 (C-6'), 104.4 (C-1''), 72.0 (C-2''), 73.5 (C-3''), 69.6 (C-4''), 76.6 (C-5''), 68.0 (C-6''), 100.8 (C-1'''), 74.0 (C-2'''), 74.3 (C-3'''), 70.2 (C-4'''), 73.6 (C-5'''), 63.1 (C-6'''), 170.5 (2'''-OCOCH$_3$), 19.3 (2''-OCOCH$_3$), 171.8 (6'''-OCOCH$_3$), 19.6 (6'''-OCOCH$_3$).

HSQC: δ 6.16 (H-6) vs δ 98.8 (C-6), δ 6.39 (H-8) vs δ 93.8 (C-8), δ 8.07 (H-2') vs δ 116.7 (C-2'), δ 6.85 (H-5') vs δ 115.4 (C-5'), δ 7.68 (H-6') vs δ 121.6 (C-6'), δ 5.08 (H-1'') vs δ 104.4 (C-1''), δ 3.82 (H-2'') vs δ 72.0 (C-2''), δ 3.55 (H-3'') vs δ 73.5 (C-3''), δ 3.70 (H-4'') vs δ 69.6 (C-4''), δ 3.52 (H-5'') vs δ 76.6 (C-5''), δ 3.64 (H-6'') vs δ 68.0 (C-6''), δ 4.16 (H-1''') vs δ 100.8 (C-1'''), δ 4.37 (H-2''') vs δ 74.0 (C-2'''), δ 2.98 (H-3''') vs δ 74.3 (C-3'''), δ 3.15 (H-4''') vs δ 70.2 (C-4'''), δ 2.84 (H-5''') vs δ 73.6 (C-5'''), δ 4.20 (H-6'''a) and 4.03 (H-6'''b) vs δ 63.1 (C-6'''), δ 1.59 (2'''-OCOCH$_3$) vs δ 19.3 (2'''-OCOCH$_3$), δ 1.98 (6'''-OCOCH$_3$) vs δ 19.6 (6'''-OCOCH$_3$).

FABMS: *m/z* 711 [M + H]$^+$

HR-FABMS: *m/z* 711.1810 ([M + H]$^+$, Calcd. for $C_{31}H_{35}O_{19}$, 711.1773).

Reference

Shang X-Y, Wang Y-H, Li C, Zhang C-Z, Yang Y-C, Shi J-G. (2006). Acetylated flavonol diglucosides from *Meconpsis quintuplinervia*. *Phytochemistry* **67**: 511–515.

Quercetin-3-*O*-[α-L-rhamnopyranosyl-(1→2)-[α-L-rhamnopyranosyl-(1→6)]-β-D-galactopyranosyl]-7-*O*-β-D-glucopyranoside

IUPAC name: 3-(((2*S*,3*R*,4*S*,5*R*,6*R*)-4,5-Dihydroxy-3-(((2*S*,3*R*,4*R*,5*R*,6*S*)-3,4,5-trihydroxy-6-methyltetrahydro-2*H*-pyran-2-yl)oxy)-6-((((2*R*,3*R*,4*R*,5*R*,6*S*)-3,4,5-trihydroxy-6-methyltetrahydro-2*H*-pyran-2-yl)oxy)methyl)tetrahydro-2*H*-pyran-2-yl)oxy)-2-(3,4-dihydroxyphenyl)-5-hydroxy-7-(((2*S*,3*R*,4*S*,5*S*,6*R*)-3,4,5-trihydroxy-6-(hydroxymethyl)tetrahydro-2*H*-pyran-2-yl)oxy)-4*H*-chromen-4-one

Sub-class: Flavonol tertaglycoside

Chemical structure

Source: *Astragalus monspessulanus* Linn. (Montpellier Milk Vetch; Family: Fabaceae); aerial parts

Molecular formula: $C_{39}H_{50}O_{25}$

Molecular weight: 918

State: Orange amorphous powder

UV (MeOH): λ_{max} (log ε) 250 (4.04), 263 (sh, 3.87), 351 (3.74) nm

^{1}H-NMR (Pyridine-d_5, 400 MHz): δ 6.74 (1H, brs, H-6), 6.87 (1H, brs, H-8), 8.32 (1H, brs, H-2'), 7.27 (1H, d, J = 8.5 Hz, H-5'), 8.36 (1H, d, J = 8.5 Hz, H-6'), 6.42 (1H, d, J = 7.8 Hz, H-1''), 4.95 (1H, m, H-2''), 4.33 (1H, m, H-3''), 4.34 (1H, m, H-4''), 4.08 (1H, m, H-5''), 4.39 (1H, m, H-6''a), 3.90 (1H, dd, J = 9.8, 5.7 Hz, H-6''b), 5.81 (1H, d, J = 7.6 Hz, H-1'''), 4.33 (1H, m, H-2'''), 4.38 (1H, m, H-3'''), 4.32 (1H, m, H-4'''), 4.13 (1H, m, H-5'''), 4.54 (1H, m, H-6'''a), 4.40 (1H, m, H-6'''b), 5.17 (1H, brs, H-1''''), 4.33 (1H, m H-2''''), 4.28 (1H, m, H-3''''), 4.12 (1H, m, H-4''''), 4.15 (1H, m, H-5''''), 1.44 (3H, d, J = 6.6 Hz, H-6''''), 6.30 (1H, brs, H-1'''''), 4.86 (1H, brs, H-2'''''), 4.83 (1H, m, H-3'''''), 4.28 (1H, m, H-4'''''), 4.96 (1H, m, H-5'''''), 1.59 (3H, d, J = 6.4 Hz, H-6''''').

^{13}C-NMR (Pyridine-d_5, 100 MHz): δ 158.1 (C-2), 134.7 (C-3), 178.6 (C-4), 162.8 (C-5), 100.2 (C-6), 163.6 (C-7), 94.9 (C-8), 156.8 (C-9), 107.0 (C-10), 122.5 (C-1'), 117.3 (C-2'), 146.9 (C-3'), 150.7 (C-4'), 116.4 (C-5'), 123.4 (C-6'), 100.9 (C-1''), 76.6 (C-2''), 75.8 (C-3''), 70.1 (C-4''), 74.8 (C-5''), 66.1 (C-6''), 101.7 (C-1'''), 74.8 (C-2'''), 78.5 (C-3'''), 71.2

(C-4‴), 79.1 (C-5‴), 62.4 (C-6‴), 101.9 (C-1⁗), 72.1 (C-2⁗), 72.6 (C-3⁗), 73.8 (C-4⁗), 69.6 (C-5⁗), 18.4 (C-6⁗), 102.5 (C-1‴‴), 72.8 (C-2‴‴), 72.8 (C-3‴‴), 74.3 (C-4‴‴), 69.9 (C-5‴‴), 18.3 (C-6‴‴).

HREIMS: m/z 963.2650 ([M + HCOO]⁻, Calcd. for $C_{40}H_{51}O_{27}$, 963.2612).

Reference

Krasteva I, Bratkov V, Bucar F, Kunert O, Kollroser M, Kondeva-Burdina M, Ionkova I. (2015). Flavoalkaloids and Flavonoids from *Astragalus monspessulanus*. *J Nat Prod* **78**: 2565–2571.

Quercetin 3-*O*-β-D-apiofuranosyl-(1→6)-β-D-glucopyranoside

IUPAC name: 3-((((2*S*,4*S*,5*S*,6*R*)-6-(((((2*R*,3*R*)-3,4-dihydroxy-4-(hydro-xymethyl)tetrahydrofuran-2-yl)oxy)methyl)-3,4,5-trihydroxytetrahydro-2*H*-pyran-2-yl)oxy)-2-(3,4-dihydroxyphenyl)-5,7-dihydroxy-4*H*-chromen-4-one

Sub-class: Flavonol glycoside

Chemical structure

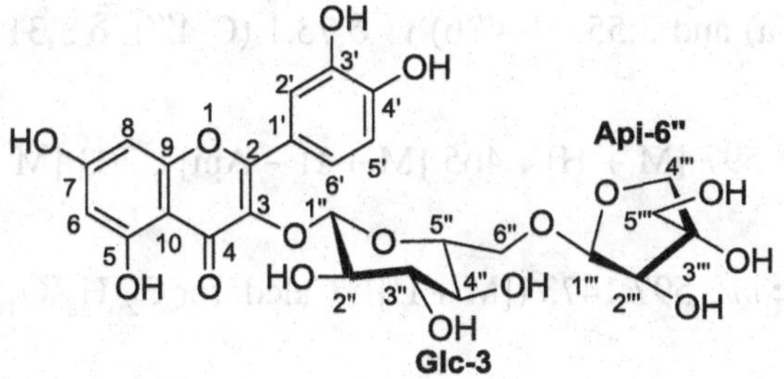

Source: *Solidago altissima* Linn. (Family: Asteraceae); leaves

Molecular formula: $C_{26}H_{28}O_{16}$

Molecular weight: 596

State: Yellow amorphous powder

Specific rotation: $[\alpha]^{25}_D$ −67.6° (MeOH, *c* 0.11)

UV (MeOH): λ_{max} 254, 263, 300 sh and 353 nm

^{1}H-NMR (DMSO-d_6, 400 MHz): δ 6.17 (1H, d, J = 2.0 Hz, H-6), 6.36 (1H, d, J = 2.0 Hz, H-8), 7.53 (2H, dd, J = 8.8, 1.6 Hz, H-2' and H-6'), 6.83 (1H, d, J = 8.8 Hz, H-5'), 5.32 (1H, d, J = 7.6 Hz, H-1''), 3.27 (1H, m, H-2''), 3.30 (1H, m, H-3''), 3.14 (1H, m, H-4''), 3.30 (1H, m, H-5''), 3.69 (1H, d, J = 12.0 Hz, H-6''a), 3.48 (1H, d, J = 12.0 Hz, H-6''b), 4.39 (1H, d, J = 1.2 Hz, H-1'''), 3.92 (1H, d, J = 2.0 Hz, H-2'''), 3.82 (1H, d, J = 10.0 Hz, H-4'''a), 3.55 (1H, d, J = 10.0 Hz, H-4'''b), 3.31 (2H, s, H-5''').

^{13}C-NMR (DMSO-d_6, 100 MHz): δ 156.3 (C-2), 133.3 (C-3), 177.2 (C-4), 160.8 (C-5), 98.8 (C-6), 166.9 (C-7), 93.7 (C-8), 156.2 (C-9), 103.6 (C-10), 121.0 (C-1'), 114.9 (C-2'), 144.7 (C-3'), 148.6 (C-4'), 116.3 (C-5'), 121.4 (C-6'), 100.9 (C-1''), 74.4 (C-2''), 75.7 (C-3''), 70.2 (C-4''), 75.9 (C-5''), 67.2 (C-6''), 109.2 (C-1'''), 76.2 (C-2'''), 78.4 (C-3'''), 73.1 (C-4'''), 63.8 (C-5''').

HMQC: δ 6.17 (H-6) *vs* δ 98.8 (C-6), δ 6.36 (H-8) *vs* δ 93.7 (C-8), δ 7.53 (H-2') *vs* δ 114.9 (C-2'), δ 6.83 (H-5') *vs* δ 116.3 (C-5'), δ 7.53 (H-6') *vs* δ 121.4 (C-6'), δ 5.32 (H-1'') *vs* δ 100.9 (C-1''), δ 3.27 (H-2'') *vs* δ 74.2 (C-2''), δ 3.30 (H-3'') *vs* δ 75.7 (C-3''), δ 3.14 (H-4'') *vs* δ 70.2 (C-4''), δ 3.30 (H-5'') *vs* δ 75.9 (C-5''), δ 3.69 (H-6''a) and 3.48 (H-6''b) *vs* δ 67.2 (C-6''), δ 4.39 (H-1''') *vs* δ 109.2 (C-1'''), δ 3.92 (H-2''') *vs* δ 76.2 (C-2'''), δ 3.82 (H-4'''a) and 3.55 (H-4'''b) *vs* δ 73.1 (C-4'''), δ 3.31 (H-5''') *vs* δ 65.8 (C-5''').

ESI-MS: *m/z* 597 [M + H]$^+$, 465 [M + H − Api]$^+$, 303 [M + H − Api − Glc]$^+$.

HR-ESI-MS: *m/z* 597.1473 ([M + H]$^+$, Calcd. for $C_{26}H_{29}O_{16}$, 597.1450).

Reference

Wu B, Takahashi T, Kashiwagi T, Tebayashi S, Kim C-S. (2007). New flavonoid glycosides from the leaves of *Solidago altissima*. *Chem Pharm Bull* **55**: 815–816.

Quercetin 3-*O*-β-D-glucopyranosyl-4″,6″-digallate [2-(3,4-dihydroxyphenyl)-5,7-dihydroxy-4-oxo-4*H*-chromen-3-yl-4,6-bis-*O*-α-D-(3,4,5-trihydroxybenzoyl) glucopyranoside]

IUPAC name: (2*R*,3*S*,4*R*,6*S*)-6-((2-(3,4-Dihydroxyphenyl)-5,7-dihydroxy-4-oxo-4*H*-chromen-3-yl)oxy)-4,5-dihydroxy-2-(((3,4,5-trihydroxybenzoyl)oxy)methyl)tetrahydro-2*H*-pyran-3-yl 3,4,5-trihydroxy benzoate

Sub-class: Flavonol glycoside

Chemical structure

Source: *Triplaris cumingiana* Fisch. & C.A. Mey. ex Mey. (Family: Polygonaceae); young leaves

Molecular formula: $C_{35}H_{28}O_{20}$

Molecular weight: 768

State: Yellow amorphous powder

Specific rotation: $[\alpha]^{28}_D$ +3.6° (MeOH, *c* 0.14)

UV (MeOH): λ_{max} (log ε) 267 (4.71), 359 (4.39) nm

FT-IR (KBr): ν_{max} 3600-3000 (OH), 1620 (α,β-unsaturated carbonyl), 1560, 1350 (aromatic unsaturaion), 1180 cm^{-1}

^{1}H-NMR (CD$_3$OD, 300 MHz): δ 6.15 (1H, d, *J* = 2.3 Hz, H-6), 6.30 (1H, d, *J* = 2.3 Hz, H-8), 7.58 (1H, d, *J* = 1.5 Hz, H-2′), 6.73 (1H, d, *J* = 9.3 Hz, H-5′), 7.60 (1H, dd, *J* = 9.0, 1.5 Hz, H-6′), 5.35 (1H, d, *J* = 7.8 Hz, H-1″), 3.86 (2H, overlapping signal, H-2″ and H-3″), 5.17 (1H, t, *J* = 9.8 Hz, H-4″), 3.70 (1H, t, *J* = 9.8 Hz, H-5″), 4.20 (2H, m, H-6″), 7.07 (2H, s, H-2‴ and H-6‴), 6.92 (2H, s, H-2⁗ and H-6⁗).

^{13}C-NMR (CD$_3$OD, 75 MHz): δ 160.1 (C-2), 136.0 (C-3), 179.9 (C-4), 163.5 (C-5), 100.7 (C-6), 166.5 (C-7), 95.7 (C-8), 159.0 (C-9), 106.3 (C-10), 124.4 (C-1′), 118.0 (C-2′), 147.2 (C-3′), 150.4 (C-4′), 116.7 (C-5′), 123.7 (C-6′), 104.9 (C-1″), 76.7 (C-2″), 74.5 (C-3″ and C-5″), 73.0 (C-4″), 64.3 (C-6″), 121.9 (C-1‴), 111.2 (C-2‴ and C-6‴), 146.7 (C-3‴ and C-5‴), 140.8 (C-4‴), 168.7 (O*C*O-1‴), 121.8 (C-1⁗), 111.1 (C-2⁗ and C-6⁗), 146.5 (C-3⁗ and C-5⁗), 140.5 (C-4⁗), 168.3 (O*C*O-1⁗).

FABMS: *m/z* (%rel) 769 [M + H]$^+$ (3), 613 (3), 460 (3), 391 (3), 307 (25), 235 (3), 219 (3), 154 (100), 136 (66).

HR-FABMS: *m/z* 769.12477 ([M + H]$^+$, Calcd. for $C_{35}H_{29}O_{20}$, 769.12522).

Reference

Hussein AA, Barberena I, Correa M, Coley PD, Solis PN, Gupta MP (2005). Cytotoxic flavonol glycosides from *Triplaris cumingiana*. *J Nat Prod* **68**: 231–233.

Quercetin 4'-*O*-α-L-arabinofuranoside [2-Hydroxy-4-*O*-α-L-(3,5,7-trihydroxy-4-oxo-4*H*-chromen-2-yl)phenylarabinofuranoside]

IUPAC name: 2-(4-(((2*S*,3*R*,4*R*,5*S*)-3,4-Dihydroxy-5-(hydroxymethyl)tetrahydrofuran-2-yl)oxy)-3-hydroxyphenyl)-3,5,7-trihydroxy-4*H*-chromen-4-one

Sub-class: Flavonol glycoside

Chemical structure

Source: *Triplaris cumingiana* Fisch. & C.A. Mey. ex Mey. (Family: Polygonaceae); young leaves

Molecular formula: $C_{20}H_{18}O_{11}$

Molecular weight: 434

State: Yellow amorphous powder

Specific rotation: $[\alpha]^{28}_D$ $-106.3°$ (MeOH, c 0.08)

UV (MeOH): λ_{max} (log ε) 256 (4.43), 360 (4.34) nm

^{1}H-NMR (CD$_3$OD, 300 MHz): δ 6.22 (1H, d, J = 2.0 Hz, H-6), 6.41 (1H, d, J = 2.0 Hz, H-8), 7.52 (1H, d, J = 2.2 Hz, H-2′), 6.92 (1H, d, J = 8.5 Hz, H-5′), 7.52 (1H, dd, J = 8.5, 2.2 Hz, H-6′), 5.48 (1H, s, H-1″), 4.35 (1H, dd, J = 3.0, 1.1 Hz, H-2″), 3.92 (1H, 1H, dd, J = 6.0, 3.0 Hz, H-3″), 3.89 (1H, m, H-4″), 3.51 (2H, m, H-5″).

^{13}C-NMR (CD$_3$OD, 75 MHz): δ 159.6 (C-2), 135.7 (C-3), 180.8 (C-4), 163.8 (C-5), 100.7 (C-6), 166.8 (C-7), 95.6 (C-8), 160.1 (C-9), 106.4 (C-10), 123.9 (C-1′), 117.7 (C-2′), 147.1 (C-3′), 150.6 (C-4′), 117.2 (C-5′), 123.8 (C-6′), 110.3 (C-1″), 84.1 (C-2″), 79.5 (C-3″), 88.7 (C-4″), 63.4 (C-5″).

FABMS: m/z (%rel) 435 [M + H]$^+$ (4), 391 (10), 303 (10), 185 (61), 149 (10), 115 (10), 93 (100).

HR-FABMS: m/z 435.09184 ([M + H]$^+$, Calcd. for C$_{20}$H$_{19}$O$_{11}$, 435.09274).

Reference

Hussein AA, Barberena I, Correa M, Coley PD, Solis PN, Gupta MP. (2005). Cytotoxic flavonol glycosides from *Triplaris cumingiana*. *J Nat Prod* **68**: 231–233.

Quercetin 4'-*O*-α-rhamnopyranosyl-3-*O*-β-D-allopyranoside

IUPAC name: 5,7-Dihydroxy-2-(3-hydroxy-4-(((2*S*,3*S*,4*S*,6*R*)-3,4,5-trihydroxy-6-methyltetrahydro-2*H*-pyran-2-yl)oxy)phenyl)-3-(((2*S*,3*R*,4*R*,5*S*,6*R*)-3,4,5-trihydroxy-6-(hydroxymethyl)tetrahydro-2*H*-pyran-2-yl)oxy)-4*H*-chromen-4-one

Sub-class: Flavone gylcoside

Chemical structure

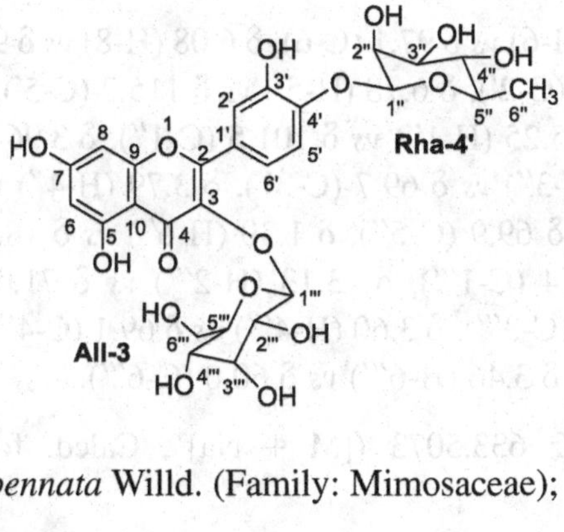

Source: *Acacia pennata* Willd. (Family: Mimosaceae); leaves

Molecular formula: $C_{27}H_{30}O_{16}$

Molecular weight: 610

State: Yellow amorphous powder

Specific rotation: $[\alpha]^{20}_{D}$ −25.1° (MeOH, *c* 0.19)

Bioactivity studied: Cyclooxygenases-1 (COX-1) inhibitory activity (IC_{50} = 11.6 µg/mL)

UV: λ_{max} (MeOH) 358, 299 sh, 267 sh, 258 nm

IR (KBr): ν_{max} 3420 (OH), 1690, 1660 (α,β-unsaturated carbonyl), 1606, 1510, 1448 (aromatic unsaturaion), 1257, 1065, 815 cm^{-1}.

^{1}H-NMR (DMSO-d_6, 600 MHz): δ 6.38 (1H, br s, H-6), 6.08 (1H, br s, H-8), 7.21 (1H, br s, H-2′), 6.78 (1H, d, J = 8.5 Hz, H-5′), 7.18 (1H, br s, H-6′), 5.25 (1H, br s, H-1″), 3.96* (1H, H-2″), 3.50* (1H, H-3″), 3.79* (1H, H-4″), 3.35* (1H, H-5″), 1.30 (3H, d, J = 6.2 Hz, H-6″), 5.92 (1H, d, J = 7.4 Hz, H-1‴), 3.13* (1H, H-2‴), 3.60* (2H, H-3‴ and H-4‴), 3.39* (1H, H-5‴), 3.46* (1H, H-6‴) [*multiplicity was not determined due to overlap of the respective peak].

^{13}C-NMR (DMSO-d_6, 150 MHz): δ 157.8 (C-2), 134.5 (C-3), 177.9 (C-4), 161.1 (C-5), 97.1 (C-6), 164.9 (C-7), 94.6 (C-8), 157.1 (C-9), 104.0 (C-10), 120.0 (C-1′), 114.5 (C-2′), 148.2 (C-3′), 149.9 (C-4′), 115.2 (C-5′), 121.0 (C-6′), 101.5 (C-1″), 69.9 (C-2″ and 5″), 69.7 (C-3″), 74.2 (C-4″), 18.5 (C-6″), 100.4 (C-1‴), 71.0 (C-2‴), 70.1 (C-3‴), 69.1 (C-4‴), 72.0 (C-5‴), 60.0 (C-6‴).

HMQC: δ 6.38 (H-6) *vs* δ 97.1 (C-6), δ 6.08 (H-8) *vs* δ 94.6 (C-8), δ 7.21 (H-2′) *vs* δ 114.5 (C-2′), δ 6.78 (H-5′) *vs* δ 115.2 (C-5′), δ 7.18 (H-6′) *vs* δ 121.0 (C-6′), δ 5.25 (H-1″) *vs* δ 101.5 (C-1″), δ 3.96 (H-2″) *vs* δ 69.9 (C-2″), δ 3.50 (H-3″) *vs* δ 69.7 (C-3″), δ 3.79 (H-4″) *vs* δ 74.2 (C-4″), δ 3.35 (H-5″) *vs* δ 69.9 (C-5″), δ 1.30 (H-6″) *vs* δ 18.5 (C-6″), δ 5.92 (H-1‴) *vs* δ 100.4 (C-1‴), δ 3.13 (H-2‴) *vs* δ 71.0 (C-2‴), δ 3.60 (H-3‴) *vs* δ 70.1 (C-3‴), δ 3.60 (H-4‴) *vs* δ 69.1 (C-4‴), δ 3.39 (H-5‴) *vs* δ 72.0 (C-5‴), δ 3.46 (H-6‴) *vs* δ 60.0 (C-6‴).

HR-ESI-MS: m/z 633.5073 ([M + Na]$^+$, Calcd. for $C_{27}H_{30}O_{16}Na$, 633.5750).

Reference

Dongmo AB, Miyamoto T, Yoshikawa K, Arihara S, Lacaille-Dubois M-A. (2007). Flavonoids from *Acacia pennata* and their cyclooxygenase (COX-1 and COX-2) inhibitory activities. *Planta Med* **73**: 1202–1207.

Rotundaflavone Ia

IUPAC name: (2*S*)-8-((*E*)-3,7-Dimethylocta-2,6-dien-4-yl)-5-hydroxy-7-methoxy-2-phenylchroman-4-one

Sub-class: Prenylated flavanone

Chemical structure

Source: *Boesenbergia rouunda* (Linn.) Mansf. [Synonym: *B. pandulata* (Roxb.) Schltr.] (Family: Zingiberaceae); rhizomes

Molecular formula: $C_{26}H_{30}O_4$

Molecular weight: 406

State: Amorphous powder

Specific rotation: $[\alpha]^{25}_D$ −47° (MeOH, *c* 0.82)

UV: λ_{max} (MeOH) (log ε): 294 (4.43), 340 (3.78) nm

IR (KBr): ν_{max} 1646 (carbonyl), 1583, 1452 (aromatic unsaturaion), 1211, 1106 cm^{-1}

^{1}H-NMR (CDCl$_3$, 500 MHz): δ 5.42 (1H, dd, J = 13.4, 2.7 Hz, H-2β), 2.79 (1H, dd, J = 17.1, 2.7 Hz, H-3$_{eq}$), 3.10 (1H, dd, J = 17.1, 13.4 Hz, H-3$_{ax}$), 6.06 (1H, s, H-6), 7.47-7.37 (5H, m, H-2'-H-6'), 1.60 (3H, d, J = 6.8 Hz, H-1''), 5.36 (1H, q, J = 6.8 Hz, H-2''), 3.80 (1H, dd-like, H-4''), 1.58 (3H, s, H-5''), 2.56 (1H, m, H-1'''a), 2.67 (1H, m, H-1'''b), 5.00 (1H, dd, J = 7.6, 7.4 Hz, H-2'''), 1.59 (3H, s, H-4'''), 1.56 (3H, s, H-5'''), 3.78 (3H, s, 7-OCH_3), 12.24 (1H, br s, 5-OH).

^{13}C-NMR (CDCl$_3$, 125 MHz): δ 79.31 (C-2), 43.60 (C-3), 195.86 (C-4), 161.54 (C-5), 91.11 (C-6), 166.59 (C-7), 111.89 (C-8), 161.37 (C-9), 102.85 (C-10), 138.57 (C-1'), 126.16 (C-2' and C-6'), 128.90 (C-3', C-4'and C-5'), 13.56 (C-1''), 117.45 (C-2''), 136.73 (C-3''), 41.97 (C-4''), 15.40 (C-5''), 28.93 (C-1'''), 124.08 (C-2'''), 131.22 (C-3'''), 25.79 (C-4'''), 17.81 (C-5'''), 55.69 (7-OCH$_3$).

HMBC: δ 5.42 (H-2) *vs* δ 195.86 (C-4), 135.57 (C-1') and 126.16 (C-2'/C-6'), δ 2.79 (H-3$_{eq}$) *vs* δ 195.86 (C-4) and 102.85 (C-10), δ 3.10 (H-3$_{ax}$) *vs* δ 135.57 (C-1'), δ 2.79 (H-6) *vs* δ 111.89 (C-8) and 102.85 (C-10), δ 3.78 (7-OCH$_3$) *vs* δ 166.59 (C-7), δ 1.60 (H-1'') *vs* δ 136.73 (C-3''), δ 5.36 (H-2'') *vs* δ 136.73 (C-3'') and 41.97 (C-4''), δ 3.80 (H-4'') *vs* δ 117.45 (C-2''), 136.73 (C-3'') and 166.59 (C-7), δ 1.58 (H-5'') *vs* δ 136.73 (C-3'') and 41.97 (C-4''), δ 2.56 (H-1'''a) and 2.67 (H-1'''b) *vs* δ 111.89 (C-8), δ 2.56 (H-2''') *vs* δ 25.79 (C-4''') and 17.81 (C-5'''), δ 1.59 (H-4''') *vs* δ 124.08 (C-2'''), 131.22 (C-3''') and 17.81 (C-5'''), δ 1.56 (H-5''') *vs* δ 124.08 (C-2'''), 131.22 (C-3''') and 25.79 (C-4''').

EIMS: *m/z* (%rel.) 406 ([M]$^+$, 2), 337 (96), 233 (100)

HR-EIMS: *m/z* 406.2147 ([M]$^+$, Calcd. for C$_{26}$H$_{30}$O$_4$, 406.2144).

Reference

Morikawa T, Funakoshi K, Ninomiya K, Yasuda D, Miyagawa K, Matsuda H, Yoshikawa M. (2008). Medicinal foodstuffs. XXXIV. Structures of new prenylchalcones and prenyl-flavanones with TNF-α and aminopeptidase N inhibitory activities from *Boesenbergia rotunda*. *Chem Pharm Bull* **56**: 956–962.

Sarcandracoumarin

IUPAC name: 3-(1-(3,4-Dihydroxyphenyl)ethyl)-7-hydroxy-6,8-dimethoxy-2*H*-chromen-2-one

Sub-class: Coumarin

Chemical structure

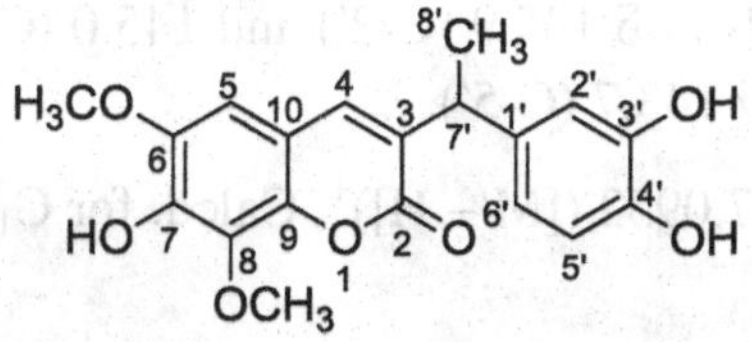

Source: *Sarcandra glabra* (Thunb.) Nakai (Family: Chloranthaceae); whole plants

Molecular formula: $C_{19}H_{18}O_7$

Molecular weight: 358

State: Brownish solid

Specific rotation: $[\alpha]^{20}_D$ +3° (MeOH, *c* 1.0)

UV (MeOH): λ_{max} (log ε) 205 (4.92), 230 (sh), 288 (4.12), 342 (4.26) nm

FT-IR (KBr): ν_{max} 3382 (OH), 1693 (lactone carbonyl), 1587, 1500, 1463 (aromatic unsaturation), 1280, 1199, 1106, 1045 cm^{-1}

^{1}H-NMR (DMSO-d_6, 400 MHz): δ 7.66 (1H, s, H-4), 7.02 (1H, s, H-5), 6.63 (1H, d, *J* = 2.0 Hz, H-2′), 6.64 (1H, d, *J* = 8.4 Hz, H-5′), 6.52 (1H, dd, J = 8.4, 2.0 Hz, H-6′), 3.97 (1H, q, J = 7.1 Hz, H-7′), 1.43 (3H, d, J = 7.1 Hz, H-8′), 3.79 (6H, s, 6-OCH_3 and 8-OCH_3), 9.73 (1H, br s, 7-OH), 8.75 (1H, br s, 3′-OH), 8.72 (1H, br s, 4′-OH).

^{13}**C-NMR (DMSO-d_6, 100 MHz):** δ 160.6 (C-2), 129.3 (C-3), 138.5 (C-4), 104.3 (C-5), 145.6 (C-6), 143.0 (C-7), 134.5 (C-8), 141.6 (C-9), 110.6 (C-10), 118.1 (C-1′), 135.9 (C-2′), 114.9 (C-3′), 145.0 (C-4′), 143.7 (C-5′), 115.5 (C-6′), 38.0 (C-7′), 20.3 (C-8′), 60.7 (6-OCH_3), 56.1 (8-OCH_3).

HMBC: δ 7.66 (H-4) vs δ 160.6 (C-2), 129.3 (C-3), 104.3 (C-5), 141.6 (C-9), 110.6 (C-10) and 38.0 (C-7′), δ 7.02 (H-5) vs δ 138.5 (C-4), 145.6 (C-6), 143.0 (C-7), 141.6 (C-9) and 110.6 (C-10), δ 6.63 (H-2′) vs δ 118.1 (C-1′), 114.9 (C-3′), 145.0 (C-4′) and 38.0 (C-7′), δ 6.64 (H-5′) vs δ 118.1 (C-1′), 114.9 (C-3′) and 145.0 (C-4′), δ 6.52 (H-5′) vs δ 118.1 (C-1′), 145.0 (C-4′) and 38.0 (C-7′), δ 1.43 (H-8′) vs δ 129.3 (C-3) and 118.1 (C-1′), δ 3.79 (6-OCH_3) vs δ 145.6 (C-6), δ 3.79 (8-OCH_3) vs δ 134.5 (C-8), δ 8.75 (3′-OH) vs δ 135.9 (C-2′) and 145.0 (C-4′), δ 8.72 (4′-OH) vs δ 114.9 (C-3′) and 143.7 (C-5′).

HR-ESIMS: m/z 357.0952 ([M − H])$^+$, Calcd. for $C_{19}H_{17}O_7$, 357.0974).

Reference

Feng S, Xu L, Wu M, Hao J, Qiu S X, Wei X. (2010). A new coumarin from *Sarcandra glabra*. *Fitoterapia* **81**: 472–474.

Sarmenoside VII [Limocitrin 3-*O*-(6-*O*-acetyl-β-D-glucopyranosyl)-7-*O*-β-D-glucopyranoside]

IUPAC name: ((2*R*,3*S*,4*S*,5*R*,6*S*)-3,4,5-Trihydroxy-6-((5-hydroxy-2-(4-hydroxy-3-methoxyphenyl)-8-methoxy-4-oxo-7-((((2*S*,3*R*,4*S*,5*S*,6*R*)-3,4,5-trihydroxy-6-(hydroxymethyl)tetrahydro-2*H*-pyran-2-yl)oxy)-4*H*-chromen-3-yl)oxy)tetrahydro-2*H*-pyran-2-yl)methyl acetate

Sub-class: Flavonol glycoside

Chemical structure

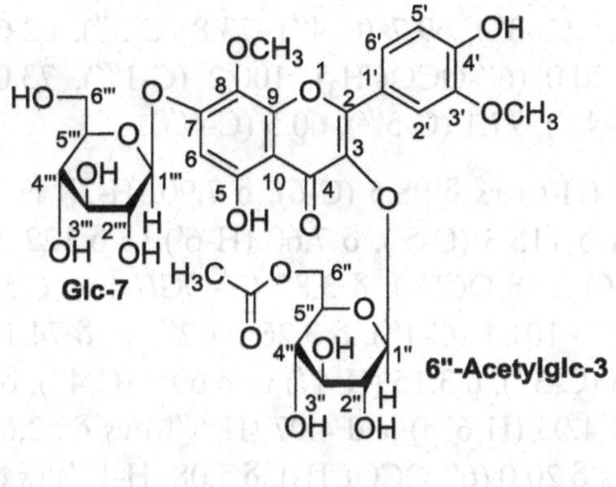

Source: *Sedum sarmentosum* (Family: Crassulaseae); whole plants

Molecular formula: $C_{31}H_{36}O_{19}$

Molecular weight: 712

State: Amorphous powder

Specific rotation: $[\alpha]^{26}_{D}$ −47.7° (MeOH, c 0.20)

Bioactivity studied: Showed inhibitory activity against oleic acid-albumin-induced lipid accumulation

UV (MeOH): λ_{max} (log ε) 257 (4.22) and 358 (4.05) nm

FT-IR (KBr): ν_{max} 3420 (OH), 2934, 1720 (ester carbonyl), 1653 (α,β-unsaturated carbonyl), 1601, 1559, 1456, 1071 cm^{-1}

^{1}H-NMR (DMSO-d_6, 600 MHz): δ 6.68 (1H, s, H-6), 7.90 (1H, d, J = 2.0 Hz, H-2'), 6.96 (1H, d, J = 8.3 Hz, H-5'), 7.60 (1H, dd, J = 8.3, 2.0 Hz, H-6'), 12.28 (1H, br s, 5-OH), 3.87 (6H, s, 8-OCH_3 and 3'-OCH_3), 5.45 (1H, d, J = 7.6 Hz, H-1''), 3.26 (1H, m, H-2''), 3.30 (1H, m, H-3''), 3.15 (1H, m, H-4''), 3.33 (1H, m, H-5''), 4.03 (1H, dd, J = 11.7, 5.5 Hz, H-6''a), 4.07 (1H, dd, J = 11.7, 2.1 Hz, H-6''b), 1.74 (3H, s, 6''-OCOCH_3), 5.08 (1H, d, J = 7.6 Hz, H-1'''), 3.32 (1H, m, H-2'''), 3.28 (1H, m, H-3'''), 3.18 (1H, m, H-4'''), 3.47 (1H, m, H-5'''), 3.45 (1H, m, H-6'''a), 3.70 (1H, br d, $ca.$ 12, H-6'''b).

^{13}C-NMR (DMSO-d_6, 125 MHz): δ 156.5 (C-2), 133.1 (C-3), 177.6 (C-4), 155.6 (C-5), 98.5 (C-6), 156.0 (C-7), 128.9 (C-8), 147.9 (C-9), 105.1 (C, C-10), 120.8 (C-1'), 113.1 (C-2'), 146.8 (C-3'), 149.8 (C-4'), 115.3 (C-5'), 122.2 (C-6'), 61.2 (8-OCH_3), 55.4 (3'-OCH_3), 101.1 (C-1''), 74.1 (C-2''), 76.5 (C-3''), 69.7 (C-4''), 73.8 (C-5''), 62.6 (C-6''), 169.7 (6''-OCOCH$_3$), 20.0 (6''-OCOCH$_3$), 100.2 (C-1'''), 73.0 (C-2'''), 76.0 (C-3'''), 69.5 (C-4'''), 77.1 (C-5'''), 60.5 (C-6''').

HMQC: δ 6.68 (H-6) vs δ 98.5 (C-6), δ 7.90 (H-2') vs δ 113.1 (C-2'), δ 6.96 (H-5') vs δ 115.3 (C-5'), δ 7.60 (H-6') vs δ 122.2 (C-6'), δ 3.87 (8-OCH_3) vs δ 61.2 (8-OCH_3), δ 3.87 (3'-OCH_3) vs δ 55.4 (3'-OCH_3), δ 5.45 (H-1'') vs δ 101.1 (C-1''), δ 3.26 (H-2'') vs δ 74.1 (C-2''), δ 3.30 (H-3'') vs δ 76.5 (C-3''), δ 3.15 (H-4'') vs δ 69.7 (C-4''), δ 3.33 (H-5'') vs δ 73.8 (C-5''), δ 4.03 (H-6''a) and 4.07 (H-6''b) vs δ 62.6 (C-6''), δ 1.74 (6''-OCOCH_3) vs δ 20.0 (6''-OCOCH_3), δ 5.08 (H-1''') vs δ 100.2 (C-1'''), δ 3.32 (H-2''') vs δ 73.0 (C-2'''), δ 3.28 (H-3''') vs δ 76.0 (C-3'''), δ 3.18 (H-4''') vs δ 69.5 (C-4'''), δ 3.47 (H-5''') vs δ 77.1 (C-5'''), δ 3.45 (H-6'''a) and 3.70 (H-6'''b) vs δ 60.5 (C-6''').

HMBC: δ 6.68 (H-6) *vs* δ 128.9 (C-8) and 105.1 (C, C-10), δ 7.90 (H-2′) *vs* δ 149.8 (C-4′) and 122.2 (C-6′), δ 6.96 (H-5′) *vs* δ 120.8 (C-1′) and 146.8 (C-3′), δ 7.60 (H-6′) *vs* δ 113.1 (C-2′), δ 3.87 (3′-OCH_3) *vs* δ 146.8 (C-3′), δ 5.45 (H-1″) *vs* δ 133.1 (C-3),″), δ 4.03 (H-6″a) and 4.07 (H-6″b) *vs* δ 169.7 (6″-OCOCH$_3$), δ 1.74 (6″-OCOCH_3) *vs* δ 169.7 (6″-OCOCH$_3$) (selected 2D-correlations were shown).

HR-FABMS: *m/z* 735.1752 ([M + Na]$^+$, Calcd. for C$_{31}$H$_{36}$O$_{19}$Na, 735.1748).

Reference

Morikawa T, Ninomiya K, Zhang Y, Yamada T, Nakamura S, Matsuda H, Muraoka O, Hayakawa T, Yoshikawa M. (2012). Flavonol glycosides with lipid accumulation inhibitory activity from *Sedum sarmentosum*. *Phytochemistry Lett* **5**: 53–58.

Siamenol A

IUPAC name: 4-Hydroxy-2-(2-hydroxypropan-2-yl)-9-((S)-1-hydroxypropyl)-5-isobutyryl-2H-furo[2,3-f]chromen-7(3H)-one

Sub-class: Coumarin

Chemical structure

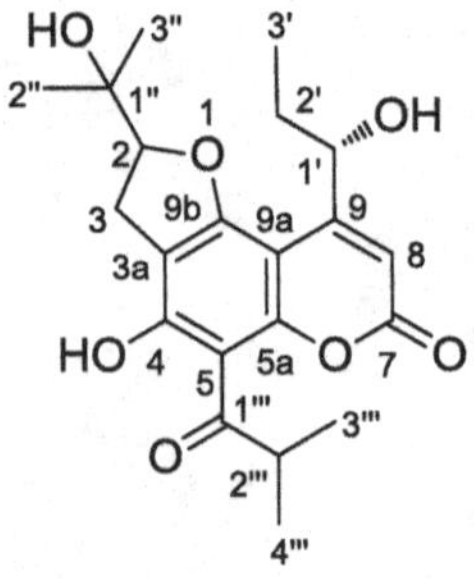

Source: *Mammea siamensis* (Family: Clusiaceae); twigs

Molecular formula: $C_{21}H_{26}O_7$

Molecular weight: 390

State: Colorless oil

Specific rotation: $[\alpha]_D^{29}$ −53.33° (CHCl$_3$, c 0.135)

UV (EtOH): λ_{max} (log ε) 221 (4.14), 295 (4.19) nm

IR (KBr): ν_{max} 3437 (OH), 2931, 1724 (C=O), 1631 (C=C), 1607, 1391 (aromatic unsaturaion), 1129 cm^{-1}

^{1}**H-NMR (CDCl$_3$, 400 MHz):** δ 4.86 (1H, t, J = 9.0 Hz, H-2), 3.13 (1H, dd, J = 15.4, 9.9 Hz, H-3a), 3.27 (1H, dd, J = 15.4, 8.7 Hz, H-3b), 6.42 (1H, s, H-8), 14.12 (1H, s, C$_4$-OH), 5.16 (1H, dd, J = 8.2, 2.0 Hz, H-1'), 1.46 (1H, m, H-2'a), 1.81 (1H, m, H-2'b), 1.09 (3H, t, J = 7.3 Hz, H-3'), 1.25 (3H, s, H-2''), 1.51 (3H, s, H-3''), 3.80 (1H, hept, J = 6.7 Hz, H-2'''), 1.14 (3H, d, J = 6.7 Hz, H-3'''), 1.20 (3H, d, J = 6.7 Hz, H-4''').

^{13}**C-NMR ((CDCl$_3$, 100 MHz):** δ 93.49 (C-2), 26.40 (C-3), 110.68 (C-3a), 163.46 (C-4), 103.99 (C-5), 156.55 (C-5a), 160.46 (C-7), 105.62 (C-8), 160.46 (C-9), 97.11 (C-9a), 161.05 (C-9b), 71.47 (C-1'), 30.52 (C-2'), 10.38 (C-3'), 71.10 (C-1''), 24.69 (C-2''), 26.96 (C-3''), 210.27 (C-1'''), 40.31 (C-2'''), 19.68 (C-3'''), 18.45 (C-4''').

HMBC: δ 3.13 (H-3a) *vs* δ 93.49 (C-2) and 71.10 (C-1''), δ 3.27 (H-3b) *vs* δ 110.68 (C-3a), δ 14.12 (C$_4$-OH) *vs* δ 110.68 (C-3a), 163.46 (C-4) and 103.99 (C-5), δ 1.25 (H-2'') *vs* δ 71.10 (C-1'') and 93.49 (C-2), δ 1.51 (H-3'') *vs* δ 71.10 (C-1'') and 93.49 (C-2), δ 3.80 (H-2''') *vs* δ 19.68 (C-3''') and 18.45 (C-4'''), δ 1.14 (H-3''') *vs* δ 210.27 (C-1'''), δ 1.20 (H-4''') *vs* δ 210.27 (C-1'''), δ 6.42 (H-8) *vs* δ 97.11 (C-9a) and 71.47 (C-1'), δ 5.16 (H-1') *vs* δ 105.62 (C-8), 97.11 (C-9a), 30.52 (C-2') and 10.38 (C-3'), δ 1.46 (H-2'a) and 1.81 (H-2'b) *vs* δ 10.38 (C-3'), δ 1.09 (H-3') *vs* δ 71.47 (C-1') and 30.52 (C-2') (selected HMC correlations were shown).

EIMS: *m/z* (%rel) 390 ([M]$^+$, 17), 364 (20), 345 (4), 332 (95), 331 (57), 329 (30), 319 (18), 303 (36), 271 (17), 257 (27).

HR-FABMS: *m/z* 391.1756 ([M + H]$^+$, Calcd. for C$_{21}$H$_{27}$O$_7$, 391.1757).

Reference

Prachyawarakorn V, Mahidol C, Ruchirawat S. (2006). Siamenols A-D, four new coumarins from *Mammea siamensis. Chem Pharm Bull* **54**: 884–886.

Sophoranone [3,5,7,3′-Tetrahydroxy-2′,4′-dimethoxy-5′-prenylisoflavanone]

IUPAC name: (S)-3,5,7-Trihydroxy-3-(3-hydroxy-2,4-dimethoxy-5-(3-methylbut-2-en-1-yl)phenyl)chroman-4-one

Sub-class: Isoflavanone

Chemical structure

Source: *Sophora mollis* subsp. *griffithii* (Stocks) Ali (Family: Leguminosae); roots

Molecular formula: $C_{22}H_{24}O_8$

Molecular weight: 416

State: Yellow gummy material

Bioactivity studied: Immunomodulatory

Specific rotation: $[\alpha]_D^{26}$ +0.4° (MeOH, c 0.09)

CD (MeOH): λ_{max} ($\Delta\varepsilon$) 288 (+29.14), 297 (+0.2287), 269 (+28.89) nm

UV (MeOH): λ_{max} 291, 330 nm

IR (KBr): ν_{max} 3371 (OH), 1639 (C=O), 1601-1400 (aromatic and aliphatic C=C) cm^{-1}

^{1}H-NMR (CD$_3$OD, 300 MHz): δ 4.69 (1H, d, J = 11.7 Hz, H-2a), 4.01 (1H, d, J = 11.7 Hz, H-2b), 5.91 (1H, d, J = 2.1 Hz, H-6), 5.88 (1H, d, J = 2.1 Hz, H-8), 7.03 (1H, s, H-6'), 3.75 (3H, s, 2'-OCH_3), 3.52 (3H, s, 4'-OCH_3), 3.23 (2H, br d, J = 6.5 Hz, H-1''), 5.17 (1H, br t, J = 6.5 Hz, H-2''), 1.72 (3H, s, H-4''), 1.67 (3H, s, H-5'').

^{13}C-NMR (CD$_3$OD, 75 MHz): δ 75.6 (C-2), 75.4 (C-3), 196.5 (C-4), 168.7 (C-5), 97.5 (C-6), 168.8 (C-7), 96.2 (C-8), 164.4 (C-9), 101.9 (C-10), 128.8 (C-1'), 147.3 (C-2'), 132.5 (C-3'), 149.5 (C-4'), 148.2 (C-5'), 114.4 (C-6'), 60.7 (2'-OCH$_3$), 61.9 (4'-OCH$_3$), 25.2 (C-1''), 124.5 (C-2''), 129.9 (C-3''), 25.8 (C-4''), 18.0 (C-5'').

HMQC: δ 4.69 (H-2a) and 4.01 (H-2b) vs δ 75.6 (C-2), δ 5.91 (H-6) vs δ 97.5 (C-6), δ 5.88 (H-8) vs δ 96.2 (C-8), δ 7.03 (H-6') vs δ 114.4 (C-6'), δ 3.75 (2'-OCH_3) vs δ 60.7 (2'-OCH_3), δ 3.52 (4'-OCH_3) vs δ 61.9 (4'-OCH$_3$), δ 3.23 (H-1'') vs δ 25.2 (C-1''), δ 5.17 (H-2'') vs δ 124.5 (C-2''), δ 1.72 (H-4'') vs δ 25.8 (C-4''), δ 1.67 (H-5'') vs δ 18.0 (C-5'').

HMBC: δ 4.69 (H-2a) vs δ 196.5 (C-4) and 164.4 (C-9), δ 4.01 (H-2b) vs δ 75.4 (C-3), δ 7.03 (H-6') vs δ 75.4 (C-3), δ 3.75 (2'-OCH_3) vs δ 147.3 (C-2'), δ 3.23 (H-1'') vs δ 114.4 (C-6') and 18.0 (C-5''), δ 5.17 (H-2'') vs δ 129.9 (C-3'') (selected HMBC correlations were shown).

EIMS: m/z (%rel) 416 (43.1), 264 (57.7), 249 (44.5), 195 (11.0), 153 (100), 124 (10.0), 68.9 (21.2), 55 (16.9).

HR-ESIMS: m/z 416.1479 ([M]$^+$), Calcd. for C$_{22}$H$_{24}$O$_8$, 416.1471).

Reference

Rahman Atta-ur, Haroone MS, Tareen RB, Mesaik MA, Jan S, Abbaskhan A, Asif M, Gulzar T, Al-Majid AM, Yousuf S, Choudhary MI. (2012). Secondary metabolites of *Sophora mollis* subsp. *griffithii* (Stocks) Ali. *Phytochemistry Lett* **5**: 613–616.

Sophoronol A

IUPAC name: (*S*)-3,5,7-Trihydroxy-3-(5-hydroxy-2,2-dimethyl-2*H*-chromen-6-yl)chroman-4-one

Sub-class: Pyranoisoflavanone

Chemical structure

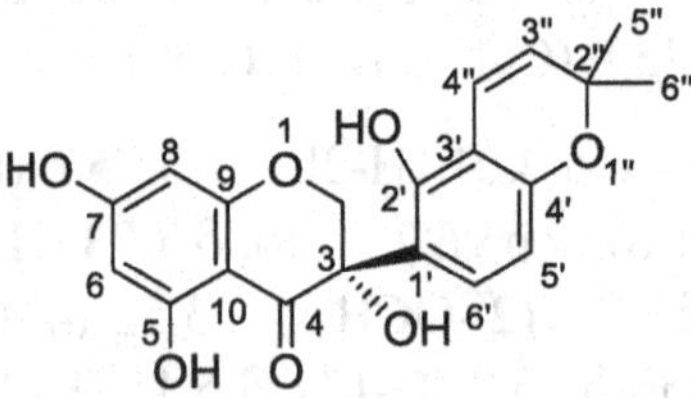

Source: *Sophora mollis* F. (Family: Liguminosae); roots

Molecular formula: $C_{20}H_{18}O_7$

Molecular weight: 370

State: White powder

Specific rotation: $[\alpha]^{25}_D$ +200° (MeOH, *c* 0.3)

UV (MeOH): λ_{max} (log ε): 228 (3.66), 290 (3.38), 331 (3.00) nm

IR (KBr): ν_{max} 3400 (OH), 1656 (C=O), 1620, 1500 cm^{-1}

^{1}H-NMR (DMSO-*d*$_6$, 400 MHz): δ 4.58 (1H, d, *J* = 11.7 Hz, H-2a), 4.06 (1H, d, *J* = 11.7 Hz, H-2b), 5.68 (2H, s, H-6 and H-8), 6.26 (1H, d, *J* = 8.4 Hz, H-5'), 7.15 (1H, d, *J* = 8.4 Hz, H-6'), 5.62 (1H, d, *J* = 10.2 Hz, H-3''),

6.62 (1H, d, J = 10.2 Hz, H-4″), 1.33 (6H, s, H-5″ and H-6″), 11.60 (1H, s, 5-OH).

^{13}C-NMR (DMSO-d_6, 100 MHz): δ 73.5 (C-2), 73.5 (C-3), 194.2 (C-4), 165.2 (C-5), 96.2 (C-6), 169.8 (C-7), 97.3 (C-8), 162.8 (C-9), 100.0 (C-10), 129.6 (C-1′), 153.9 (C-2′), 110.6 (C-3′), 150.8 (C-4′), 107.8 (C-5′), 128.0 (C-6′), 75.9 (C-2″), 129.6 (C-3″), 117.4 (C-4″), 28.0 (C-5″), 28.1 (C-6″).

HR-ESIMS: m/z 393.0955 ([M + Na]$^+$, Calcd. for $C_{20}H_{18}O_5Na$, 393.0950).

Reference

Zhang G-P, Xiao Z-Y, Rafique J, Arfan M, Smith P J, Lategan C A, Hu L-H. (2009). Antiplasmodial isoflavanones from the roots of *Sophora mollis*. *J Nat Prod* **72**: 1265–1268.

Sophoronol C

IUPAC name: (*S*)-3,5-Dihydroxy-3-(5-methoxy-2,2-dimethylchroman-6-yl)-8,8-dimethyl-2,3,9,10-tetrahydropyrano[2,3-*f*]chromen-4(8*H*)-one

Sub-class: Pyranoisoflavanone

Chemical structure

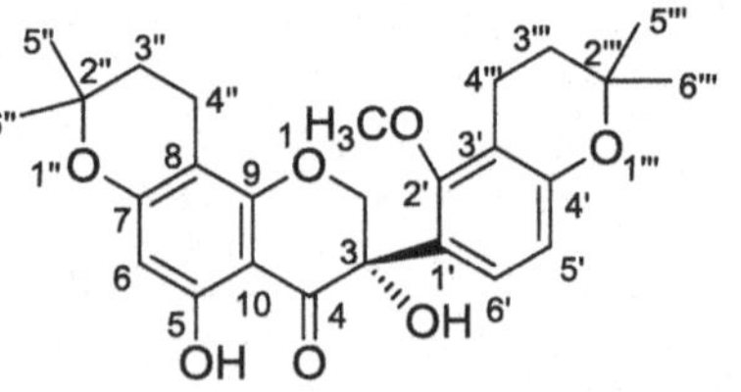

Source: *Sophora mollis* F. (Family: Liguminosae); roots

Molecular formula: $C_{26}H_{30}O_7$

Molecular weight: 454

State: White powder

Specific rotation: $[\alpha]^{25}_D$ +113° (MeOH, *c* 0.3)

UV (MeOH): λ_{max} (log ε): 297 (3.43) nm

IR (KBr): ν_{max} 3465 (OH), 2975, 1645 (C=O), 1585, 1480 cm^{-1}

¹H-NMR (CDCl₃, 400 MHz): δ 4.84 (1H, dd, *J* = 11.7 Hz, H-2a), 4.28 (1H, dd, *J* = 11.7 Hz, H-2b), 6.00 (1H, s, H-6), 6.58 (1H, d, *J* = 8.4 Hz, H-5'), 7.26 (1H, d, *J* = 8.4 Hz, H-6'), 1.75 (4H, m, H-3'' and H-3'''), 2.56 (2H, t, *J* = 6.9 Hz, H-4''), 1.33 (3H, s, H-5''), 1.32 (3H, s, H-6''), 2.72

(2H, t, J = 6.0 Hz, H-4'''), 1.35 (3H, s, H-5'''), 1.33 (3H, s, H-6'''), 3.72 (1H, s, 3-OH), 11.57 (1H, s, 5-OH), 3.71 (3H, s, 2'-OCH_3).

^{13}**C-NMR (CDCl$_3$, 100 MHz):** δ 74.3 (C-2), 73.7 (C-3), 196.1 (C-4), 161.9 (C-5), 97.8 (C-6), 163.2 (C-7), 101.0 (C-8), 159.6 (C-9), 101.1 (C-10), 121.8 (C-1'), 156.5 (C-2'), 114.9 (C-3'), 155.8 (C-4'), 112.8 (C-5'), 125.7 (C-6'), 76.2 (C-2''), 31.8 (C-3''), 16.0 (C-4''), 26.6 (C-5''), 26.1 (C-6''), 74.0 (C-2'''), 32.1 (C-3'''), 18.3 (C-4'''), 27.4 (C-5'''), 26.08 (C-6'''), 60.6 (2'-OCH_3).

HR-ESIMS: m/z 477.1891 ([M+Na]$^+$, Calcd. for C$_{26}$H$_{30}$O$_7$Na, 477.1891).

Reference

Zhang G-P, Xiao Z-Y, Rafique J, Arfan M, Smith P J, Lategan C A, Hu L-H. (2009). Antiplasmodial isoflavanones from the roots of *Sophora mollis*. *J Nat Prod* **72**: 1265–1268.

Terpurinflavone [7-Acetoxy-8-(3″-acetoxy-2″,2″-dimethyltetrahydro-4″-furanyl)flavone]

IUPAC name: (3*S*,4*S*)-4-(7-Acetoxy-4-oxo-2-phenyl-4*H*-chromen-8-yl)-2,2-dimethyltetrahydrofuran-3-yl acetate

Sub-class: Flavone

Chemical structure

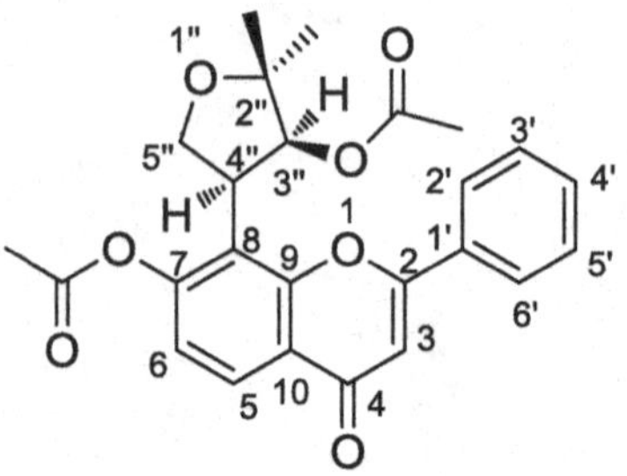

Source: *Tephrosia purpurea* (Family: Leguminosae-Papilionoideae); stems

Molecular formula: $C_{25}H_{24}O_7$

Molecular weight: 436

State: White amorphous powder

Melting point: 144–145 °C

Bioactivity studied: Antimalarial [IC_{50} values ($\mu M \pm SD$) of 3.12 ± 0.28 and 6.26 ± 2.66, respectively, against D6 and W2 strains of *Plasmodium falciparum*)

UV (CHCl$_3$): λ_{max} 295, 325 nm

^{1}H-NMR (acetone-d_6, 500 MHz): δ 6.79 (1H, s, H-3), 8.00 (1H, d, J = 8.5 Hz, H-5), 6.94 (1H, d, J = 8.5 Hz, H-6), 8.18 (2H, m, H-2' and H-6'), 7.63 (3H, m, H-3', H-4', H-5'), 5.35 (1H, d, J = 8.5 Hz, H-3''), 4.44 (1H, ddd, J = 8.5, 8.0, 2.0 Hz, H-4''), 4.84 (1H, dd, J = 9.5, 8.0 Hz, H-5''a), 5.02 (1H, dd, J = 9.5, 2.0 Hz, H-5''b), 2.00 (3H, s, C$_7$-OCOCH_3), 1.76 (3H, s, C$_{2'''}$-CH_3), 1.61 (6H, s, C$_{2'''}$-CH_3 and C$_{3'''}$-OCOCH_3).

^{13}C-NMR (acetone-d_6, 125 MHz): δ 163.9 (C-2), 108.5 (C-3), 177.6 (C-4), 129.3 (C-5), 110.1 (C-6), 167.9 (C-7), 116.0 (C-8), 155.9 (C-9), 119.9 (C-10), 133.6 (C-1'), 127.9 (C-2' and C-6'), 130.7 (C-3' and C-5'), 133.1 (C-4'), 84.0 (C-2''), 78.7 (C-3''), 42.3 (C-4''), 79.1 (C-5''), 170.3 (C$_7$-OCOCH$_3$), 23.0 (C$_7$-OCOCH$_3$), 22.4 and 24.4 (2 × C$_{2'''}$-CH$_3$), 170.3 (C$_{3'''}$-OCOCH$_3$), 20.9 (C$_{3'''}$-OCOCH$_3$).

HMBC: δ 6.79 (H-3) *vs* δ 163.9 (C-2), 177.6 (C-4), 119.9 (C-10) and 13.6 (C-1'), δ 8.00 (H-5) *vs* δ 177.6 (C-4), 167.9 (C-7) and 155.9 (C-9), δ 6.94 (H-6) *vs* δ 116.0 (C-8) and 119.9 (C-10), δ 8.18 (H-2'/H-6') *vs* δ 163.9 (C-2) and 133.6 (C-1'), δ 7.63 (H-3'/H-5') *vs* δ 127.9 (C-2'/C-6') and 133.1 (C-4'), δ 5.35 (H-3'') *vs* δ 84.0 (C-2''), 79.1 (C-5'') and 170.3 (C$_{3'''}$-OCOCH$_3$), δ 4.44 (H-4'') *vs* δ 167.9 (C-7), 116.0 (C-8), 155.9 (C-9), 84.0 (C-2''), 78.7 (C-3'') and 79.1 (C-5''), δ 4.84 (H-5''a) *vs* δ 78.7 (C-3''), δ 5.02 (H-5''b) *vs* δ 116.0 (C-8) and 42.3 (C-4''), δ 1.61 (C$_{3'''}$-OCOCH$_3$) *vs* δ 170.3 (C$_{3'''}$-OCOCH$_3$), δ 1.61 (C$_{2'''}$-CH$_3$) *vs* δ 84.0 (C-2''), 78.7 (C-3'') and 22.4 (C$_{2'''}$-CH$_3$), δ 1.76 (C$_{2'''}$-CH$_3$) *vs* δ 84.0 (C-2''), 78.7 (C-3'') and 24.4 (C$_{2'''}$-CH$_3$), δ 2.00 (C$_7$-OCOCH$_3$) *vs* δ 170.3 (C$_7$-OCOCH$_3$).

EIMS: *m/z* (%rel.) 437 ([M + 1]$^+$, 8), 376 (15), 317 (55), 316 (100), 264 (27), 265 (75).

ESI-TOF-MS: *m/z* 437.1593 ([M + H]$^+$, Calcd. for C$_{25}$H$_{25}$O$_7$, 437.1600).

Reference

Juma W P, Akala H M, Eyase F L, Muiva L M, Heydenreich M, Okalebo F A, Gitu P M, Peter M G, Walsh DS, Imbuga M, Yenesew A. (2011). Terpurinflavone: An antiplasmodial flavone from the stem of *Tephrosia purpurea. Phytochemistry Lett* **4**: 176–178.

2,3′,4,4′-Tetrahydroxy-3, 5′-diprenylchalcone

IUPAC name: (*E*)-1-(2,4-dihydroxy-3-(3-methylbut-2-en-1-yl)phenyl)-3-(3,4-dihydroxy-5-(3-methylbut-2-en-1-yl)phenyl)prop-2-en-1-one

Sub-class: Prenylated chalcone

Chemical structure

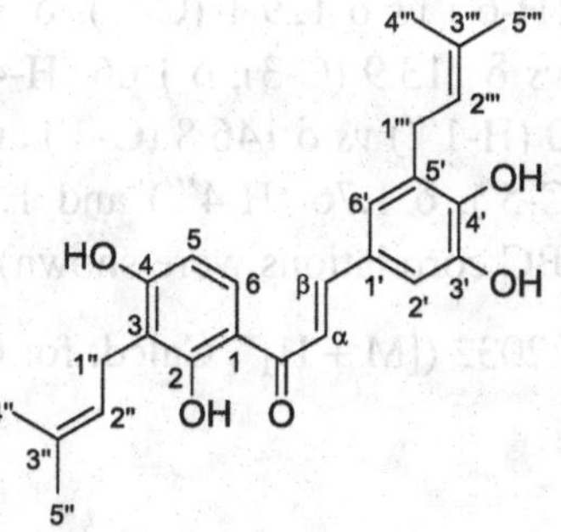

Source: *Glycyrrhiza glabra* F. (Licorice; Family: Fabaceae); roots

Molecular formula: $C_{25}H_{28}O_5$

Molecular weight: 408

State: Yellow powder

Bioactivity studied: Peroxisome proliferator-activated receptor-γ (PPAR-γ) ligand-binding activity

UV (MeOH): λ_{max} (log ε): 383 (4.39) nm

IR (KBr): v_{max} 3375 (OH), 1702, 1614 (C=O), 1566, 1494 cm^{-1}

^{1}H-NMR (Acetone-d_6, 500 MHz): δ 6.56 (1H, d, J = 8.9 Hz, H-5), 7.90 (1H, d, J = 8.9 Hz, H-6), 12.31 (1H, s, 2-OH), 7.20 (1H, br s, H-2'), 7.19 (1H, br s, H-6'), 7.64 (1H, d, J = 15.3 Hz, H-α), 7.75 (1H, d, J = 15.3 Hz, H-β), 3.39 (2H, d, J = 7.1 Hz, H-1''), 5.29 (1H, m, H-2''), 1.66 (3H, br s, H-4''), 1.79 (3H, br s, H-5''), 3.40 (2H, d, J = 7.1 Hz, H-1'''), 5.38 (1H, m, H-2'''), 1.76 (3H, br s, H-4'''), 1.74 (3H, br s, H-5''').

^{13}C-NMR (Acetone-d_6, 125 MHz): δ 115.7 (C-1), 164.6 (C-2), 113.9 (C-3), 162.1 (C-4), 107.5 (C-5), 129.6 (C-6), 126.8 (C-1'), 113.3 (C-2'), 145.1 (C-3'), 146.8 (C-4'), 129.1 (C-5'), 123.4 (C-6'), 117.8 (C-α), 145.2 (C-β), 192.5 (C=O), 21.8 (C-1''), 122.8 (C-2''), 131.0 (C-3''), 25.4 (C-4''), 17.4 (C-5''), 28.5 (C-1'''), 122.9 (C-2'''), 132.2 (C-3'''), 25.4 (C-4'''), 17.4 (C-5''').

HMBC: δ 7.90 (H-6) *vs* δ 162.1 (C-4) and 192.5 (C=O), δ 12.31 (2-OH) *vs* δ 164.6 (C-2), δ 7.64 (H-α) *vs* δ 192.5 (C=O) and 126.8 (C-1'), δ 7.75 (H-β) *vs* δ 192.5 (C=O), δ 7.20 (H-2') *vs* δ 145.2 (C-β), 145.1 (C-3') and 146.8 (C-4'), δ 7.19 (H-6') *vs* δ 129.1 (C-5'), δ 3.39 (H-1'') *vs* δ 162.1 (C-4), δ 5.29 (H-2'') *vs* δ 113.9 (C-3), δ 1.66 (H-4'') and 1.79 (H-5'') *vs* δ 122.8 (C-2''), δ 3.40 (H-1''') *vs* δ 146.8 (C-4') and 123.4 (C-6'), δ 5.38 (H-2''') *vs* δ 129.1 (C-5'), δ 1.76 (H-4''') and 1.74 (H-5''') *vs* δ 122.9 (C-2''') (selected HMBC correlations were shown).

HR-ESIMS: m/z 409.2032 ([M + H]$^+$, Calcd. for $C_{25}H_{29}O_5$, 409.2015).

Reference

Kuroda M, Mimaki Y, Honda S, Tanaka H, Yokota S, Mae T. (2010). Phenolics from *Glycyrrhiza glabra* roots and their PPAR-γ ligand-binding activity. *Bioorg Med Chem* **18**: 962–970.

5,7,3′,4′-Tetrahydroxy-3, 6-diprenylflavone

IUPAC name: 2-(3,4-Dihydroxyphenyl)-5,7-dihydroxy-3,6-bis(3-methylbut-2-en-1-yl)-4*H*-chromen-4-one

Sub-class: Prenylated flavone

Chemical structure

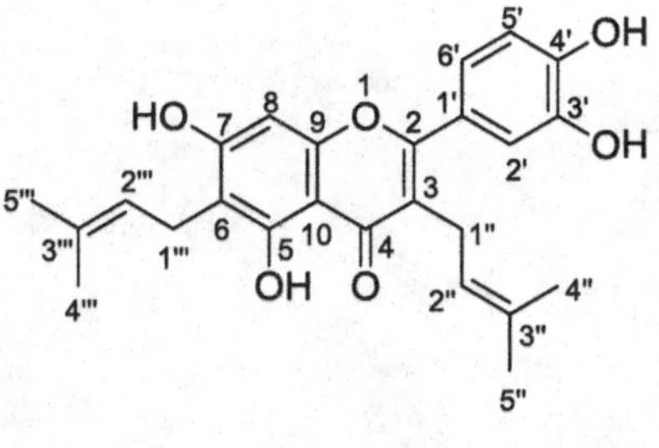

Source: *Macaranga gigantifolia* Merr. (Family: Euphorbiaceae); leaves

Molecular formula: $C_{25}H_{26}O_6$

Molecular weight: 422

State: Yellow powder

Bioactivity studied: Cytotoxic

^{1}H-NMR (Acetone-d_6, 500 MHz): δ 6.58 (1H, s, H-8), 8.05 (1H, d, *J* = 1.95 Hz, H-2′), 8.14 (1H, d, *J* = 8.43 Hz, H-5′), 7.97 (1H, dd, *J* = 8.43, 1.95 Hz, H-6′), 3.39 (2H, d, *J* = 7.14 Hz, H-1″), 5.28 (1H, t, *J* = 7.14 Hz,

H-2″), 1.74 (6H, s, H-4″ and H-5″), 3.36 (2H, d, J = 7.14 Hz, H-1‴), 5.38 (1H, t, J = 7.14 Hz, H-2‴), 1.78 (3H, s, H-4‴), 1.65 (3H, s, H-5‴).

^{13}C-NMR (Acetone-d_6, 125 MHz): δ 157.8 (C-2), 129.1 (C-3), 176.6 (C-4), 158.9 (C-5), 111.7 (C-6), 162.7 (C-7), 93.8 (C-8), 155.6 (C-9), 104.1 (C-10), 123.5 (C-1′), 130.3 (C-2′), 160.2 (C-3′), 147.0 (C-4′), 130.5 (C-5′), 128.0 (C-6′), 29.1 (C-1″), 123.3 (C-2″), 133.2 (C-3″), 17.9 (C-4″), 25.9 (C-5″), 22.0 (C-1‴), 123.2 (C-2‴), 131.7 (C-3‴), 17.9 (C-4‴), 25.9 (C-5‴).

HR-ESIMS: m/z 421.1635 [M − H]⁻ (Calcd. for $C_{25}H_{25}O_6$, 421.1631).

Reference

Darmawan A, Megawati, Lotulung P D N, Fajriah S, Primahana G, Meiliawai L. (2015). A new flavonoid derivative as cytotoxic compound isolated from ethyl acetate extract of *Macaranga gigantifolia* Merr. Leaves. *Procedia Chem* **16**: 53–57.

5,7,2',2'-Tetrahydroxyflavone 7-O-β-D-glucuronopyranoside

IUPAC name: 5(2S,3S,4S,5R,6S)-6-((2-(2,5-dihydroxyphenyl)-5-hydroxy-4-oxo-4H-chromen-7-yl)oxy)-3,4,5-trihydroxytetrahydro-2H-pyran-2-carboxylic acid

Sub-class: Flavone glucuronopyranoside

Chemical structure

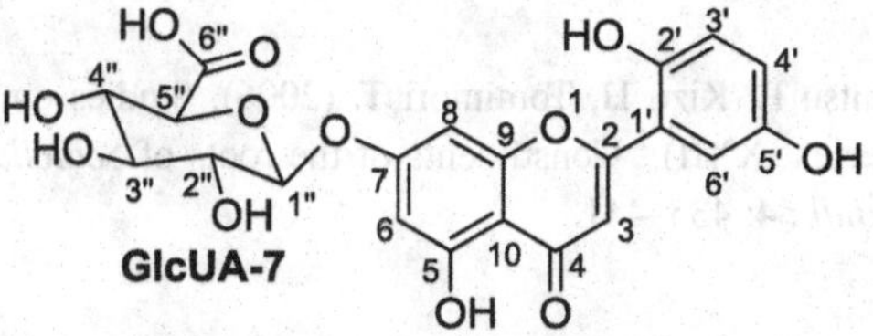

GlcUA-7

Source: *Scutellaria amabilis* Hara (Family: Labiatae); roots

Molecular formula: $C_{21}H_{18}O_{12}$

Molecular weight: 462

State: Yellow needles

Meting point: 226–227 °C (dec.)

Specific rotation: $[\alpha]^{25}_D$ −69.1° (MeOH, c 0.039)

UV (MeOH): λ_{max} (log ε): 247 (4.01), 254 (415), 260 (4.26), 267 (4.37), 300 (4.00), 362 (4.03) nm

IR (KBr): ν_{max} 3432 (OH), 1744, 1654, 1616, 1574 (aromatic unstauration) cm^{-1}

^{1}H-NMR (DMSO-d_6, 400 MHz): δ 7.14 (1H, s, H-3), 6.44 (1H, d, J = 2.2 Hz, H-6), 6.78 (1H, d, J = 2.2 Hz, H-8), 6.90 (1H, d, J = 8.8 Hz, H-3'), 6.85 (1H, dd, J = 8.8, 2.4 Hz, H-4'), 7.28 (1H, d, J = 2.4 Hz, H-6'), 5.18 (1H, d, J = 7.4 Hz, H-1''), 3.16–3.50 (3H, m, H-2'', H-3'' and H-4''), 3.80 (1H, d, J = 9.0 Hz, H-5''), 12.91 (1H, s, 5-OH), 10.22 (1H, s, 2'-OH), 9.24 (1H, s, 5'-OH), 5.31 (1H, d, J = 4.0 Hz, 2''-OH), 5.48 (1H, d, J = 5.2 Hz, 3''-OH).

^{13}C-NMR (DMSO-d_6, 100 MHz): δ 161.0 (C-2), 109.2 (C-3), 182.1 (C-4), 161.8 (C-5), 99.4 (C-6), 162.9 (C-7), 94.4 (C-8), 157.1 (C-9), 105.4 (C-10), 116.9 (C-1'), 149.7 (C-2'), 118.0 (C-3'), 120.5 (C-4'), 149.9 (C-5'), 113.4 (C-6'), 99.5 (C-1''), 73.0 (C-2''), 75.2 (C-3''), 71.4 (C-4''), 76.0 (C-5''), 170.6 (C-6'').

HR-FAB-MS: m/z 463.0871 [M + H]$^+$, Calcd. for $C_{22}H_{23}O_{11}$, 463.0877.

Reference

Miyaichi Y, Hanamitsu E, Kizu H, Tomimori T. (2006). Studies on the constituents of *Scutellaria* species (XXII).[1] Constituents of the roots of *Scutellaria amabilis* Hara. *Chem Pharm Bull* **54**: 435–441.

(2*S*)-5,7,3',4',-
Tetramethoxyflavanone

IUPAC name: (*S*)-2-(3,4-Dimethoxyphenyl)-5,7-dimethoxychroman-4-one

Sub-class: Flavanone

Chemical structure

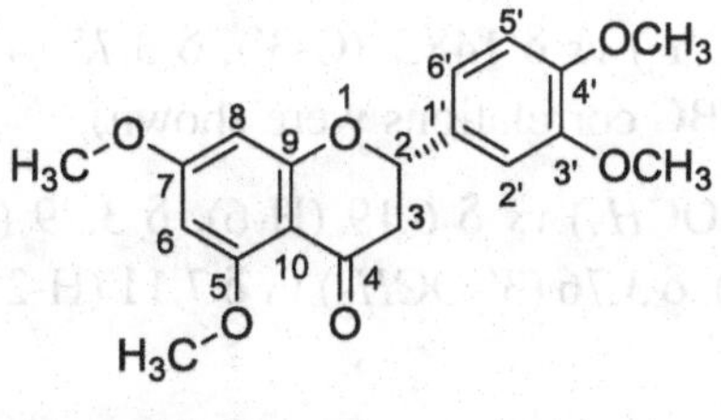

Source: *Limnophila indica* Linn. (Family: Scrophulariaceae); whole plants

Molecular formula: $C_{19}H_{20}O_6$

Molecular weight: 344

State: Colorless crystalline solid

Melting point: 119–121 °C

Specific rotation: $[\alpha]^{25}_D$ −18.2° (MeOH, *c* 0.16)

UV (MeOH): λ_{max} (log ε): 283 (4.22), 324 (sh, 3.67) nm

CD (MeOH, c 0.16): $\Delta\varepsilon_{283}$ −0.21, $\Delta\varepsilon_{324}$ +0.07

IR (KBr): ν_{max} 2823 (OCH$_3$), 1669 (C=O), 1619, 1515, 1460, 1390 cm^{-1}

^{1}H-NMR (DMSO-d_6, 400 MHz): δ 5.42 (1H, dd, J = 12.8, 2.9 Hz, H-2β), 3.09 (1H, dd, J = 16.4, 12.8 Hz, H-3$_{ax}$), 2.58 (1H, dd, J = 16.4, 2.9 Hz, H-3$_{eq}$), 6.19 (1H, d, J = 2.3 Hz, H-6), 6.21 (1H, d, J = 2.3 Hz, H-8), 7.11 (1H, d, J = 2.0 Hz, H-2'), 6.95 (1H, d, J = 8.5 Hz, H-5'), 7.01 (1H, dd, J = 8.5, 2.0 Hz, H-6'), 3.77 (3H, s, 5-OCH_3), 3.79 (3H, s, 7-OCH_3), 3.76 (3H, s, 3'-OCH_3), 3.75 (3H, s, 4'-OCH_3).

^{13}C-NMR (DMSO-d_6, 100 MHz): δ 78.5 (C-2), 44.8 (C-3), 188.0 (C-4), 161.7 (C-5), 92.7 (C-6), 165.4 (C-7), 93.6 (C-8), 164.3 (C-9), 105.4 (C-10), 131.1 (C-1'), 110.4 (C-2'), 148.7 (C-3'), 149.0 (C-4'), 111.5 (C-5'), 119.0 (C-6'), 55.8 (5-OCH_3), 55.7 (7-OCH_3), 55.6 (3'-OCH_3), 55.5 (4'-OCH_3).

HMBC: δ 5.42 (H-2β) *vs* δ 188.0 (C-4) and 110.4 (C-2'), δ 3.09 (H-3$_{ax}$) and 2.58 (H-3$_{eq}$) *vs* δ 188.0 (C-4), δ 6.19 (H-6) *vs* δ 161.7 (C-5) and 165.4 (C-7), δ 6.21 (H-8) *vs* δ 165.4 (C-7), δ 7.11 (H-2') *vs* δ 131.1 (C-1'), δ 6.95 (H-5') *vs* δ 119.0 (C-6'), δ 7.01 (H-6') *vs* δ 131.1 (C-1') and 78.5 (C-2), δ 3.77 (5-OCH_3) *vs* δ 161.7 (C-5), δ 3.79 (7-OCH_3) *vs* δ 165.4 (C-7), δ 3.76 (3'-OCH_3) *vs* δ 148.7 (C-3'), δ 3.75 (4'-OCH_3) *vs* δ 149.0 (C-4') (selected HMBC correlations were shown).

NOESY: δ 3.77 (5-OCH_3) *vs* δ 6.19 (H-6), δ 3.79 (7-OCH_3) *vs* δ 6.19 (H-6) and 6.21 (H-8), δ 3.76 (3'-OCH_3) *vs* δ 7.11 (H-2'), δ 3.75 (4'-OCH_3) *vs* δ 6.95 (H-5').

ESI-TOFMS: m/z 711.2000 [2M + Na]$^+$, 367.1238 [M + Na]+, 689.2305 [2M + H]$^+$, 345.1284 [M + H]$^+$, (Calcd. for $C_{19}H_{21}O_6$, 345.1332 [M + H]$^+$).

Reference

Reddy N P, Reddy B A K, Gunasekar D, Blond A, Bodo B, Murthy M M. (2007). Flavonoids from *Limnophila indica*. *Phytochemistry* **68**: 636–639.

Tomentodiplacol

IUPAC name: (2*R*,3*R*)-3,5,7-Trihydroxy-6-((*E*)-6-hydroxy-3,7-dimethylocta-2,7-dien-1-yl)-2-(4-hydroxy-3-methoxyphenyl)chroman-4-one

Sub-class: *C*-Geranyl flavanone

Chemical structure

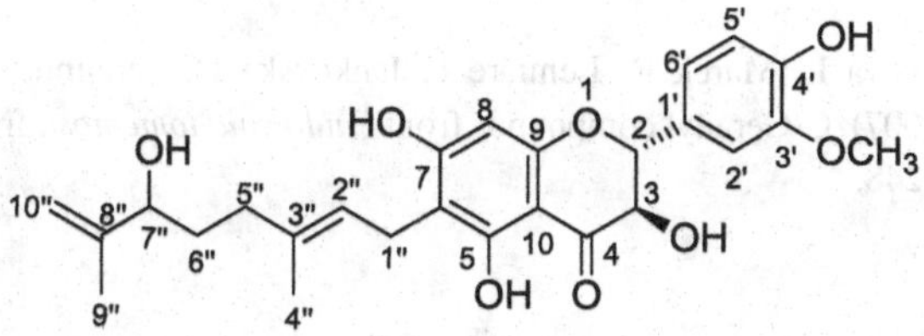

Source: *Paulownia tomentosa* Steud. (Family: Scrophulariaceae); fruit

Molecular formula: $C_{26}H_{30}O_8$

Molecular weight: 470

State: Yellow powder

UV (MeOH): λ_{max} (log ε) 215 (4.12), 236 (sh), 295 (3.52), 341 (sh) nm

FT-IR (KBr): 3419 (OH), 2929, 2833, 1636 (aromatic unsaturation), 1517, 1463, 1276, 1164, 1118 cm^{-1}

^{1}H-NMR (DMSO-*d$_6$*, 500 MHz): δ 5.00 (1H, d, *J* = 11.5 Hz, H-2β), 4.62 (1H, d, *J* = 11.5 Hz, H-3α), 5.94 (1H, s, H-8), 7.08 (1H, d, *J* = 1.8 Hz, H-2'), 6.78 (1H, d, *J* = 8.1 Hz, H-5'), 6.89 (1H, dd, *J* = 8.1, 1.8 Hz, H-6'),

293

3.13 (2H, d, H-1″), 5.13 (1H, t, H-2″), 1.70 (3H, s, H-4″), 1.85 (1H, m, H-5″a), 1.94 (1H, m, H-5″b), 1.45 (2H, m, H-6″), 3.80 (1H, q, H-7″), 1.61 (3H, s, H-9″), 4.70 (1H, s, H-10″a), 4.81 (1H, s, H-10″b), 3.78 (3H, s, 3′-OCH_3), 5.71 (1H, d, 3-OH), 12.19 (1H, s, 5-OH), 10.75 (1H, br s, 7-OH), 9.07 (1H, s, 4′-OH), 4.60 (1H, s, 7″-OH).

^{13}C-NMR (DMSO-d_6, 125 MHz): δ 83.1 (C-2), 71.4 (C-3), 198.0 (C-4), 160.3 (C-5), 107.8 (C-6), 164.4 (C-7), 94.4 (C-8), 160.1 (C-9), 100.2 (C-10), 128.2 (C-1′), 112.2 (C-2′), 147.3 (C-3′), 147.0 (C-4′), 114.9 (C-5′), 121.1 (C-6′), 20.5 (C-1″), 122.0 (C-2″), 134.1 (C-3″), 16.0 (C-4″), 35.2 (C-5″), 33.4 (C-6″), 73.4 (C-7″), 148.2 (C-8″), 17.5 (C-9″), 109.8 (C-10″), 55.7 (3′-OCH_3).

ESIMS: m/z 469 [M − H]$^-$

HRMS-QTOF: m/z 469.1851 [M − H]$^-$ (Calcd. for $C_{26}H_{29}O_8$, 469.2019).

Reference

Šmejkal K, Grycová L, Marek R, Lemiáre F, Jankovská D, Forejtníková H, Vančo J, Suchý V. (2007). C-Geranyl compounds from *Paulownia tomentosa* fruits. *J Nat Prod* **70**: 1244–1248.

(2*R*,3*R*)-3,4′,7-Trihydroxy-3′-prenylflavanone

IUPAC name: (2*R*,3*R*)-3,7-Dihydroxy-2-(4-hydroxy-3-(3-methylbut-2-en-1-yl)phenyl)chroman-4-one

Sub-class: Prenylated flavanone

Chemical structure

Source: *Glycyrrhiza glabra* F. (Licorice; Family: Fabaceae); roots

Molecular formula: $C_{20}H_{20}O_5$

Molecular weight: 340

State: Yellow powder

Bioactivity studied: Peroxisome proliferator-activated receptor-γ (PPAR-γ) ligand-binding activity

Specific rotation: $[\alpha]^{25}_D$ –38.0° (MeOH, *c* 0.01)

UV (MeOH): λ_{max} (log ε): 276 (4.10), 313 (3.84) nm

IR (KBr): ν_{max} 3374 (OH), 1673 (C=O), 1608, 1502 cm^{-1}

¹H-NMR (Acetone-d_6, 500 MHz): δ 5.03 (1H, d, J = 11.9 Hz, H-2β), 4.59 (1H, d, J = 11.9 Hz, H-3α), 7.74 (1H, d, J = 8.6 Hz, H-5), 6.64 (1H, dd, J = 8.6, 2.2 Hz, H-6), 6.41 (1H, d, J = 2.2 Hz, H-8), 7.35 (1H, d, J = 2.0 Hz,, H-2'), 6.90 (1H, d, J = 8.2 Hz, H-5'), 7.27 (1H, dd, J = 8.2, 2.2 Hz, H-6'), 3.38 (2H, d, J = 7.3 Hz, H-1''), 5.39 (1H, m, H-2''), 1.72 (3H, br s, H-4''), 1.74 (3H, br s, H-5'').

¹³C-NMR (Acetone-d_6, 125 MHz): δ 84.6 (C-2), 73.4 (C-3), 192.7 (C-4), 129.3 (C-5), 111.2 (C-6), 165.4 (C-7), 103.1 (C-8), 164.1 (C-9), 112.5 (C-10), 128.9 (C-1'), 130.0 (C-2'), 128.1 (C-3'), 155.8 (C-4'), 115.0 (C-5'), 127.1 (C-6'), 28.7 (C-1''), 123.1 (C-2''), 132.0 (C-3''), 25.4 (C-4''), 17.4 (C-5'').

HMBC: δ 5.03 (H-2) *vs* δ 128.9 (C-1'), δ 4.59 (H-3) *vs* δ 128.9 (C-1'), δ 7.74 (H-5) *vs* δ 192.7 (C-4) and 164.1 (C-9), δ 6.64 (H-6) *vs* δ 165.4 (C-7), δ 6.41 (H-8) *vs* δ 165.4 (C-7) and 164.1 (C-9), δ 7.35 (H-2') *vs* δ 84.6 (C-2) and 155.8 (C-4'), δ 6.90 (H-5') *vs* δ 155.8 (C-4'), δ 7.27 (H-6') *vs* δ 84.6 (C-2) and 155.8 (C-4'), δ 3.38 (H-1'') *vs* δ 130.0 (C-2'), 128.1 (C-3'), 155.8 (C-4'), δ 5.39 (H-2'') *vs* δ 128.1 (C-3'), δ 1.72 (H-4'') and 1.74 (H-5'') *vs* δ 123.1 (C-2'') (selected HMBC correlations were shown).

HR-ESIMS: m/z 341.1411 ([M + H]⁺, Calcd. for $C_{20}H_{21}O_5$, 341.1389).

Reference

Kuroda M, Mimaki Y, Honda S, Tanaka H, Yokota S, Mae T. (2010). Phenolics from *Glycyrrhiza glabra* roots and their PPAR-γ ligand-binding activity. *Bioorg Med Chem* **18**: 962–970.

(3*S*)-2',4',5-Trihydroxy-[5''-(1,2-dihydroxy-1-methylethyl)-dihydrofurano(2'',3'':7,8)]-isoflavanone

IUPAC name: (3*S*)-3-(2,4-dihydroxyphenyl)-8-(1,2-dihydroxypropan-2-yl)-5-hydroxy-8,9-dihydro-2*H*-furo[2,3-*h*]chromen-4(3*H*)-one

Sub-class: Isoflavanone (dihydrofuranoisoflavanone)

Chemical structure

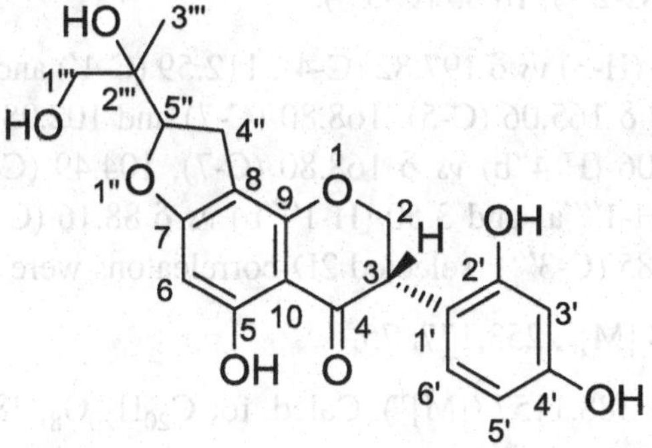

Source: *Lespedeza maximowiczi* Schneid (Family: Liguminosae); leaves

Molecular formula: $C_{20}H_{20}O_8$

Molecular weight: 388

State: Brownish powder

Melting point: 149–151 °C

Bioactivity studied: Inhibitor against formation of advanced glycation end products (IC_{50} 20.6 μM)

Specific rotation: $[\alpha]_D^{18}$ +126.3° (MeOH, c 0.025)

UV (MeOH): λ_{max} (log ε) 241 (sh, 3.14), 295 (3.46), 336 (sh, 2.73) nm

FTIR (ATR): 3284 (OH), 1644 (C=O), 1602, 1519, 1471, 1375 (aromatic ring), 1272, 1209, 1172, 1136, 1098 cm^{-1}

^{1}H-NMR (CD$_3$OD, 600 MHz): δ 4.59 (1H, t, J = 10.3 Hz, H-2a), 4.43 (1H, dd, J = 10.3, 5.4 Hz, H-2b), 4.18 (1H, dd, J = 10.3, 5.4 Hz, H-3β), 5.94 (1H, s, H-6), 6.33 (1H, d, J = 2.3 Hz, H-3′), 6.25 (1H, dd, J = 8.6, 2.3 Hz, H-5′), 6.82 (1H, d, J = 8.6 Hz, H-6′), 3.07 (1H, dd, J = 15.0, 8.4 Hz, H-4″a), 3.06 (1H, dd, J = 15.0, 9.5 Hz, H-4″b), 4.92 (1H, br t, J = 9.0 Hz, H-5″), 3.68 (1H, d, J = 11.0 Hz, H-1⁗a), 3.50 (1H, d, J = 11.0 Hz, H-1⁗b), 1.16 (3H, s, H-3⁗).

^{13}C-NMR (CD$_3$OD, 150 MHz): δ 70.26 (C-2), 46.58 (C-3), 197.82 (C-4), 165.06 (C-5), 90.66 (C-6), 168.80 (C-7), 104.49 (C-8), 157.81 (C-9), 102.98 (C-10), 112.59 (C-1′), 157.63 (C-2′), 102.38 (C-3′), 156.22 (C-4′), 106.37 (C-5′), 130.46 (C-6′), 25.37 (C-4″), 88.16 (C-5″), 66.30 (C-1‴), 73.32 (C-2‴), 18.85 (C-3‴).

HMBC: δ 4.18 (H-3) *vs* δ 197.82 (C-4), 112.59 (C-1′) and 130.46 (C-6′), δ 5.94 (H-6) *vs* δ 165.06 (C-5), 168.80 (C-7) and 102.98 (C-10), δ 3.07 (H-4″a) and 3.06 (H-4″b) *vs* δ 168.80 (C-7), 104.49 (C-8) and 157.81 (C-9), δ 3.68 (H-1⁗a) and 3.50 (H-1⁗b) *vs* δ 88.16 (C-5″), and 73.32 (C-2‴) and 18.85 (C-3‴) (selected 2D-correleatons were shown).

EIMS: m/z 388 [M]$^+$, 253,177, 75.

HR-EIMS: m/z 388.1157 ([M]$^+$), Calcd. for $C_{20}H_{20}O_8$, 388.1158).

Reference

Park H Y, Kim G B, Kwon Y S. (2010). Two new dihydrofuranoisoflavanones from the leaves of *Lespedeza maximowiczi* and their inhibitory effect on the formation of advanced glycation end products. *Arch Pharm Res* **33**: 1159–1163.

(3*S*)-7,2′,4′-Trihydroxy-5, 5′-dimethoxy-6-(3-methylbut-2-enyl)-isoflavan

IUPAC name: (*S*)-4-(7-Hydroxy-5-methoxy-6-(3-methylbut-2-en-1-yl) chroman-3-yl)-6-methoxybenzene-1,3-diol

Sub-class: Prenylated isoflavan

Chemical structure

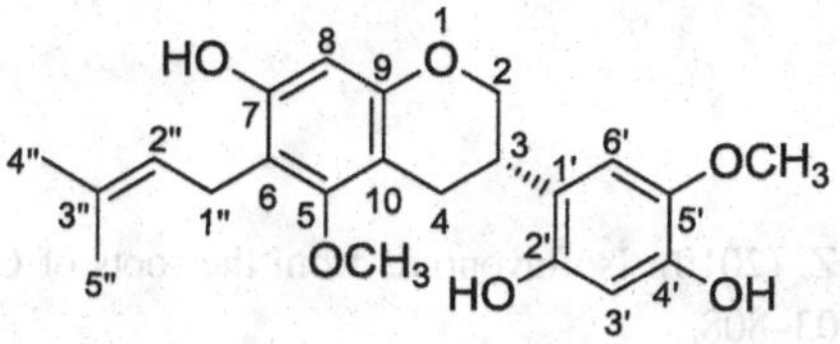

Source: *Campylotropis hirtella* (Franch.) Schindl. (Family: Leguminosae); roots

Molecular formula: $C_{22}H_{26}O_6$

Molecular weight: 386

State: Pale yellow oil

Bioactivity studied: *In vitro* immunosuppressive activity

Specific rotation: $[\alpha]^{25}_D$ −33.1° (MeOH, *c* 1.01)

UV: λ_{max} (MeOH): 235, 285 nm

IR (KBr): v_{max} 3385 (OH), 2935, 1697, 1616, 1518, 1450, 1198, 1165, 1124, 1113, 1076, 860, 835 cm^{-1}

^{1}H-NMR (CD$_3$COCD$_3$, 400 MHz): δ 4.18 (1H, m, H-2a), 4.01 (1H, m, H-2b), 3.42 (1H, m, H-3), 2.89 (1H, m, H-4a), 2.82 (1H, m, H-4b), 6.17 (1H, s, H-8), 6.48 (1H, s, H-3′), 6.479 (1H, s, H-6′), 3.28 (2H, m, H-1″), 5.26 (1H, m, H-2″), 1.75 (3H, s, H-4″), 1.64 (3H, s, H-5″), 3.71 (3H, s, 5-OCH$_3$), 3.74 (3H, s, 5′-OCH$_3$).

^{13}C-NMR (CD$_3$COCD$_3$, 100 MHz): δ 69.6 (C-2), 31.9 (C-3), 26.0 (C-4), 157.7 (C-5), 113.8 (C-6), 154.8 (C-7), 99.2 (C-8), 153.8(C-9), 107.7 (C-10), 118.3 (C-1′), 149.6 (C-2′), 103.7 (C-3′), 146.2 (C-4′), 141.3 (C-5′), 112.4 (C-6′), 22.7 (C-1″), 124.7 (C-2″), 129.6 (C-3″), 17.3 (C-4″), 25.2 (C-5″), 60.0 (5-OCH$_3$), 56.7 (5′-OCH$_3$).

EIMS: m/z (%rel.) 386 ([M]$^+$, 85), 369 (17), 331 (10), 222 (14), 221 (100), 178 (11), 177 (18), 167 (10), 166 (84), 165 (83), 153 (26), 151 (23), 137 (11).

ESIMS: m/z 387 [M + H]$^+$

HR-EIMS: m/z 386.1737 ([M$^+$], Calcd. for C$_{22}$H$_{26}$O$_6$, 386.1729).

Reference

Shou Q, Tan Q, Shen Z. (2010). Isoflavonoids from the roots of *Campylotropis hirtella*. *Planta Med* **76**: 803–808.

5,2',4'-Trihydroxy-7-methoxy-5"-(3-methylbuten-2-yl)isoflavone

IUPAC name: 3-(2,4-Dihydroxy-5-(3-methylbut-2-en-1-yl)phenyl)-5-hydroxy-7-methoxy-4H-chromen-4-one

Sub-class: Prenylatedisoflavone

Chemical structure

Source: *Erythrina poeppigiana* (Walp.) O.F. Cook (Family: Fabaceae); stem barks

Molecular formula: $C_{21}H_{20}O_6$

Molecular weight: 368

Bioactivity studied: Binding affinity for estrogen receptor

State: Yellow amorphous solid

UV (MeOH): λ_{max} (log ε): 266 (4.24), 365 (3.86) nm

IR (KBr): ν_{max} 3340 (OH), 1658 (C=O), 1618, 1580, 1513 cm^{-1}

^{1}**H-NMR (CDCl$_3$, 600 MHz):** δ 7.98 (1H, s, H-2), 6.45 (1H, d, J = 2.0 Hz, H-6), 6.48 (1H, d, J = 2.0 Hz, H-8), 3.90 (3H, s, 7-OCH_3), 12.31 (1H, s, 5-OH), 6.56 (1H, s, H-3′), 6.86 (1H, s, H-6′), 3.31 (2H, d, J = 7.3 Hz, H-1″), 5.30 (1H, t, J = 7.3 Hz, H-2″), 1.77 (6H, s, H-4″ and H-5″).

^{13}**C-NMR (CDCl$_3$, 150 MHz):** δ 155.0 (C-2), 123.4 (C-3), 182.0 (C-4), 162.3 (C-5), 98.9 (C-6), 166.2 (C-7), 92.5 (C-8), 157.8 (C-9), 105.6 (C-10), 111.9 (C-1′), 155.5 (C-2′), 106.8 (C-3′), 156.5 (C-4′), 120.0 (C-5′), 130.7 (C-6′), 28.9 (C-1″), 121.8 (C-2″), 134.8 (C-3″), 25.8 (C-4″), 17.9 (C-5″), 55.9 (7-OCH$_3$).

ESIMS (+ve): m/z 369 [M + 1]$^+$ (100), 337 (28), 301 (14).

HR-ESIMS: m/z 368.3855 ([M]$^+$, Calcd. for C$_{21}$H$_{20}$O$_6$, 368.3857).

Reference

Djiogue S, Halabalaki M, Alexi X, Njamen D, Fomum Z T, Alexis M N, Skaltsounis A-L. (2009). Isoflavonoids from *Erythrina poeppigiana*: evaluation of their binding affinity for the estrogen receptor. *J Nat Prod* **72**: 1603–1607.

5,7,2'-Trihydroxy-8-methoxyflavone 7-*O*-β-D-glucopyranoside

IUPAC name: 5-Hydroxy-2-(2-hydroxyphenyl)-8-methoxy-7-(((2S,3R,4S,5S,6R)-3,4,5-trihydroxy-6-(hydroxymethyl)tetrahydro-2H-pyran-2-yl)oxy)-4H-chromen-4-one

Sub-class: Flavone glycoside

Chemical structure

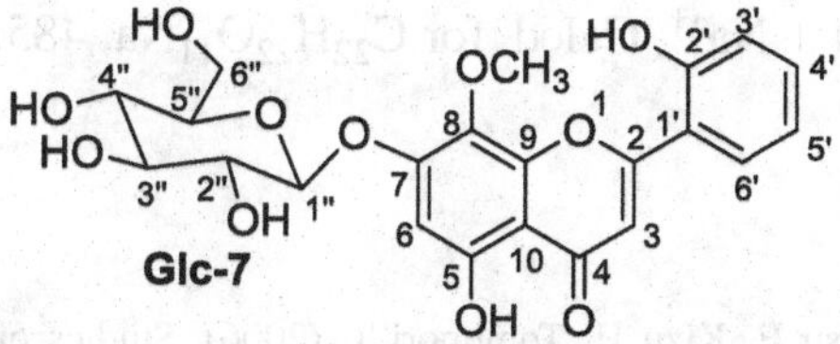

Source: *Scutellaria amabilis* Hara (Family: Labiatae); roots

Molecular formula: $C_{22}H_{22}O_{11}$

Molecular weight: 462

State: Yellow needles

Meting point: 263–264 °C (dec.)

Specific rotation: $[\alpha]^{25}_D$ −77.5° (MeOH, *c* 0.035)

UV (MeOH): λ_{max} (log ε): 246 (3.60), 254 (3.82), 260 (4.14), 272 (4.40), 337 (4.06) nm

303

IR (KBr): v_{max} 3404 (OH), 1654 (α,β-unsaturated carbonyl), 1616, 1574 (aromatic unstauration) cm^{-1}

^{1}H-NMR (DMSO-d_6, 400 MHz): δ 7.13 (1H, s, H-3), 6.66 (1H, s, H-6), 7.08 (1H, br d, J = 8.0 Hz, H-3'), 7.43 (1H, ddd, J = 8.0, 2.0 Hz, H-4'), 7.05 (1H, br dd, J = 8.0 Hz, H-5'), 7.89 (1H, dd, J = 8.0, 2.0 Hz, H-6'), 5.07 (1H, d, J = 7.6 Hz, H-1''), 3.18-3.35 (3H, m, H-2'', H-3'' and H-4''), 3.42 (1H, m, H-5''), 3.84 (1H, m, H-6''a), 3.71 (1H, dd, J = 11.0, 5.8 Hz, H-6''b), 3.87 (3H, s, 8-OCH_3), 12.59 (1H, s, 5-OH), 10.91 (1H, s, 2'-OH), 5.06 (1H, d, J = 5.0 Hz, 2''-OH), 5.14 (1H, d, J = 4.5 Hz, 3''-OH), 5.40 (1H, d, J = 5.0 Hz, 4''-OH), 4.61 (1H, t, J = 5.8 Hz, 6''-OH).

^{13}C-NMR (DMSO-d_6, 100 MHz): δ 161.7 (C-2), 109.0 (C-3), 182.4 (C-4), 156.0 (C-5), 98.6 (C-6), 156.3 (C-7), 129.1 (C-8), 149.1 (C-9), 104.9 (C-10), 117.2 (C-1'), 157.0 (C-2'), 117.2 (C-3'), 133.1 (C-4'), 119.6 (C-5'), 128.3 (C-6'), 100.3 (C-1''), 73.2 (C-2''), 76.6 (C-3''), 69.6 (C-4''), 77.2 (C-5''), 60.6 (C-6''), 61.3 (8-OCH_3).

EIMS: m/z (%rel): 300 (45), 285 (76), 270 (100), 257 (25).

HR-FAB-MS: m/z 463.1231 ([M + H]$^+$, Calcd. for $C_{22}H_{23}O_{11}$, 463.1240); m/z 485.1054 ([M + Na]$^+$, Calcd. for $C_{22}H_{22}O_{11}Na$, 485.1060)

Reference

Miyaichi Y, Hanamitsu E, Kizu H, Tomimori T. (2006). Studies on the constituents of *Scutellaria* species (XXII).[1] Constituents of the roots of *Scutellaria amabilis* Hara. *Chem Pharm Bull* **54**: 435–441.

5,7,4'-Trihydroxy-8-*p*-hydroxybenzyldihydroflavonol

IUPAC name: (2*R*,3*R*)-3,5,7-trihydroxy-8-(4-hydroxybenzyl)-2-(4-hydroxyphenyl)chroman-4-one

Sub-class: Benzylated dihydroflavonol

Chemical structure

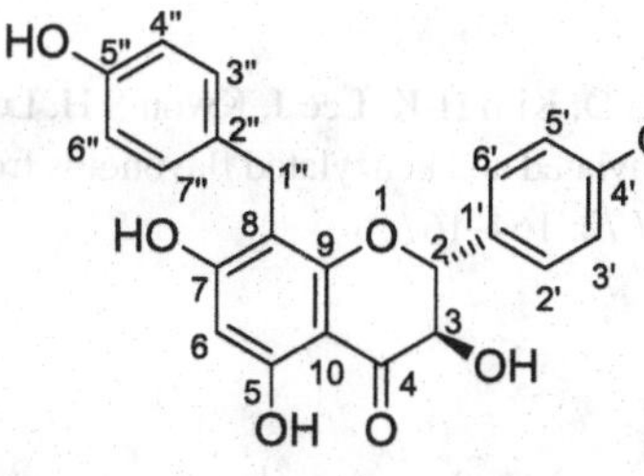

Source: *Cudrania tricuspidata* (carr.) Bur. (Family: Moraceae); fruits

Molecular formula: $C_{22}H_{18}O_7$

Molecular weight: 394

State: Yellow powder

Melting point: 126–128 °C

Specific rotation: $[\alpha]^{25}_D$ +30.81° (MeOH, *c* 0.03)

UV (MeOH): λ_{max} (log ε) 296.2 (4.3) nm

IR (dried film): v_{max} 3439 (OH), 1635 (α,β-unsaturated carbonyl), 609 cm^{-1}

^{1}H-NMR (Acetone-d_6, 500 MHz): δ 5.23 (1H, d, J = 11.5 Hz, H-2β), 4.75 (1H, d, J = 11.5 Hz, H-3α), 6.26 (1H, s, H-6), 7.57 (2H, d, J = 8.4 Hz, H-2′ and H-6′), 7.05 (2H, d, J = 8.4 Hz, H-3′ and H-5′), 3.91 (1H, d, J = 14.0 Hz, H-1″a), 3.84 (1H, d, J = 14.0 Hz, H-1″b), 7.21 (2H, d, J = 8.3 Hz, H-3″ and H-7″), 6.80 (2H, d, J = 8.3 Hz, H-4″ and H-6″), 11.81 (1H, s, 5-OH).

^{13}C-NMR (Acetone-d_6, 125 MHz): δ 84.0 (C-2), 73.0 (C-3), 198.3 (C-4), 162.6 (C-5), 96.5 (C-6), 165.3 (C-7), 109.1 (C-8), 160.8 (C-9), 101.3 (C-10), 129.1 (C-1′), 129.7 (C-2′), 115.6 (C-3′), 158.5 (C-4′), 115.6 (C-5′), 129.7 (C-6′), 27.5 (C-1″), 132.7 (C-2″), 130.1 (C-3″), 115.3 (C-4″), 156.0 (C-5″), 115.3 (C-6″), 130.1 (C-7″).

HR-FABMS: m/z 395.1136 ([M + H]$^+$, Calcd. for $C_{22}H_{19}O_7$, 395.1131).

Reference

Han X H, Hong S S, Jin Q, Li D, Kim H-K, Lee J, Kwon S H, Lee D, Lee C-K, Lee M K, Hwang B Y. (2009). Prenylated and benzylated flavonoids from the fruits of *Cudrania tricuspidata*. *J Nat Prod* **72**: 164–167.

(2*S*)-6,3′,4′-Trimethoxy-[2″,3″:7,8]-furanoflavanone

IUPAC name: (*S*)-2-(3,4-Dimethoxyphenyl)-6-methoxy-2*H*-furo[2,3-*h*] chromen-4(3*H*)-one

Sub-class: Furanoflavanone

Chemical structure

Source: *Millettia erythrocalyx* Gagnep. (Family: Ligumonisae); pods

Molecular formula: $C_{20}H_{18}O_6$

Molecular weight: 354

State: White powder

Specific rotation: $[\alpha]^{28}_D$ −1.53° (MeOH, *c* 0.2)

UV (MeOH): λ_{max} (log ε): 233 (4.46), 348 (2.78) nm

IR (film): ν_{max} 1678 (C=O), 1265, 1217 cm^{-1}

^{1}H-NMR (CDCl$_3$, 500 MHz): δ 5.51 (1H, dd, *J* = 13.4, 3.1 Hz, H-2β), 3.16 (1H, dd, *J* = 17.1, 13.4 Hz, H-3$_{ax}$), 2.89 (1H, dd, *J* = 17.1, 3.1 Hz,

H-3$_{eq}$), 7.28 (1H, s, H-5), 7.03 (1H, d, J = 2.1 Hz, H-2′), 6.91 (1H, d, J = 8.8 Hz, H-5′), 7.04 (1H, dd, J = 8.8, 2.1 Hz, H-6′), 6.92 (1H, d, J = 2.1 Hz, H-4″), 7.61 (1H, d, J = 2.1 Hz, H-5″), 3.99 (3H, s, 6-OCH_3), 3.91 (3H, s, 3′-OCH_3), 3.90 (3H, s, 4′-OCH_3).

^{13}C-NMR (CDCl$_3$, 125 MHz): δ 80.5 (C-2), 44.3 (C-3), 191.4 (C-4), 102.1 (C-5), 141.5 (C-6), 149.8 (C-7), 119.2 (C-8), 151.6 (C-9), 115.4 (C-10), 131.2 (C-1′), 109.5 (C-2′), 149.5 (C-3′), 149.2 (C-4′), 111.1 (C-5′), 118.9 (C-6′), 105.3 (C-4″), 145.3 (C-5″), 56.3 (6-OCH_3), 56.0 (3′-OCH_3), 56.0 (4′-OCH_3).

HR-ESI-TOFMS: m/z 355.1190 ([M + H])$^+$, Calcd. for C$_{20}$H$_{19}$O$_6$, 355.1181).

Reference

Sritularak B, Likhitwitayawuid K. (2006). Flavonoids from the pods of *Millettia erythrocalyx*. *Phytochemistry* **67**: 812–817.

4′,5,6-Trimethoxyisoflavone-7-*O*-β-D-glucopyranosyl(1→4)-α-L-rhamnopyranosyl(1→6)-β-D-glucopyranosyl(1→3)-[α-L-rhamnopyranosyl(1→6)]-β-D-glucopyranoside

IUPAC name: 7-((((2*R*,3*S*,4*R*,5*S*,6*S*)-4-(((2*R*,3*S*,4*R*,5*R*,6*S*)-6-(((((2*R*,3*R*,4*S*,5*R*,6*S*)-3,4-dihydroxy-6-methyl-5-(((2*R*,3*S*,4*R*,5*R*,6*S*)-3,4,5-trihydroxy-6-(hydroxymethyl)tetrahydro-2*H*-pyran-2-yl)oxy)tetrahydro-2*H*-pyran-2-yl)oxy)methyl)-3,4,5-trihydroxytetrahydro-2*H*-pyran-2-yl)oxy)-3,5-dihydroxy-6-(((((2*R*,3*R*,4*R*,5*R*,6*R*)-3,4,5-trihydroxy-6-methyltetrahydro-2*H*-pyran-2-yl)oxy)methyl)tetrahydro-2*H*-pyran-2-yl)oxy)-5,6-dimethoxy-3-(4-methoxyphenyl)-4*H*-chromen-4-one

Sub-class: Isoflavonoid glycoside

Chemical structure

Source: *Baphia bancoensis* (Family: Fabaceae); roots

Molecular formula: $C_{48}H_{66}O_{29}$

Molecular weight: 1107

State: Brown lacquer

Specific rotation: $[\alpha]^{25}{}_D$ −8.0° (MeOH, *c* 0.23)

UV (MeOH): λ_{max} (log ε) 260 (4.35), 317 (3.82) nm

FT-IR (KBr): ν_{max} 3450 (OH), 2959, 1658 (α,β-unsaturated carbonyl), 1623, 1575, 1460, 1371 (aromatic unsaturation), 1295, 1231, 1153, 1067, 1036, 812 cm^{-1}

^{1}H-NMR (CD$_3$OD, 500 MHz): δ 8.25 (1H, s, H-2), 7.22 (1H, s, H-8), 7.50 (2H, d, *J* = 8.7 Hz,, H-2′ and H-4′), 7.01 (2H, d, *J* = 8.7 Hz,, H-3′ and H-5′), 3.96 (3H, s, 5-OC*H₃*), 3.94 (3H, s, 6-OC*H₃*), 3.85 (3H, s, 4′-OC*H₃*), 5.21 (1H, d, *J* = 8.0 Hz, H-1″), 3.84 (1H, t, *J* = 8.0 Hz, H-2″), 3.69 (1H, t, *J* = 8.7 Hz, H-3″), 3.50 (1H, dd, *J* = 9.6, 8.7 Hz, H-4″), 3.89 (1H, m, H-5″), 3.59 (1H, dd, *J* = 10.4, 7.3 Hz, H-6″a), 4.16 (1H, dd, *J* = 10.6, 2.7 Hz, H-6″b), 4.52 (1H, d, *J* = 7.8 Hz, H-1‴), 3.37 (1H, dd, *J* = 8.9, 7.7 Hz, H-2‴), 3.45 (1H, t, *J* = 9.1 Hz, H-3‴), 3.27 (1H, t, *J* = 9.1 Hz, H-4‴), 3.64 (1H, m, H-5‴), 3.71 (1H, dd, *J* = 11.0, 7.8 Hz, H-6‴a), 4.06 (1H, dd, *J* = 11.1, 2.2 Hz, H-6‴b), 4.74 (1H, d, *J* = 1.6 Hz, H-1⁗), 3.95 (1H, dd, *J* = 3.4, 1.6 Hz, H-2⁗), 3.90 (1H, dd, *J* = 9.4, 3.4 Hz, H-3⁗), 3.40 (1H, t, *J* = 9.6 Hz, H-4⁗), 3.74 (1H, dq, *J* = 9.5, 6.2 Hz, H-5⁗), 1.27 (3H, d, *J* = 6.2 Hz, H-6⁗), 4.78 (1H, d, *J* = 1.7 Hz, H-1‴‴), 3.95 (1H, dd,

$J = 3.4$, 1.6 Hz, H-2‴‴′), 4.06 (1H, dd, $J = 9.3$, 3.4 Hz, H-3‴‴′), 3.63 (1H, t, $J = 9.2$ Hz, H-4‴‴′), 3.84 (1H, m, H-5‴‴′), 1.42 (3H, d, $J = 6.2$ Hz, H-6‴‴′), 4.56 (1H, d, $J = 7.7$ Hz, H-1‴‴″), 3.35 (1H, dd, $J = 8.5$, 7.7 Hz, H-2‴‴″), 3.41 (1H, t, $J = 9.1$ Hz, H-3‴‴″), 3.18 (1H, t, $J = 9.5$ Hz, H-4‴‴″), 3.28 (1H, m, H-5‴‴″), 3.55 (1H, dd, $J = 12.0$, 6.0 Hz, H-6‴‴″a), 3.80 (1H, dd, $J = 12.1$, 2.4 Hz, H-6‴‴″b).

^{13}C-NMR (CD$_3$OD, 125 MHz): δ 154.0 (C-2), 126.1 (C-3), 177.5 (C-4), 154.1 (C-5), 142.1 (C-6), 157.1 (C-7), 102.1 (C-8), 155.7 (C-9), 115.3 (C-10), 125.6 (C-1′), 131.6 (C-2′), 114.7 (C-3′), 161.1 (C-4′), 114.7 (C-5′), 131.6 (C-6′), 62.7 (5-OCH$_3$), 62.2 (6-OCH$_3$), 55.7 (4′-OCH$_3$), 101.6 (C-1″), 73.6 (C-2″), 89.4 (C-3″), 70.3 (C-4″), 76.6 (C-5″), 67.5 (C-6″), 105.5 (C-1‴), 75.2 (C-2‴), 77.7 (C-3‴), 72.3 (C-4‴), 76.8 (C-5‴), 70.4 (C-6‴), 101.9 (C-1‴′), 71.9 (C-2‴′), 72.3 (C-3‴′), 74.4 (C-4‴′), 69.9 (C-5‴′), 18.1 (C-6‴′), 103.3 (C-1‴″), 71.9 (C-2‴″), 72.0 (C-3‴″), 85.0 (C-4‴″), 68.6 (C-5‴″), 18.1 (C-6‴″), 106.4 (C-1‴‴′), 76.3 (C-2‴‴′), 78.2 (C-3‴‴′), 71.6 (C-4‴‴′), 78.4 (C-5‴‴′), 62.8 (C-6‴‴′).

HR-ESI-MS: m/z 1129.3580 ([M + Na]$^+$, Calcd. for C$_{48}$H$_{66}$O$_{29}$Na, 1129.3587).

Reference

Yao-Kouassi P A, Magid A A, Richard B, Martinez A, Jacquier M-J, Caron C, Debar E L M, Gangloff S C, Coffy A A, Zèches-Hanrot M. (2008). Isoflavonoid glycosides from the roots of *Baphia bancoensis*. *J Nat Prod* **71**: 2073–2076.

Tupichinol A [(2*R*,3*R*)-3,4′-Dihydroxy-7-methoxy-8-methylflavan]

IUPAC name: (2*R*,3*R*)-2-(4-Hydroxyphenyl)-7-methoxy-8-methylchroman-3-ol

Sub-class: Flavanol

Chemical structure

Source: *Tupistra chinensis* Baker (Family: Liliaceae); rhizomes

Molecular formula: $C_{17}H_{18}O_4$

Molecular weight: 286

State: Colorless prisms (EtOAc)

Melting point: 141–142 °C

Specific rotation: $[\alpha]^{24}_D$ −40.1° (MeOH, *c* 0.034)

Bioactivity studied: Cytotoxic (showed 30% inhibition against human gastric tumor cells)

¹H-NMR (Acetone-d_6, 400 MHz): δ 5.03 (1H, s, H-2), 4.23 (1H, br s, H-3$_\beta$), 2.78 (1 H, dd, J = 16.4, 3.6 Hz, H-4$_{\alpha\text{-eq}}$), 3.16 (1 H, dd, J = 16.4, 3.6 Hz, H-4$_{\beta\text{-ax}}$), 6.86 (1H, d, J = 8.4 Hz, H-5), 6.51 (1H, d, J = 8.4 Hz, H-6), 7.39 (2H, d, J = 8.0 Hz, H-2′ and H-6′), 6.85 (2H, d, J = 8.0 Hz, H-3′ and H-5′), 3.72 (1H, br s, C$_3$-OH), 3.78 (3H, s, 7-OCH_3), 2.10 (3H, s, C$_8$-CH_3), 8.37 (1H, br s, C$_4$-OH).

¹³C-NMR (Acetone-d_6, 100 MHz): δ 80.1 (C-2), 67.6 (C-3), 34.9 (C-4), 128.6 (C-5), 104.7 (C-6), 158.3 (C-7), 114.0 (C-8), 154.4 (C-9), 113.6 (C-10), 132.1 (C-1′), 129.5 (C-2′, C-6′), 116.2 (C-3′, C-5′), 158.2 (C-4′), 56.6 (C$_7$-OCH_3), 9.2 (C$_8$-CH_3).

HMQC: δ 5.03 (H-2) vs δ 80.1 (C-2), δ 4.23 (H-3) vs δ 67.6 (C-3), δ 2.78 (H-4$_{\alpha\text{-eq}}$) and 3.16 (H-4$_{\beta\text{-ax}}$) vs δ 34.9 (C-4), δ 6.86 (H-5) vs δ 128.6 (C-5), 7.39 (H-2′/H-6′) vs δ 129.5 (C-2′/C-6′), δ 6.85 (H-3′/ H-5′) vs δ 116.2 (C-3′/C-5′), δ 3.78 (7-OCH_3) vs δ 56.6 (C$_7$-OCH_3), 2.10 (C$_8$-CH_3) vs δ 9.2 (C$_8$-CH_3).

HMBC: δ 5.03 (H-2) vs δ 132.1 (C-1′), 129.5 (C-2′/C-6′) and 116.2 (C-3′/C-5′), δ 4.23 (1H, br s, H-3$_\beta$) vs δ 113.6 (C-10), δ 2.78 (H-4$_{\lambda\text{-eq}}$) vs δ 80.1 (C-2), 67.6 (C-3), 128.6 (C-5), 154.4 (C-9) and 113.6 (C-10), 3.16 (H-4$_{\beta\text{-ax}}$) vs δ 67.6 (C-3), 154.4 (C-9) and 113.6 (C-10), δ 6.86 (H-5) vs δ 104.7 (C-6), 158.3 (C-7), and 154.4 (C-9), δ 6.51 (H-6) vs δ 128.6 (C-5), 158.3 (C-7), 114.0 (C-8) and 113.6 (C-10), δ 3.78 (7-OCH_3) vs δ 158.3 (C-7), δ 2.10 (C$_8$-CH_3) vs δ 158.3 (C-7), 114.0 (C-8) and 154.4 (C-9).

¹H-¹H COSY: δ 5.03 (H-2) vs δ 4.23 (H-3$_\beta$), δ 4.23 (H-3$_\beta$) vs δ 5.03 (H-2), 2.78 (H-4$_{\alpha\text{-eq}}$) and 3.16 (H-4$_{\beta\text{-ax}}$), δ 6.86 (H-5) vs δ 6.51 (H-6), δ 7.39 (H-2′/H-6′) vs δ 6.85 (H-3′/H-5′).

NOESY: δ 5.03 (H-2) vs δ 4.23 (H-3$_\beta$), 3.16 (H-4$_{\beta\text{-ax}}$) and 7.39 (H-2′/H-6′), δ 4.23 (H-3$_\beta$) vs δ 4.23 (H-3$_\beta$) vs δ 5.03 (H-2), 2.78 (H-4$_{\alpha\text{-eq}}$) and 3.16 (H-4$_{\beta\text{-ax}}$), δ 2.78 (H-4$_{\alpha\text{-eq}}$) vs δ 4.23 (H-3$_\beta$) and 6.86 (H-5), δ 3.16 (H-4$_{\beta\text{-ax}}$) vs δ 5.03 (H-2), 4.23 (H-3$_\beta$) and 6.86 (H-5), δ 6.86 (H-5) vs δ 2.78 (H-4$_{\alpha\text{-eq}}$), 3.16 (H-4$_{\beta\text{-ax}}$) and 6.51 (H-6), δ 6.51 (H-6) vs δ 6.86 (H-5) and 3.78 (7-OCH_3), δ 3.78 (7-OCH_3) vs δ 6.51 (H-6) and 2.10 (C$_8$-CH_3), δ 2.10 (C$_8$-CH_3) vs δ 3.78 (7-OCH_3) and 7.39 (H-2′/H-6′), δ 7.39 (H-2′/H-6′) vs δ 5.03 (H-2), 2.10 (C$_8$-CH_3) and 6.85 (H-3′ and H-5′), δ 6.85 (H-3′/H-5′) vs δ 7.39 (H-2′/H-6′).

EIMS: m/z (%rel.) 286 ([M]$^+$, 32), 151 (100), 136 (42), 107 (59).

HR-EIMS: m/z 286.1205 ([M]$^+$, Calcd. for $C_{17}H_{18}O_4$, 286.1205).

Reference

Pan W-B, Chang F.-R, Wei L-M, Wu Y-C. (2003). New flavans, spirostanol sapogenins, and a pregnane genin from *Tupistra chinensis* and their cytotoxicity. *J Nat Prod* **66**: 161–168.

Tupichinol E [3,5,4′-Trihydroxy-7-methoxy-8-methylflavone]

IUPAC name: 3,5-Dihydroxy-2-(4-hydroxyphenyl)-7-methoxy-8-methyl-4*H*-chromen-4-one

Sub-class: Flavonol

Chemical structure

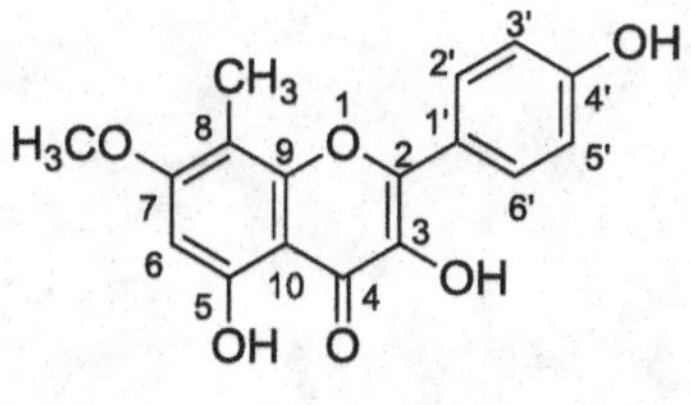

Source: *Tupistra chinensis* Baker (Family: Liliaceae); rhizomes

Molecular formula: $C_{17}H_{14}O_6$

Molecular weight: 314

State: Yellow oil

¹H-NMR (Acetone-*d₆*, 400 MHz): δ 6.49 (1H, s, H-6), 8.21 (2H, d, *J* = 9.2 Hz, H-2′ and H-6′), 7.04 (2H, d, *J* = 9.2 Hz, H-3′ and H-5′), 9.24 (1H, br s, C$_3$-O*H*), 12.19 (1H, s, C$_5$-O*H*), 3.97 (1H, s, C$_7$-OC*H₃*), 2.28 (3H, s, C$_8$-C*H₃*), 8.11 (1H, br s, C$_4'$-O*H*).

¹³C-NMR (Acetone-*d₆*, 100 MHz): δ 144.0 (C-2), 132.7 (C-3), 177.8 (C-4), 160.9 (C-5), 95.7 (C-6), 164.7 (C-7), 104.2 (C-8), 147.9 (C-9),

105.1 (C-10), 124.3 (C-1′), 131.2 (C-2′, C-6′), 117.1 (C-3′, C-5′), 148.2 (C-4′), 57.4 (C$_3$-OCH$_3$), 8.4 (C$_8$-CH$_3$).

HMQC: δ 6.49 (H-6) *vs* δ 95.7 (C-6), δ 8.21 (H-2′/H-6′) *vs* δ 131.2 (C-2′/C-6′), δ 7.04 (H-3′/H-5′) *vs* δ 117.1 (C-3′/C-5′), δ 3.97 (C$_7$-OCH_3) *vs* δ 57.4 (C$_7$-OCH_3), δ 2.28 (C$_8$-CH_3) *vs* δ 8.11 (C$_8$-CH_3).

¹H-¹H COSY: δ 8.21 (H-2′/H-6′) *vs* δ 7.04 (H-3′/H-5′).

NOESY: δ 6.49 (H-6) *vs* δ 3.88 (C$_3$-OCH_3), δ 2.28 (C$_8$-CH_3) *vs* δ 8.21 (H-2′/H-6′).

EIMS: *m/z* (%rel.) 314 ([M]$^+$, 69), 271 (19), 121 (44), 105 (100).

HR-EIMS: *m/z* 314.0795 ([M]$^+$, Calcd. for C$_{17}$H$_{14}$O$_6$, 314.0790).

Reference

Pan W-B, Wei L-M, Wei L-L, Wu Y-C. (2006). Chemical constituents of *Tupistra chinensis* rhizomes. *Chem Pharm Bull* **54**: 954–958.

Tupichinol F [5,7,4′-Trihydroxy-3-methoxy-8-methylflavone]

IUPAC name: 5,7-Dihydroxy-2-(4-hydroxyphenyl)-3-methoxy-8-methyl-4H-chromen-4-one

Sub-class: Flavone

Chemical structure

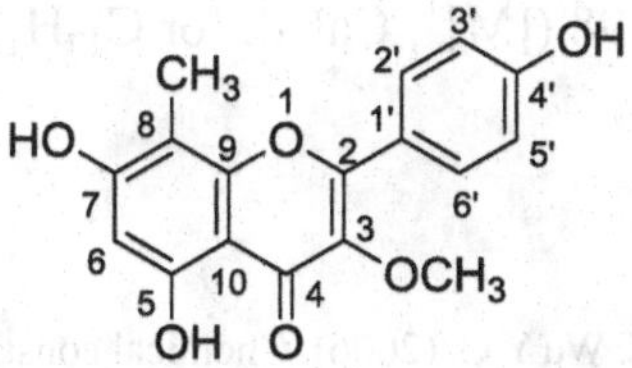

Source: *Tupistra chinensis* Baker (Family: Liliaceae); rhizomes

Molecular formula: $C_{17}H_{14}O_6$

Molecular weight: 314

State: Yellow oil

^{1}H-NMR (Acetone-d_6, 400 MHz): δ 6.34 (1H, s, H-6), 8.09 (2H, d, J = 8.8 Hz, H-2′ and H-6′), 7.05 (2H, d, J = 8.8 Hz, H-3′ and H-5′), 3.88 (3H, s, 3-OCH_3), 12.69 (1H, s, C_5-OH), 9.21 (1H, br s, C_7-OH), 2.27 (3H, s, C_8-CH_3), 8.01 (1H, br s, $C_{4'}$-OH).

^{13}C-NMR (Acetone-d_6, 100 MHz): δ 155.4 (C-2), 139.5 (C-3), 179.9 (C-4), 160.7 (C-5), 99.0 (C-6), 161.1 (C-7), 102.0 (C-8), 156.7 (C-9),

105.0 (C-10), 123.1 (C-1′), 131.3 (C-2′, C-6′), 116.6 (C-3′, C-5′), 151.8 (C-4′), 60.3 (C_3-OCH_3), 7.9 (C_8-CH_3).

HMQC: δ 6.34 (H-6) *vs* δ 99.0 (C-6), δ 8.09 (H-2′/H-6′) *vs* δ 131.3 (C-2′/C-6′), δ 7.05 (H-3′/H-5′) *vs* δ 116.6 (C-3′/C-5′), δ 3.88 (C_3-OCH_3) *vs* δ 60.3 (C_3-OCH_3), 2.27 (C_8-CH_3) *vs* δ 7.9 (C_8-CH_3).

HMBC: δ 6.34 (H-6) *vs* δ 160.7 (C-5), 161.1 (C-7) and 105.0 (C-10), δ 12.69 (C_5-OH) *vs* δ 160.7 (C-5), 99.0 (C-6) and 105.0 (C-10), δ 2.27 (C_8-CH_3) *vs* δ 161.1 (C-7), 102.0 (C-8) and 156.7 (C-9), δ 3.88 (3-OCH_3) *vs* δ 139.5 (C-3), δ 8.09 (H-2′/H-6′) *vs* δ 155.4 (C-2) and 151.8 (C-4′), δ 7.05 (H-3′/H-5′) *vs* δ 151.8 (C-4′).

^{1}H-^{1}H COSY: δ 8.09 (H-2′/H-6′) *vs* δ 7.05 (H-3′/H-5′).

NOESY: δ 2.27 (C_8-CH_3) *vs* δ 8.09 (H-2′/H-6′), δ 3.88 (C_3-OCH_3) *vs* δ 8.09 (H-2′/H-6′).

EIMS: *m/z* (%rel.) 314 ([M]$^+$, 45), 313 (47), 285 (18), 271 (28), 121 (59), 57 (100).

HR-EIMS: *m/z* 314.0798 ([M]$^+$, Calcd. for $C_{17}H_{14}O_6$, 314.0790).

Reference

Pan W-B, Wei L-M, Wei L-L, Wu Y-C. (2006). Chemical constituents of *Tupistra chinensis* rhizomes. *Chem Pharm Bull* **54**: 954–958.

Vogelin H [7,4′-Dihydroxy-8-(γ, γ-dimethylallyl)-2″ξ-(4″-hydroxyisopropyl)dihydrofurano[1″,3″:5,6]isoflavone]

IUPAC name: 4-Hydroxy-8-(4-hydroxyphenyl)-2-(2-hydroxypropan-2-yl)-5-(3-methylbut-2-en-1-yl)-2*H*-furo[2,3-*f*]chromen-9(3*H*)-one

Sub-class: Isoflavone

Chemical structure

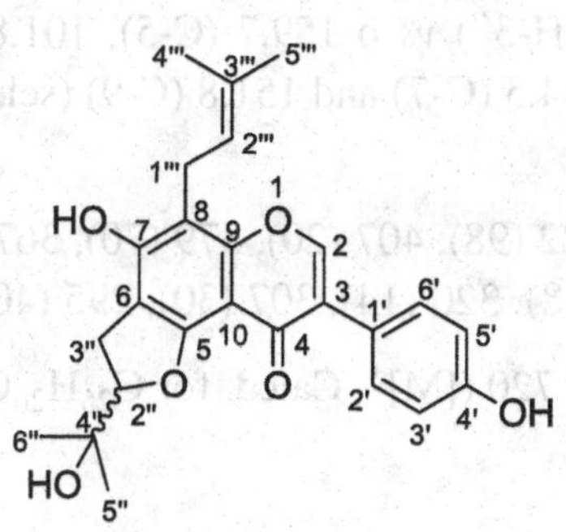

Source: *Erythrina vogelii* (Family: Fabaceae); stem bark

Molecular formula: $C_{25}H_{26}O_6$

Molecular weight: 422

State: Pale yellow powder

Melting point: 195–197 °C

Specific rotation: $[\alpha]^{20}_D$ −38° (CHCl$_3$, *c* 0.0015)

UV (MeOH): λ_{max} (log ε) 203 (4.48), 216 (4.37), 271 (4.48) nm

IR (NaCl): ν_{max} 3545 (OH), 1642 (α,β-unsaturated carbonyl), 1610, 1512, 1425 (aromatic unsaturaion), 1382, 1270, 1215, 1172, 1075, 836 cm^{-1}

^{1}H-NMR (CD$_3$OD, 400 MHz): δ 7.75 (1H, s, H-2), 7.28 (2H, d, *J* = 8.5 Hz, H-2′ and H-6′), 6.83 (2H, d, *J* = 8.5 Hz, H-3′ and H-5′), 7.74 (1H, d, *J* = 8.3 Hz, H-2″), 3.20 (2H, br d, *J* = 8.1 Hz, H-3″), 1.30 (3H, s, H-5″), 1.20 (3H, s, H-6″), 3.29 (2H, d, *J* = 7.3 Hz, H-1‴), 5.23 (1H, m, H-2‴), 1.64 (3H, s, H-4‴), 1.73 (3H, s, H-5‴).

^{13}C-NMR (CD$_3$OD, 100 MHz): δ 152.5 (C-2), 123.0 (C-3), 180.7 (C-4), 159.7 (C-5), 101.8 (C-6), 164.5 (C-7), 107.0 (C-8), 151.8 (C-9), 104.0 (C-10), 121.4 (C-1′), 130.2 (C-2′), 115.4 (C-3′), 156.0 (C-4′), 115.4 (C-5′), 130.2 (C-6′), 91.0 (C-2″), 27.1 (C-3″), 72.2 (C-4″), 25.5 (C-5″), 24.0 (C-6″), 21.8 (C-1‴), 121.4 (C-2‴), 132.2 (C-3‴), 17.7 (C-4‴), 25.5 (C-5‴).

HMBC: δ 7.75 (H-2) *vs* δ 151.8 (C-9), δ 7.74 (H-2″) *vs* δ 159.7 (C-5) and 101.8 (C-6), δ 3.20 (H-3″) *vs* δ 159.7 (C-5), 101.8 (C-6), 164.5 (C-7), δ 3.29 (H-1‴) *vs* δ 164.5 (C-7) and 151.8 (C-9) (selected 2D-correlations were shown).

EIMS: *m/z* (%rel.) 422 (98), 407 (20), 379 (70), 367 (100), 363 (25), 352 (33), 349 (28), 335 (18), 320 (14), 307 (30), 295 (40), 118 (35), 59 (10)

HR-EIMS: *m/z* 422.1720 ([M]$^+$, Calcd. for C$_{25}$H$_{26}$O$_6$, 422.1729).

Reference

Waffo A F K, Coombes P H, Mulholland D A, Nkengfack A E, Fomum Z T. (2006). Flavones and isoflavones from the west African Fabaceae *Erythrina vogelii*. *Phytochemistry* **67**:459–463.

Vogelin I [7,4'-Dihydroxy-8-[(2'''ξ,3'''-dihydroxy-3'''-methyl) butyl]-2'',2''-dimethyl-3'',4''-dehydropyrano[1'',4'':5,6]isoflavone]

IUPAC name: 6-(2,3-Dihydroxy-3-methylbutyl)-5-hydroxy-9-(4-hydroxyphenyl)-2,2-dimethylpyrano[2,3-*f*]chromen-10(2*H*)-one

Sub-class: Isoflavone

Chemical structure

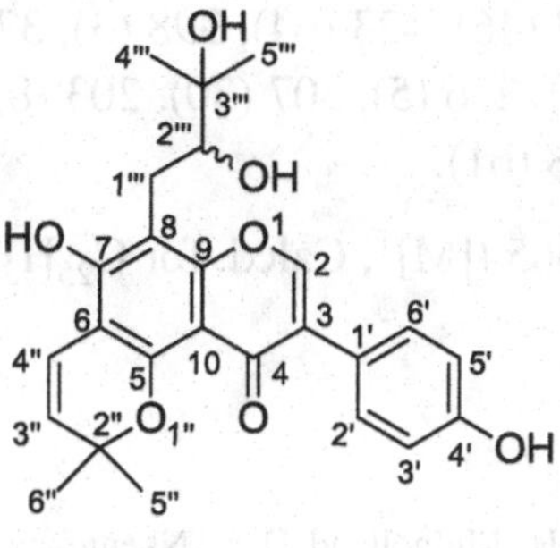

Source: *Erythrina vogelii* (Family: Fabaceae); stem bark

Molecular formula: $C_{25}H_{26}O_7$

Molecular weight: 438

State: Pale yellow powder

Melting point: 248–249 °C

Specific rotation: $[\alpha]^{20}_D$ +42° (CHCl$_3$, c 0.0050)

UV (MeOH): λ_{max} (log ε) 215 (4.41), 268 (4.40), 294 (3.93) nm

IR (NaCl): v_{max} 3525 (OH), 1640 (α,β-unsaturated carbonyl), 1610 (aromatic unsaturaion), 1085, 840 cm^{-1}

^{1}H-NMR (CD$_3$OD, 400 MHz): δ 8.01 (1H, s, H-2), 7.32 (2H, d, J = 8.6 Hz, H-2′ and H-6′), 6.84 (2H, d, J = 8.5 Hz, H-3′ and H-5′), 5.64 (1H, d, J = 10.0 Hz, H-3″), 6.67 (1H, br d, J = 10.0 Hz, H-4″), 1.46 (6H, s, H-5″ and H-6″), 2.88 (1H, m, H-1‴a), 2.85 (1H, m, H-1‴b), 3.60 (1H, m, H-2‴), 1.28 (6H, s, H-4‴ and H-5‴).

^{13}C-NMR (CD$_3$OD, 100 MHz): δ 154.2 (C-2), 124.0 (C-3), 182.4 (C-4), 155.5 (C-5), 105.5 (C-6), 158.1 (C-7), 106.5 (C-8), 156.4 (C-9), 105.8 (C-10), 122.7 (C-1′), 130.9 (C-2′), 116.0 (C-3′), 158.0 (C-4′), 116.0 (C-5′), 130.9 (C-6′), 78.9 (C-2″), 128.7 (C-3″), 116.1 (C-4″), 28.5 (C-5″ and C-6″), 25.5 (C-1‴), 78.6 (C-2‴), 73.6 (C-3‴), 26.0 (C-4‴), 24.4 (C-5‴).

HMBC: δ 8.01 (H-2) vs δ 156.4 (C-9), δ 6.67 (H-4″) vs δ 155.5 (C-5), 105.5 (C-6) and 158.1 (C-7) (selected 2D-correlations were shown).

EIMS: m/z (%rel.) 438 (46), 423 (64), 398 (3), 379 (29), 349 (100), 335 (20), 321 (60), 295 (13), 236 (5), 307 (30), 203 (8), 166 (9), 152 (7), 118 (4), 91 (18), 57 (27), 28 (61).

HR-EIMS: m/z 438.1665 ([M]$^+$, Calcd. for C$_{25}$H$_{26}$O$_7$, 438.1678).

Reference

Waffo A F K, Coombes P H, Mulholland D A, Nkengfack A E, Fomum Z T. (2006). Flavones and isoflavones from the west African Fabaceae *Erythrina vogelii*. *Phytochemistry* **67**: 459–463.

Yunngnin A

IUPAC name: 8-Formyl-7-hydroxy-5,6-dimethoxycoumarin

Sub-class: Coumarin

Chemical structure

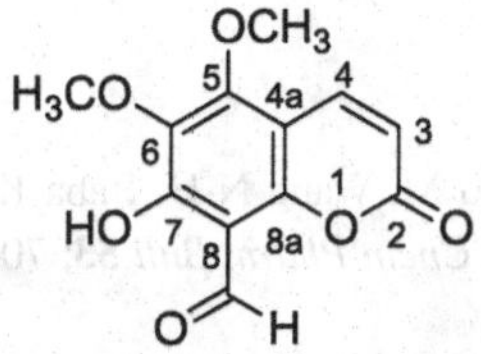

Source: *Heracleum yunngningense* HAND.-MASS. (Family: Umbelliferae); roots

Molecular formula: $C_{12}H_{10}O_6$

Molecular weight: 250

State: Colorless crystalline powder

Melting point: 158–160 °C

UV (MeOH): λ_{max} (log ε) 202.6 (4.01), 216.2 (4.02), 293.7 (3.94) nm

IR (KBr): ν_{max} 2963, 1731 (C=O), 1648 (CHO), 1594, 1475, 1424 cm^{-1}

^{1}H-NMR (CDCl$_3$, 500 MHz): δ 6.27 (1H, d, J = 9.8 Hz, H-3), 7.95 (1H, d, J = 9.5 Hz, H-4), 4.24 (3H, s, C$_5$-OCH_3), 3.90 (3H, s, C$_6$-OCH_3), 12.74 (1H, s, C$_7$-OH), 10.43 (1H, s, C$_8$-CHO).

^{13}**C-NMR (CDCl$_3$, 125 MHz):** δ 159.33 (C-2), 112.50 (C-3), 139.06 (C-4), 105.37 (C-4a), 154.87 (C-5), 134.86 (C-6), 161.53 (C-7), 105.30 (C-8), 153.32 (C-8a), 61.73 (C$_5$-OCH$_3$), 61.39 (C$_6$-OCH$_3$), 191.82 (1H, s, C$_8$-CHO).

^{1}**H-^{1}H COSY:** δ 6.27 (H-3) *vs* δ 7.95 (H-4) and *vice-versa*

HMQC: δ 6.27 (H-3) *vs* δ 112.50 (C-3), δ 7.95 (H-4) *vs* δ 139.06 (C-4), δ 4.24 (C$_5$-OCH$_3$) *vs* δ 61.73 (C$_5$-OCH$_3$), δ 3.90 (C$_6$-OCH$_3$) *vs* δ 61.39 (C$_6$-OCH$_3$), δ 10.43 (C$_8$-CHO) *vs* δ 191.82 (C$_8$-CHO).

NOE: δ 7.95 (H-4) *vs* δ 3.90 (C$_6$-OCH$_3$), δ 3.90 (C$_6$-OCH$_3$) *vs* δ 4.24 (C$_5$-OCH$_3$), δ 4.24 (C$_5$-OCH$_3$) *vs* δ 12.74 (C$_7$-O*H*), δ 10.43 (C$_8$-C*H*O) *vs* δ 12.74 (C$_7$-O*H*).

HR-EIMS: *m/z* 250.0465 ([M]$^+$, Calcd. for C$_{12}$H$_{10}$O$_6$, 265.0477).

Reference

Taniguchi M, Yokota O, Shibano M, Wang N-H, Baba K. (2005). Four coumarins from *Heracleum yunngningense*. *Chem Pharm Bull* **53**: 701–704.

Yunngnoside A
[Apterin monoacetate]

IUPAC name: ((2R,3S,5R,6S)-3,4,5-Trihydroxy-6-((2-((8S,9R)-9-hydroxy-2-oxo-8,9-dihydro-2H-furo[2,3-h]chromen-8-yl)propan-2-yl)oxy)tetrahydro-2H-pyran-2-yl)methyl acetate

Sub-class: Dihydrofuranocoumarin glycoside

Chemical structure

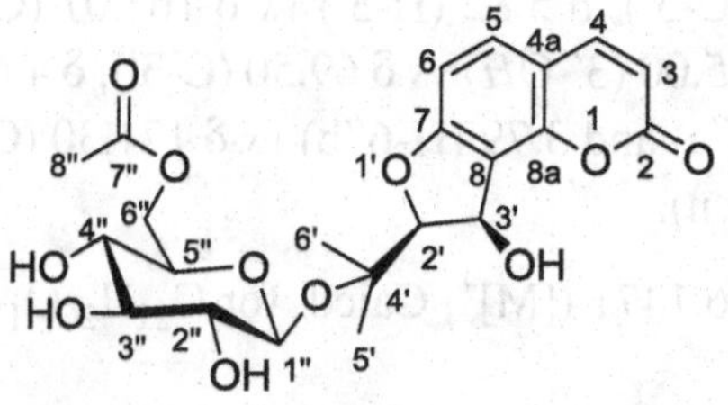

Source: *Heracleum yunngningense* HAND.-MASS. (Family: Umbelliferae); roots

Molecular formula: $C_{22}H_{26}O_{11}$

Molecular weight: 466

State: Colorless crystalline powder

Melting point: 240–242 °C

UV (MeOH): λ_{max} (log ε) 204.5 (4.62), 322 (4.15) nm

IR (KBr): ν_{max} 3447 (OH), 2926, 1737 (C=O), 1617, 1577, 1490, 1451, 1407 cm^{-1}

¹H-NMR (CDCl₃, 500 MHz): δ 6.22 (1H, d, J = 9.6 Hz, H-3), 7.66 (1H, d, J = 9.6 Hz, H-4), 7.37 (1H, d, J = 8.5 Hz, H-5), 6.81 (1H, d, J = 8.5 Hz, H-6), 4.38 (1H, d, J = 7.0 Hz, H-2′), 5.82 (1H, dd, J = 8.2, 7.0 Hz, H-3′), 5.00 (1H, d, J = 8.5 Hz, 3′-OH), 1.57 (3H, s, H-5′), 1.58 (3H, s, H-6′), 4.66 (1H, d, J = 7.8 Hz, H-1″), 3.24 (1H, dd, J = 9.0, 7.8 Hz, H-2″), 3.49 (1H, t, J = 9.0 Hz, H-3″), 3.32 (1H, dd, J = 9.6, 9.0 Hz, H-4″), 3.36 (1H, ddd, J = 9.6, 5.1, 2.2 Hz, H-5″), 4.11 (1H, dd, J = 12.0, 5.1 Hz, H-6″a), 3.79 (1H, dd, J = 12.0, 2.2 Hz, H-6″b), 2.00 (3H, s, H-8″).

¹³C-NMR (CDCl₃, 125 MHz): δ 161.05 (C-2), 112.49 (C-3), 144.18 (C-4), 113.28 (C-4a), 130.64 (C-5), 107.87 (C-6), 163.07 (C-7), 116.66 (C-8), 151.92 (C-8a), 91.19 (C-2′), 69.50 (C-3′), 78.36 (C-4′), 25.95 (C-5′), 23.30 (C-6′), 97.20 (C-1″), 73.28 (C-2″), 76.37 (C-3″), 69.82 (C-4″), 73.51 (C-5″), 62.76 (C-6″), 171.30 (C-7″), 20.74 (C-8″).

HMBC: δ 7.66 (H-4) vs δ 116.66 (C-8) and 151.92 (C-8a), δ 7.37 (H-5) vs δ 116.66 (C-8), δ 6.81 (H-6) vs δ 163.07 (C-7) and 116.66 (C-8), δ 4.38 (H-2′) vs δ 69.50 (C-3′), δ 5.82 (H-3′) vs δ 163.07 (C-7), 151.92 (C-8a) and 91.19 (C-2′), δ 5.00 (3′-OH) vs δ 69.50 (C-3′), δ 4.66 H-1″) vs δ 78.36 (C-4′), δ 4.11 (H-6″a) and 3.79 (H-6″b) vs δ 171.30 (C-7″) (selected correlations were shown).

HR-EIMS: m/z 466.1471 ([M]$^+$, Calcd. for $C_{22}H_{26}O_{11}$, 466.1475).

Reference

Taniguchi M, Yokota O, Shibano M, Wang N-H, Baba K. (2005). Four coumarins from *Heracleum yunngningense*. *Chem Pharm Bull* **53**: 701–704.

Printed in the United States
By Bookmasters